ENVIRONMENTAL RESEARCH ADVANCES

FLOODPLAINS

ENVIRONMENTAL MANAGEMENT, RESTORATION AND ECOLOGICAL IMPLICATIONS

ENVIRONMENTAL RESEARCH ADVANCES

Additional books in this series can be found on Nova's website under the Series tab.

Additional e-books in this series can be found on Nova's website under the e-book tab.

ENVIRONMENTAL RESEARCH ADVANCES

FLOODPLAINS

ENVIRONMENTAL MANAGEMENT, RESTORATION AND ECOLOGICAL IMPLICATIONS

ENNER HERENIO ALCÂNTARA, PH.D.
EDITOR

nova publishers
New York

For permission to use material from this book please contact us:
Telephone 631-231-7269; Fax 631-231-8175
Web Site: http://www.novapublishers.com

Additional color graphics may be available in the e-book version of this book.

Library of Congress Cataloging-in-Publication Data

Floodplains : environmental management, restoration, and ecological implications / edtior, Enner Herenio Alcbntara (Department of Cartography, Sco Paulo State University--UNESP, Brazil).
pages cm
Includes bibliographical references and index.
ISBN: 978-1-62808-998-1 (hardcover)
1. Floodplain management. I. Alcbntara, Enner Herenio de, editor of compilation.
TC411.F57 2013
333.75--dc23
2013033671

Published by Nova Science Publishers, Inc. † New York

CONTENTS

PREFACE

A floodplain is a flat or nearly flat land adjacent to a stream or river that experiences occasional flooding (naturally occurring process). Floodplains are of interest to most of people, because some of them live near rivers and floodplains; that are used historically for waste disposal, supply, power generation, transport and food.

This book tries to show some relevant studies in floodplains and lakes in sorted themes, such as: mercury speciation and bioavailability, plant-soil interactions, indicators for evaluating floodplain restoration, floodplain formation, phytoplankton community, storage and release of water, hydrogenic heavy metals and hydrodynamics. It is separated into nine chapters.

Chapter 1 - Mercury speciation and bioavailability in floodplain soils of east fork poplar creek, OAK Ridge, Tennessee, USA. This chapter summarizes the current status of mercury distribution, speciation and bioavailability to native Earthworms and plants in the floodplain soils of East Fork Poplar Creek after decades of US Department of Energy's remediation efforts.

Chapter 2 - Floodplains of regulated rivers: plant-soil interactions in changing ecosystems. This chapter discuss about the patters and associated factors related to plant-soil interactions in regulated rivers' floodplain.

Chapter 3 - Earthworm communities as indicators for evaluating floodplain restoration success. this chapter highlights biological and pedological features as indicator tools to evaluate the success of floodplain restoration, focusing on earthworm communities that are assumed to increase in terms of diversity, typicity and evolution along a gradient of naturality from the embanked system to the near-natural reference.

Chapter 4 - Three main stages of floodplain evolution in northern Eurasia and their ecological significance. This chapter discuss about floodplain formation by the rivers larger than modern ones, floodplain formation by the rivers smaller than modern ones, and floodplain formation by the rivers of the modern morphological type.

Chapter 5 - The importance of spatial and local environmental factors to structuring phytoplankton community in floodplain lakes of Cuiabá River (Northern Pantanal, Brazil). This chapter aimed to evaluate the effects of spatial and local environmental factors structuring the phytoplankton abundance and species diversity in nine lakes along the Cuiaba River, Northern Pantanal, Mato Grosso, during high waters and low waters.

Chapter 6 - Survey-based influence of storage and release of water from floodplain lakes on the main river in sparse-data area, The Lower Amur, Russia. This chapter shows how to use hydrometrical data to study the interrelation between water level dynamics and the water pathway in floodplains.

Chapter 7 - Hydrogenic heavy metals pollution of alluvial soils in the city of Perm (The middle cisurals region, Russia). This chapter discuss about the metals pollutions and their consequences.

Chapter 8 - The role of overbank deposition in floodplain formation and catastrophic floods in floodplain destruction in south eastern Australia. This chapter discuss about the geomorphic and sedimentologic variability in the floodplains.

Finally, the Chapter 9 - Effect of hydrodynamics of the water quality of lakes and reservoirs. The aim of this chapter is to show how different hydrodynamic processes which have their origin in the interaction of meteorological variables (wind and solar radiation) with a stratified water body of a lake determine the circulation patterns of the water and can significantly influence its quality.

Enner Herenio Alcântara, Ph.D.
Editor

In: Floodplains
Editor: Enner Herenio Alcântara
ISBN: 978-1-62808-998-1

Chapter 1

MERCURY SPECIATION AND BIOAVAILABILITY IN FLOODPLAIN SOILS OF EAST FORK POPLAR CREEK, OAK RIDGE, TENNESSEE, USA

Fengxiang X. Han[1], Yi Su[2] and David L. Monts[3]
[1]Department of Chemistry and Biochemistry, Jackson State University, Jackson, MS, US
[2]Institute for Clean Energy Technology, Mississippi State University, Starkville, MS
[3]Department of Physics and Astronomy, Mississippi State University, Mississippi State, MS, US

ABSTRACT

Historically as part of its national security mission, the U.S. Department of Energy's Y-12 National Security Facility in Oak Ridge, TN acquired a significant fraction of the world's supply of elemental mercury. During the 1950s and 1960s, a large amount of elemental mercury escaped confinement and is still present in the watershed surrounding the Y-12 facility as a medium-term sink, but long-term source. A series of remediation efforts have been deployed in the watersheds of the Oak Ridge site. Earthworms play an important role in carbon recycling and maintenance of soil fertility and soil health. Earthworms are key components in natural food chains, providing a food source for many small mammals and important food sources for small birds. This chapter summarizes our previous studies on the current status of mercury distribution, mercury speciation and bioavailability to native earthworms (*Diplocardia* spp) and plants in floodplain soils of East Fork Poplar Creek (EFPC) after decades of U.S. Department of Energy's remediation. The present study clearly shows that the total mercury in floodplain soils of EFPC significantly decreased after a series of remediation. This study confirmed the long-term effectiveness of these remediation efforts, especially after excavation of highly contaminated floodplain soils. The major mercury form in the current floodplain soils of EFPC after these decades of remediation is mainly the non-cinnabar mercury bound form in clay minerals (4M HNO_3-extractable residual fraction). The results show strong linear relationships between mercury concentrations in native earthworms (both mature and immature groups) and the non-cinnabar mercury form, while cinnabar mercury is less bioavailable to native earthworms. Native earthworms may be used as a potential mercury ecological bio-indicator (bio-marker) for demonstrating mercury bioavailability and ecotoxicity in the ecosystem. Extractability and bioavailability of HgS in Oak Ridge soils were also discussed.

1. INTRODUCTION

The Y-12 National Security Facility near the city of Oak Ridge, Tennessee, USA is a manufacturing and developmental engineering facility that formerly produced components for various nuclear weapons systems. Mercury has been identified as a key contaminant in soil, sediment, surface water, groundwater, buildings, drains, and sumps in the Y-12 watershed (USDOE, 1998; Han et al., 2006). The source of the mercury is from elemental mercury used during the 1950s and early 1960s for the manufacture of nuclear weapons. Mercury was used to capture enriched lithium by separating the lithium isotopes. The estimates of the total mercury released to the environment range from about 75 to 150 metric tons (Turner et al., 1985; USEPA, 1989). Most mercury has been accumulated in the upper 3 m of floodplain soils and the sediments of a 24-km length of the East Fork Poplar Creek (EFPC). Lower East Fork Poplar Creek (LEFPC) flows north from the Y-12 plant, off site into the city of Oak Ridge through a gap in Pine Ridge. Lower EFPC flows through residential and business sections of Oak Ridge to join Poplar Creek, which flows into the Clinch River. The concentrations of mercury in the Upper EFPC watershed (soil) ranged from 0.01 to 7700 mg/kg (Philips, 2004). Mercury has been detected at higher than background levels in sediments of the Clinch River and the Tennessee River near Chattanooga, some 190 km downstream of Oak Ridge (USEPA, 1989).

A series of remediation efforts (pollution abatement and management actions) have been employed in the Oak Ridge watersheds associated with Upper EFPC (UEFPC) from 1985 through 2004.

These include central pollution control facility, source collection, elimination of untreated discharges, central and east end mercury treatment systems, relining of sanitary and storm sewers, permanent bypass of Lake Reality (a concrete-lined catch basin located on Y-12 plants), dechlorination of cooling water discharges, and bank stabilization project (Loar, 2004). The U.S. Department of Energy has removed highly mercury contaminated floodplain soil at several locations along the creek where mercury concentrations were higher than 400 mg/kg (USEPA, 1989; USDHHS, 2004). Remediation field activities with excavation began in 1996. All floodplain soils with mercury above this level have been removed (USEPA, 1989; USDHHS, 2004). All these remediation efforts have significantly reduced mercury concentrations in water of Upper EFPC from 1.6-1.8 μg/L in 1988 to 0.3-0.5 μg/L in 2003-2004 (Loar, 2004).

Mercury speciation in UEFPC floodplain soil strongly controls solubility, mobility, and bioavailability of mercury in both terrestrial and aquatic ecosystems and downstream mercury concentrations. Floodplain soils can act both as a medium-term sink and long-term mercury source. Our previous study was to mimics the initial stage of transformation and redistribution of mercury (as nitrate mercury and other forms), which actually occurred in contaminated field soils (floodplain) in Oak Ridge (Han et al., 2006). It is essential to assess the current status of mercury concentrations, speciation, distribution and its ecological bioavailability in floodplain soils after this series of remediation efforts.

This chapter summarizes our previous studies on the current status of mercury distribution, speciation and bioavailability to native earthworms and plants in the floodplain soils of East Fork Poplar Creek after decades of US Department of Energy's remediation efforts (Han et al., 2008, 2012). This study may confirm the long-term effectiveness of the remediation, especially after excavation of highly contaminated floodplain soils.

2. Spatial and Soil Profile Distribution of Total Hg

The sampling site was located in a floodplain field of LEFPC of Oak Ridge, TN, USA (Figure 1). Surface soils, soil profile, creek bank soil and sediment were taken from the site. Two sampling strategies were used in this study: surface sampling (0-20 cm) for examining the mercury spatial distribution in floodplain field while soil profile sampling (three layers, 0-10, 50-60 and 100-110 cm) for Hg vertical distribution along soil profiles and creek bank. A total 23 surface soils (0-20 cm) were randomly sampled from both woodland area and wetland/grassland. Two 8x8 m sites were selected in the woodland for profile sampling (the total 10 soil profiles).

Five profiles consisting of three layers were randomly taken from each field. The three layers were the surface (0-10cm), subsurface (50-60 cm), and bottom layer (100-110 cm). In addition, the total six bank soils were taken from the three sites along the creek bank (0-10 cm and 50-60 cm two exposed soil depths at each site) and one sediment sample was taken from the bottom of the creek. Soil in both woodland and wetland areas was well developed with a clear B horizon. Fresh field samples were stored in a refrigerator for mercury level and speciation analyses. The subsamples were air-dried and ground through 1mm sieve for determination of soil physicochemical properties (Table 1).

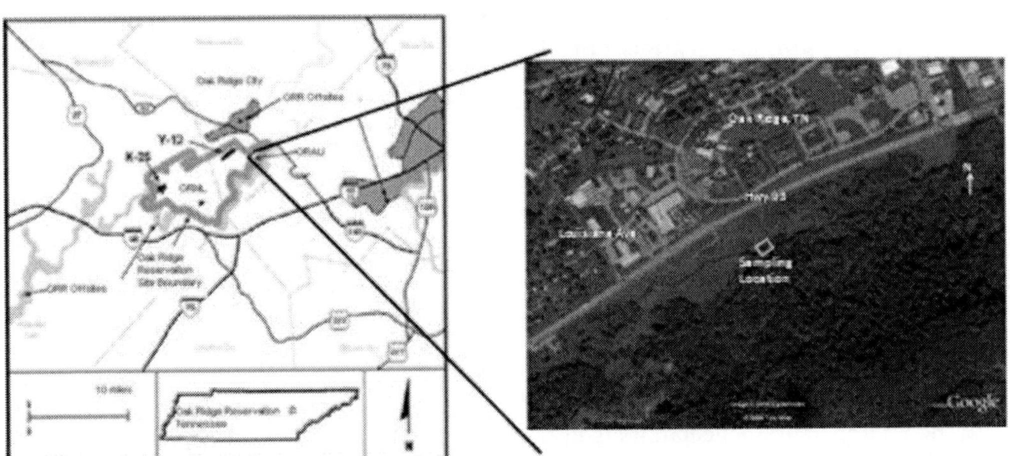

Figure 1. Location of field sampling (Han et al., 2012).

Table 1. Some soil physicochemical properties in surface and subsurface field soils, creek bank soils and creek sediment from Oak Ridge, TN. (Han et al., 2012)

Soil Types	Depth cm	Land Use		Fe2O3% %	Mn% %	CEC cmol/kg	Carbon %	pH (range)
Surface soil	0-20	Grassland	Average	2.45	0.11	18.50	4.44	7.04-7.46
			Standard Deviation	0.29	0.02	7.73	0.93	
		Woodland	Average	2.20	0.10	18.89	4.07	7.35-7.54
			Standard Deviation	0.86	0.04	9.80	1.73	
Bank soil	0-10		Average	1.99	0.09	13.31	2.09	7.32-7.46
			Standard Deviation	0.26	0.01	7.36	0.97	
	50-60		Average	2.91	0.08	6.96	1.29	7.40-7.84
			Standard Deviation	0.07	0.01	0.24	0.48	
Sediment				2.50	0.14	7.30	2.78	7.52
Soil Profile	0-10	Woodland	Average	2.85	0.11	12.33	3.10	7.38-7.51
			Standard Deviation	0.89	0.04	5.12	1.24	
	50-60		Average	2.15	0.08	11.33	1.34	7.37-7.99
			Standard Deviation	0.59	0.03	4.47	0.84	
	100-110		Average	2.79	0.08	10.20	2.48	7.47-7.64
			Standard Deviation	0.30	0.05	3.33	2.54	

Table 2. Concentrations of cinnabar mercury, noncinnabar mercury and total mercury in surface floodplain soils (0-20 cm) of Oak Ridge, TN. (Han et al., 2012)

Land use		Noncinnabar-Hg mg/kg	Cinnabar-Hg mg/kg	Total Hg mg/kg	Noncinnabar Hg %	Cinnabar-Hg %
Wetland/Grassland	Average	51.6	10.9	62.5	97.3	2.7
	Standard Deviation	19.5	25.4	34.0	2.1	2.1
	Maximum	77.8	68.4	125.4	99.0	6.8
	Minimum	29.0	0.4	31.2	93.2	1.0
Woodland	Average	39.4	0.6	40.0	98.5	1.5
	Standard Deviation	9.4	0.2	9.6	0.4	0.4
	Maximum	52.4	0.8	53.0	98.8	2.1
	Minimum	27.8	0.3	28.1	97.9	1.2

Table 3. Concentrations of cinnabar mercury, noncinnabar mercury and total mercury in bank soils and sediment of LEFPC, Oak Ridge, TN. (Han et al., 2012)

Soil type	Depth Cm		Noncinnabar-Hg mg/kg	Cinnabar-Hg mg/kg	Total Hg mg/kg	Noncinnabar Hg %	Cinnabar-Hg %
Bank soil	0-10	Average	51.4	1.0	52.3	98.2	1.8
		Standard Deviation	4.7	0.7	5.3	1.1	1.1
		Maximum	56.3	1.7	58.1	98.9	3.0
		Minimum	47.0	0.5	47.5	97.0	1.1
	50-60	Average	23.2	17.9	41.1	63.0	37.0
		Standard Deviation	26.9	28.5	20.0	53.9	53.9
		Maximum	52.9	50.8	53.9	98.2	99.0
		Minimum	0.5	1.0	18.1	1.0	1.8
Sediment			72.5	1.7	74.2	97.7	2.3

The total mercury concentrations in floodplain surface soils (0-20 cm) from Lower East Fork Poplar Creek (LEFPC) of Oak Ridge, TN were in the range of 28 to 125.4 mg/kg with an average of 53.2 mg/kg ±28.3 mg/kg (Table 2) (Han et al., 2012). Concentrations of total mercury at both depths of soil profiles (around 50-60 cm and 100-110 cm) ranged from 0.4 to 239 mg/kg and averaged 44-88 mg/kg (Table 3). There are larger variations in the total mercury in subsurface and deep subsurface soils. The total mercury was distributed among soil profiles irregularly (Figure 2). Some profiles had much higher mercury in surface soils, some profiles had higher mercury concentrations in the middle layer around 50-60 cm, while other profiles had the highest mercury in the deep down soil profiles.

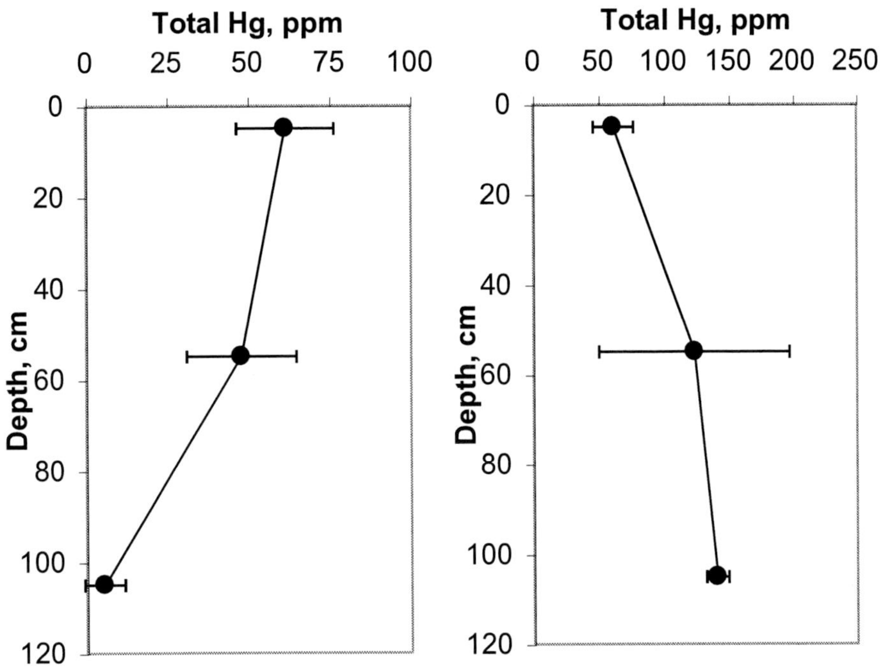

Figure 2. Mercury distribution along soil profiles in the woodland. (Han et al., 2012).

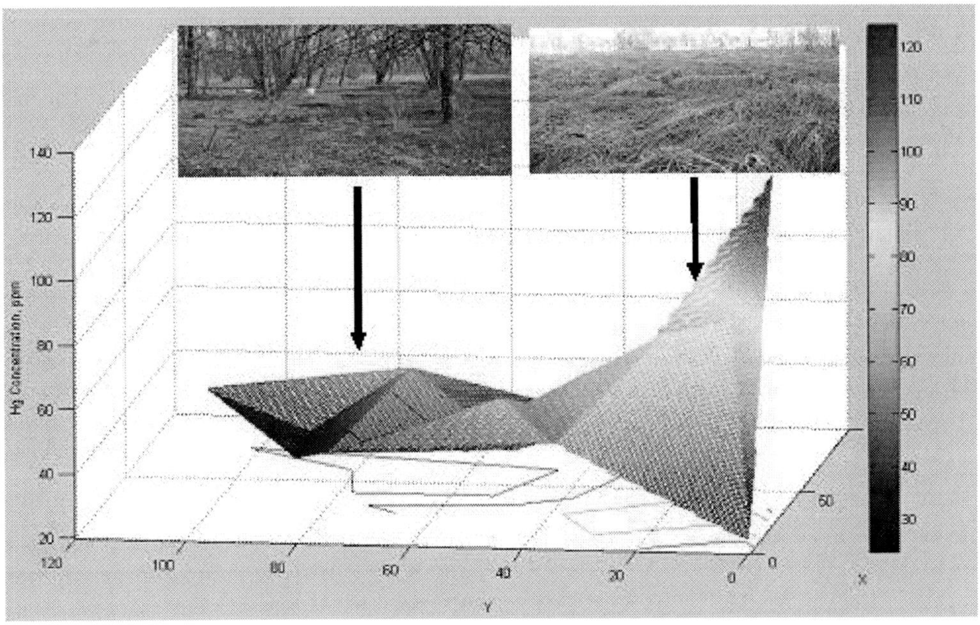

Figure 3. Spatial distribution of the total Hg in a floodplain field (wetland/grassland had higher mercury concentrations than woodland) (X, Y: in meter). (Han et al., 2012).

The total mercury was strongly affected by topography and land use. Spatially the total mercury concentrations were higher in wetlands (with depressions) than in soils of woodland (Figure 3). Wetlands usually had lower elevation than woodlands. Wetland along Lower East Fork Poplar Creek was overflown by surface/runoff water after storms and from the creek. Wetland was flourished with dense grass Tall fescue (*Festuca arundinacea*). On the other hand, woodland had relatively high elevator with less dense grasses. Wetland/grassland surface soils had an average of 62.5 ± 34 mg/kg the total mercury, while woodland surface soils had 40 ± 9.6 mg/kg total mercury. Woodland soils had less variation (CV%: 24%) than those in wetland/grassland (CV%: 54.4%). This also indicates more heterogeneous distribution of mercury in wetlands due to frequent inputs from the creek than the woodlands.

On the other hand, surface creek bank soils had similar total mercury (47.5-58.1mg/kg) with an average 52.3±5.3 mg/kg (Table 4). However, mercury concentrations decreased to 18.1-53.9 mg/kg (average: 41.1±20mg/kg) in the deep exposed bank soil around 50-60 cm. The creek sediment contained total mercury 74.2 mg/kg, which was a little higher than bank soils.

No any significant correlation between total mercury and soil physicochemical properties (CEC, total C, iron oxide, manganese oxide, pH etc). The surface soils had 2.45±0.29%, 0.11±0.02%, 3.7±1.7% iron oxide, manganese and organic carbon as well as 18.5 ± 7.73 $cmol_c/kg$, respectively. Basically, the soils down the soil profiles had less iron oxide, CEC, and less organic carbon contents, but similar Mn contents (Table 1). Soils from banks of the creek had relatively less iron oxide, CEC and organic carbon in the surface bank soils. This phenomenon on soil-property-independent-distribution of the total mercury and irregular distribution along soil profiles may be explained by a history of a series of remediation activities, including digging and filling. In the past decades, US Department of Energy has excavated all floodplain soils along the East Pork Poplar Creek with the total mercury 400 mg/kg (USEPA, 1989; USDHHS, 2004), filled with guest clean soils. The present analyses confirmed the effect of cleaning-up and remediation efforts, which indeed lowered the total mercury well below the DOE target, 400 mg/kg. As indicated by this study, the total mercury concentrations in the all soil samples investigated in the present study were well below that target.

Table 4. Concentrations of cinnabar mercury, noncinnabar mercury and total mercury in soil profiles (woodland) of floodplain of Oak Ridge, TN. (Han et al., 2012)

Depth Cm		Noncinnabar-Hg mg/kg	Cinnabar-Hg mg/kg	Total Hg mg/kg	Noncinnabar Hg %	Cinnabar-Hg %
0-10	Average	65.3	5.3	70.6	92.3	7.7
	Standard Deviation	45.1	9.3	45.4	11.1	11.1
	Maximum	197	33.1	201	98.7	39.7
	Minimum	36.4	0.9	38.2	60.3	1.3
50-60	Average	74.6	13.8	88.4	82.8	17.2
	Standard Deviation	62.8	22.8	79.8	16.8	16.8
	Maximum	180	76.1	239	97.3	49.1
	Minimum	0.3	0.3	0.6	50.9	2.7
100-110	Average	38.2	5.5	43.7	63.4	36.6
	Standard Deviation	59.9	8.7	62.4	31.2	31.2
	Maximum	142	29.1	151	95.0	98.7
	Minimum	0.1	0.2	0.4	1.3	5.0

It is clearly demonstrated by the present study that the total mercury in floodplain soils of EFPC decreased after a series of remediation, especially excavation of the highly contaminated soils/sediments along the creek. The concentrations of mercury in Upper EFPC watershed soil were earlier reported to range from 0.01 to 7700 mg/kg (Philips, 2004). Some sediment cores contained 460 mg/kg mercury at depths of 80-84 cm (Olsen et al., 1992). Mercury has been detected at higher than background levels in sediments of the Clinch River and the Tennessee River near Chattanooga, some 118 miles (190 km) downstream of Oak Ridge (USEPA, 1989). Another recent study by Pant and Allen (2007) indicated the total mercury in floodplain soils (54 samples) along EFPC at Oak Ridge, TN ranged from 0.11 to 103.3 mg/kg with a mean of 24.3 mg/kg. All these imply the overall effectiveness of DOE remediation efforts that achieved the cleaning up targets.

3. Cinnabar and Non-Cinnabar Mercury and Mercury Speciation in Floodplain Soils

We firstly divided the total mercury in floodplain soils into non-cinnabar and cinnabar form. Non-cinnabar mercury was the major form (97.8%) of mercury in the floodplain surface soils (46.5 ± 16.7 mg/kg non-cinnabar mercury) (Table 2) (Han et al., 2012). Cinnabar mercury was 2.2% (6.6±19.5 mg/kg) of the total mercury. In addition, wetland/grassland soils contained 51.6 ±19.5 mg/kg non-cinnabar mercury and 10.9 ±25.4 mg/kg cinnabar mercury, respectively. Woodland soils had 39.4±9.4 mg/kg and 0.6±0.2 mg/kg for non-cinnabar and cinnabar mercury, respectively (Figure 4). In the deeper soil profiles, non-cinnabar mercury decreased with depth (surface: 92%, middle subsurface: 83 and deep subsurface: 63 %), while cinnabar mercury increased with depth (Figure 5).

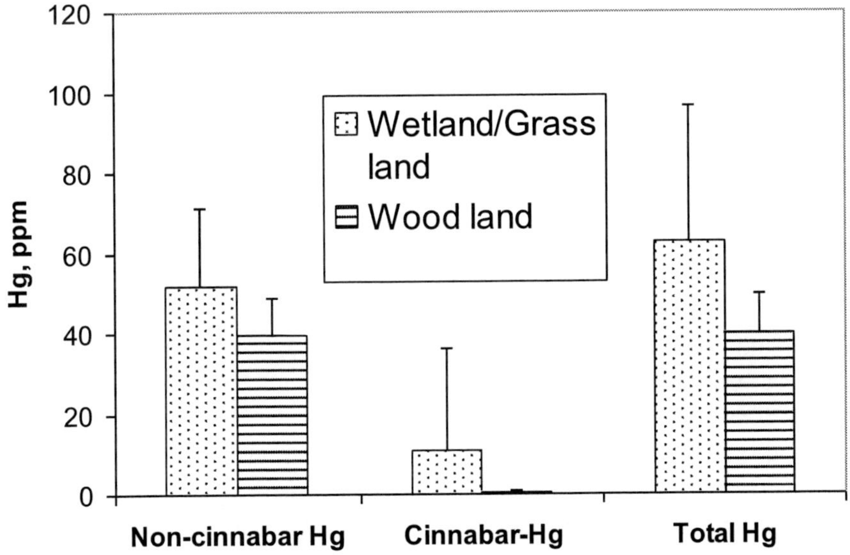

Figure 4. Concentrations of total mercury, cinnabar-mercury and non-cinnabar mercury in the surface soils (0-20 cm) of floodplain fields under wetland/grassland and woodland uses. (Han et al., 2012).

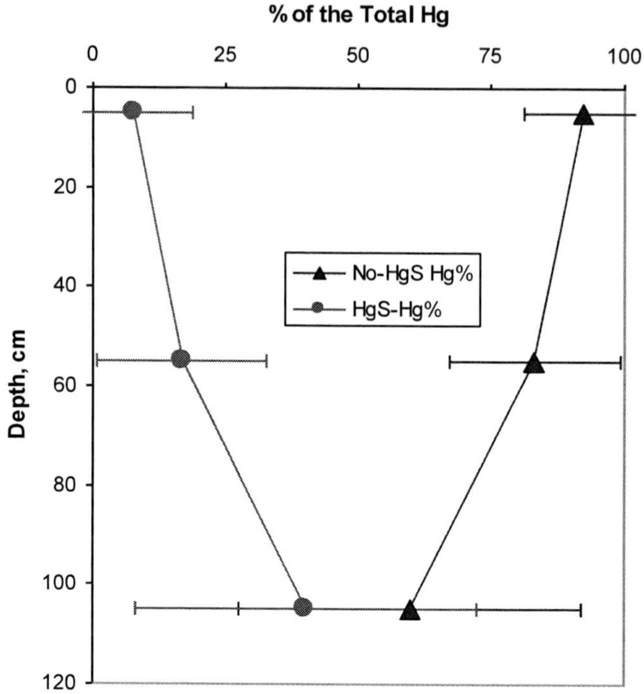

Figure 5. Soil profile distribution of cinnabar-Hg and non-cinnabar Hg in a floodplain field (averages with standard deviation).

On the other hand, creek bank soils had a similar percentage of non-cinnabar mercury and cinnabar mercury (Table 4). Surface creek bank soils contained 98±1.1% non-cinnabar mercury while cinnabar form was 1.8 ± 1.1%. However the bank soils around 50-60 cm had a significantly higher percentage cinnabar mercury (37%) compared to surface bank soil.

Furthermore, non-cinnabar mercury in the solid-phase components was further divided into six operationally defined fractions (exchangeable, easily reducible oxide bound, organic matter bound, amorphous iron oxide bound, crystalline iron oxide bound, and non-cinnabar residual forms). In general, a majority of non-cinnabar mercury was residual mercury bound with clay minerals (91-99.8%), followed by organic matter bound and crystalline iron oxide bound (Figures 6-8). Mercury in the exchangeable and easily oxide bound fractions was the lowest with most below the detection limit.

In the surface soils of woodlands, non-cinnabar mercury was mainly in a clay mineral-bound residual fraction (92%±7.0%), followed by the organic matter-bound fraction (3.8%±3.1%) and crystalline iron oxide-bound fraction (2.7%±3.9%) (Figure 6). Mercury in the amorphous iron oxide-bound, easily reducible oxide-bound, and the exchangeable fractions was 0.8±2.3%, 0.18±0.68%, and 0.4±1.5%, respectively. Compared to woodland soil, surface soils in the wetland/grassland had higher mercury concentrations in the organic matter-bound fraction (7.5±4.3%) and lower in non-cinnabar bound residual fraction (90.9±5.4%) (Figure 6). On the other hand, bank surface and subsurface soils contained slightly higher non-cinnabar clay mineral-bound residual fraction (88-100%), followed by organic bound fraction and amorphous and crystalline iron oxide-bound fractions (Figure 7).

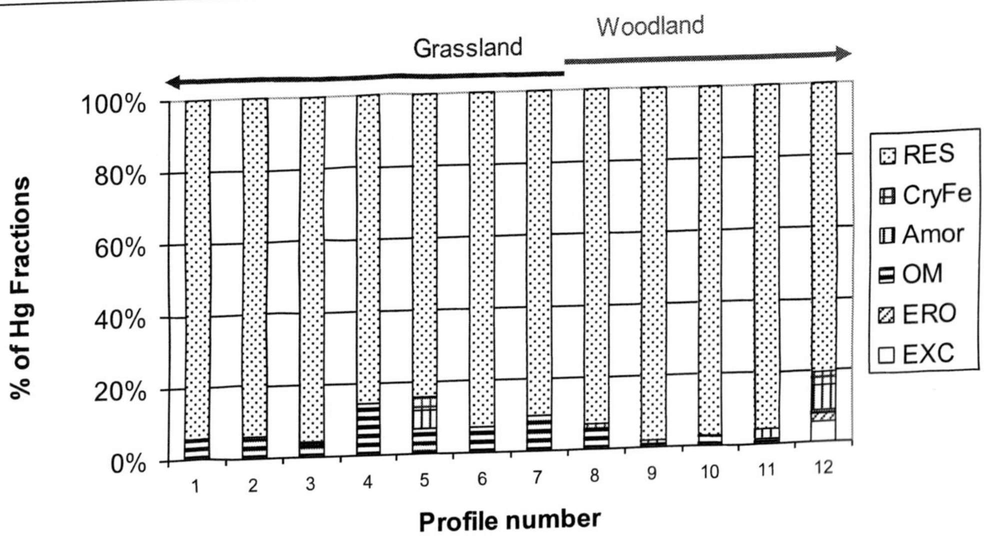

Figure 6. Mercury in solid-phase components of surface soils of wetland/grassland and woodlands (EXC: NH_4OAc-extractable mercury, soluble plus exchangeable mercury; ERO: $NH_2OH{\cdot}HCl$-extractable mercury, easily reducible oxides-bound mercury; OM: H_2O_2-oxidizable mercury, organic matter-bound mercury; AmorFe: oxalate-extractable mercury, amorphous iron oxides-bound mercury; CryFe: hot $NH_2OH{\cdot}HCl$ and HOAc-extractable mercury, crystalline iron oxides- bound mercury; RES: HNO_3-extractable mercury, residual non-cinnabar mercury; and HgS: cinnabar mercury. (Han et al., 2012).

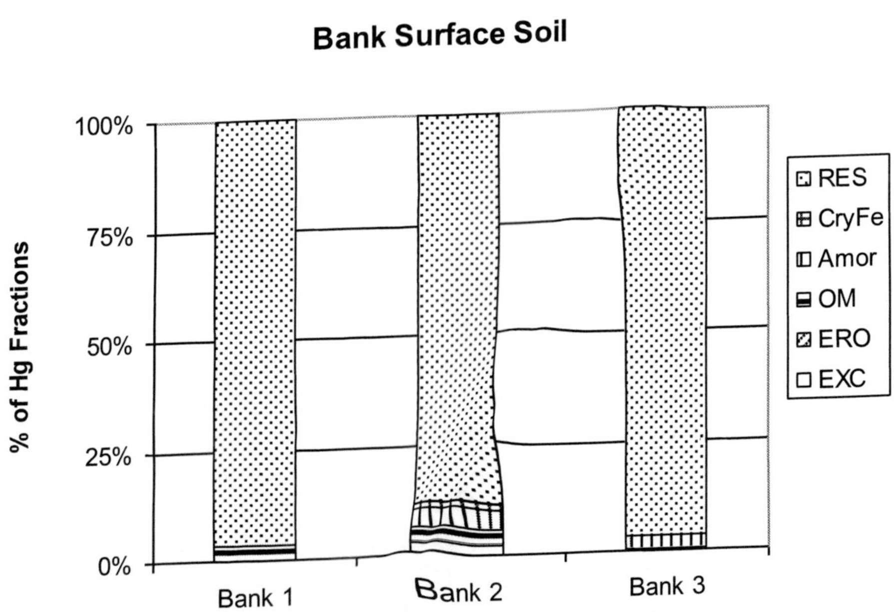

Figure 7. Mercury in solid-phase components in bank surface soils of LEFPC of Oak Ridge, TN.

Compared to surface soils, subsurface soils (both middle layers around 50-60 cm and the bottom layer around 100-110cm) from the woodland had higher mercury concentrations in the non-cinnabar clay mineral-bound residual fraction (>99.5%), less mercury in the organic fraction (0.12%), crystalline iron oxide-bound, amorphous iron oxide-bound, easily reducible oxide and exchangeable fractions (Figure 8). Mercury in exchangeable and easily oxide reducible oxide-bound fractions was below the detection limit.

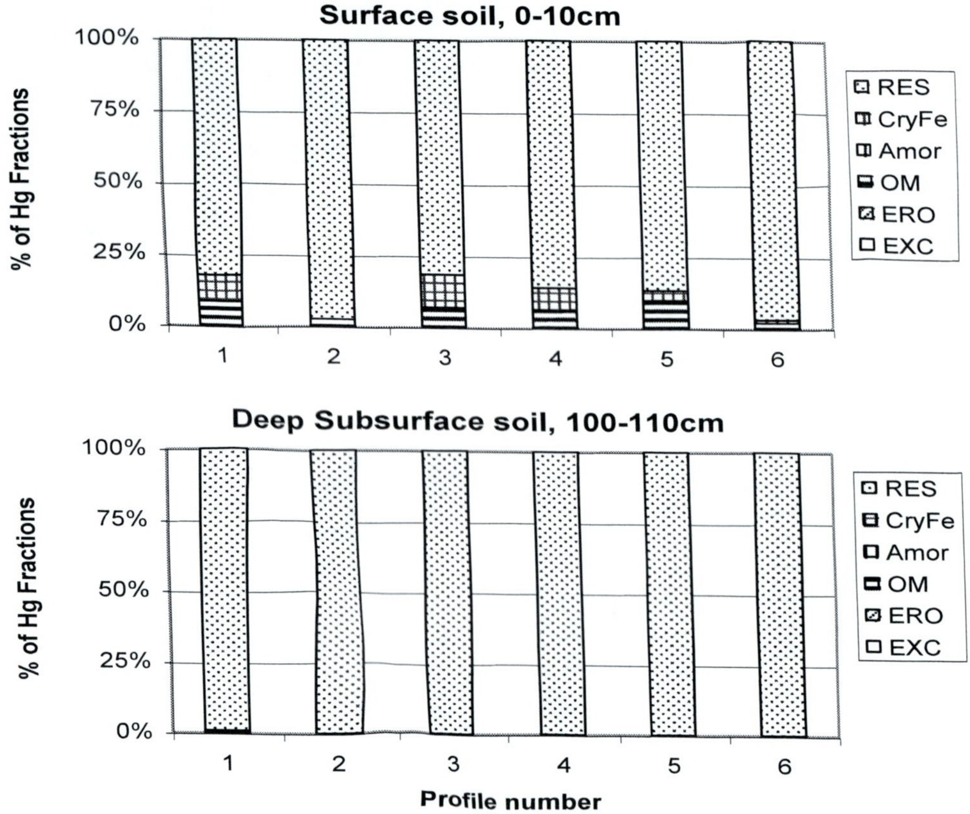

Figure 8. Mercury in solid-phase components of surface, subsurface and deep subsurface soils of woodland.

Mercury can occur as various species (e.g., HgS, Hg(II), methyl-Hg, Hg(O)), but the predominant form of mercury in the floodplain soils of the region was previously reported to be mercuric sulfide (cinnabar and meta-cinnabar forms) (NCEDR, 1996; Barnett et al., 1995). Barnett et al. (1995) used three different sequential extraction speciation schemes and found that mercury at the site was predominantly relatively insoluble mercuric sulfide or metallic Hg. They further confirmed the presence of meta-cinnabar, a form of mercuric sulfide, in site soils using X-ray and electron beam studies; this confirmed the first known evidence of authigenic mercuric sulfide formation in floodplain soils. Han et al. (2006) simulated the solubility and initial distribution /transformation of mercury in solid-phase components in floodplain soils freshly spiked with various forms of mercury. They found that mercury in floodplain soils freshly spiked with soluble mercury chloride and nitrate was dominated by the organic matter bound fraction, while mercury in soils spiked with mercury sulfide was dominated by the cinnabar form.

However, the present study indicates that the major mercury in the floodplain soils of EFPC after decades of remediation is mainly in non-cinnabar mercury bound in clay minerals. This form is not so stable as the cinnabar form, but it is more stable and less bioavailable than other solid-phase mercury components such as organic bound and Fe/Mn oxides bound mercury as discussed earlier.

Liu et al. (2006) reported that some floodplain soils from EFPC of Oak Ridge were dominated by the organic matter bound mercury fraction (50%). They further reported that the redoximorphic concentration component had higher Hg concentrations than the redoximorphic depletion component in the soil and Hg retained in the redoximorphic concentrations was less volatile and labile than Hg in the redoximorphic depletions possibly due to the strong binding affinity of Fe/Mn oxides and organic matter to Hg. As indicated by the present study, mercury bound in both iron oxide and organic matter bound fractions, which are more sensitive to environmental changes such as redox changes than other solid phase compounds, play an important role in controlling mercury solubility and bioavailability in the floodplain ecosystem.

4. Bioavailability of Mercury in Floodplain Soils to Native Earthworms

The total mercury in earthworms collected from floodplain soils of Oak Ridge, TN was in the range of 30-120 and 14-89 mg/kg for the mature and immature groups, respectively (Figure 9) (Han et al., 2012). The average mercury concentrations in both mature and immature groups were 68± 33 and 39 ±29 mg/kg, respectively. Concentrations of mercury in earthworms in contaminated floodplain soils of East Fork Poplar Creek were significantly higher than those in grassland where there is no mercury contamination (1.97± 0.46 mg/kg).

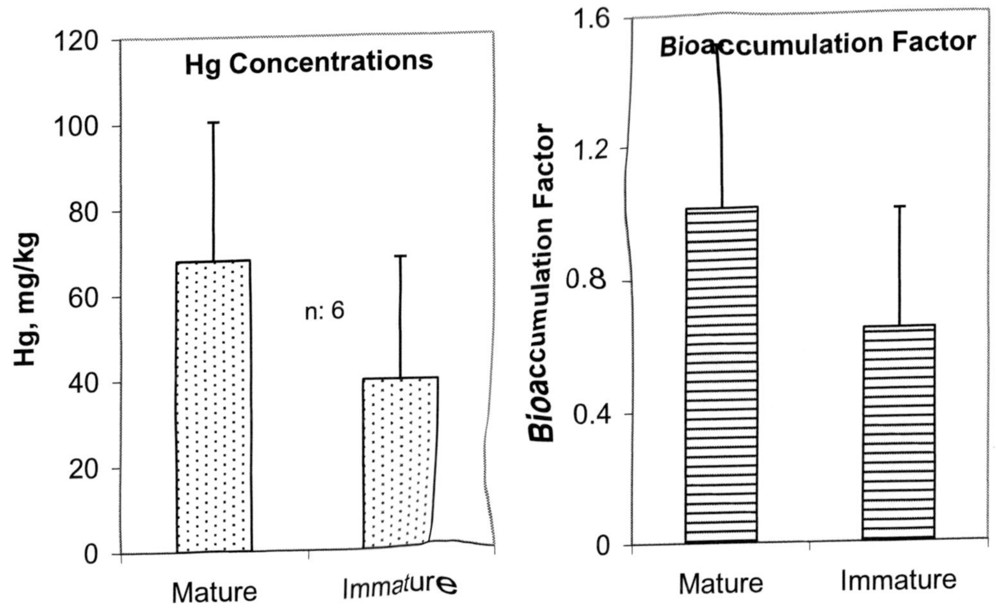

Figure 9. Concentrations and bioaccumulation factors of mercury in both mature and immature earthworms grown in contaminated floodplain field soils from Oak Ridge, TN. (Han et al., 2012).

The bioaccumulation factors (BF) (ratios of concentrations of mercury in earthworm to those in soil) of mercury in the mature group ranged from 0.47 to 1.75 with an average of 1.0 ±0.49 (Figure 9). However in the immature group of earthworm, BF factors were from 0.32 to 1.11 and averaged 0.64±0.36.

This indicates that mercury has been accumulated in the mature group and mature earthworms had almost double BF as the immature group. Burton et al. (2006) reported bioaccumulation factors for inorganic matter in earthworms were 0.6 to 3.3 and were usually < 1, but BFs for monomethylmercury ranged from 175 to 249. This may imply that most mercury in soils of floodplain of Oak Ridge site may be mainly inorganic mercury with limited amount of monomethylmercury.

Further we examined the contribution of mercury species in solid phases to bioavailability of mercury in earthworms (Figure 10). A strong linear relationship was found between mercury concentrations in earthworms (both mature and immature groups) and those in non-cinnabar mercury while no correlations between cinnabar mercury and mercury in earthworm were found (Figure 10). This clearly indicates that 4 M HNO_3 extractable non-cinnabar mercury posed higher bioavailability of mercury to earthworms than the stable cinnabar form in floodplain soils. It is interesting that no correlation between mercury concentrations in earthworm and mercury concentrations in other solid phase components was found.

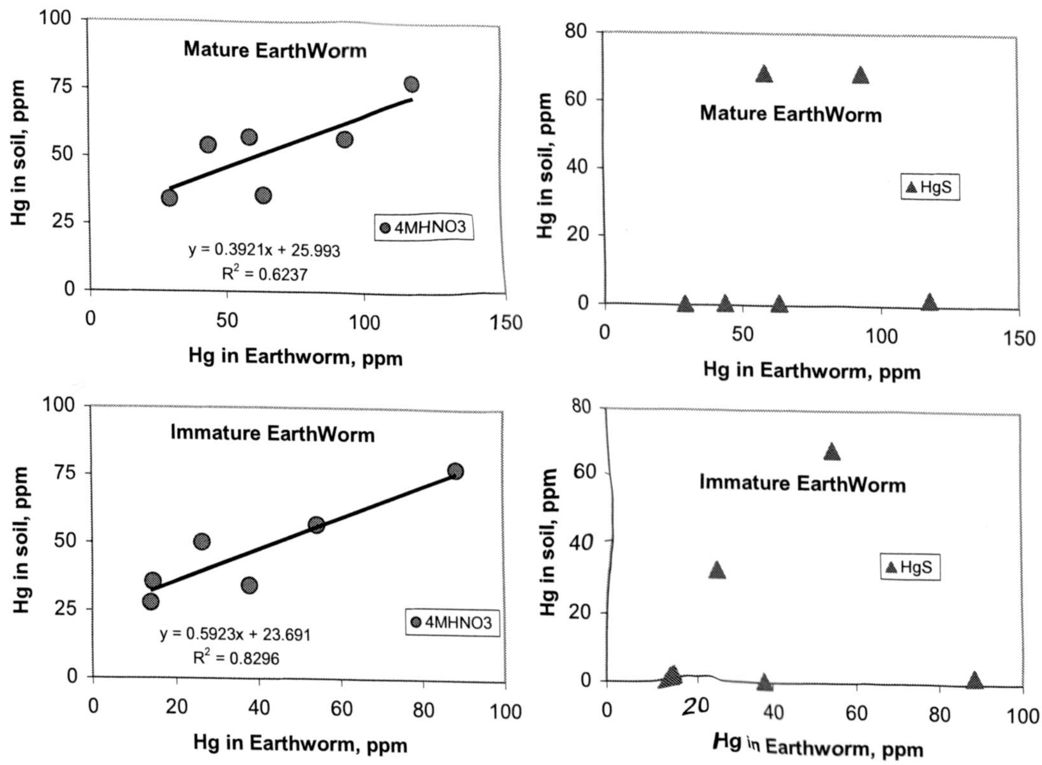

Figure 10. Relationship between concentrations of mercury in earthworms (both mature and immature) and non-cinnabar (with 4M HNO_3) and cinnabar mercury (with $HCl:HNO_3:H_2O$, 1:6:7) in surface floodplain soils from Oak Ridge, TN. (Han et al., 2012).

Earthworms play an important role in carbon recycling and maintenance of soil fertility and texture through decomposing organic residuals and mechanical mixing. They are essential to maintain good physical soil characteristics and balance soil aeration, water permeability, and mineral turnover. In addition, earthworms are key components in natural food chains, providing a food source for many small mammals and important food sources for small birds. The present study indicates that the native earthworms grown in the field mercury contaminated floodplain soils is a better estimate of mercury bioavailability and bioaccumulation in risk assessment than using freshly spiked contaminated soils as reported in the literature (Burton et al., 2006).

5. Extractability of Mercury in Cinnabar Spiked Floodplain Soil

Mercury sulfide is stable in soils under normal conditions. The solubility product constants of HgS is $10^{(-52)-(-54)}$ (Schuster, 1991). Thus, many common chemical extractants are not able to solubilize and extract mercury from pure HgS. Strong acids, such as 4M HNO_3 and 12 M HNO_3, only extracted <0.033% and <0.086% of Hg from commercial chemical HgS, respectively (Table 5, Figure 11).

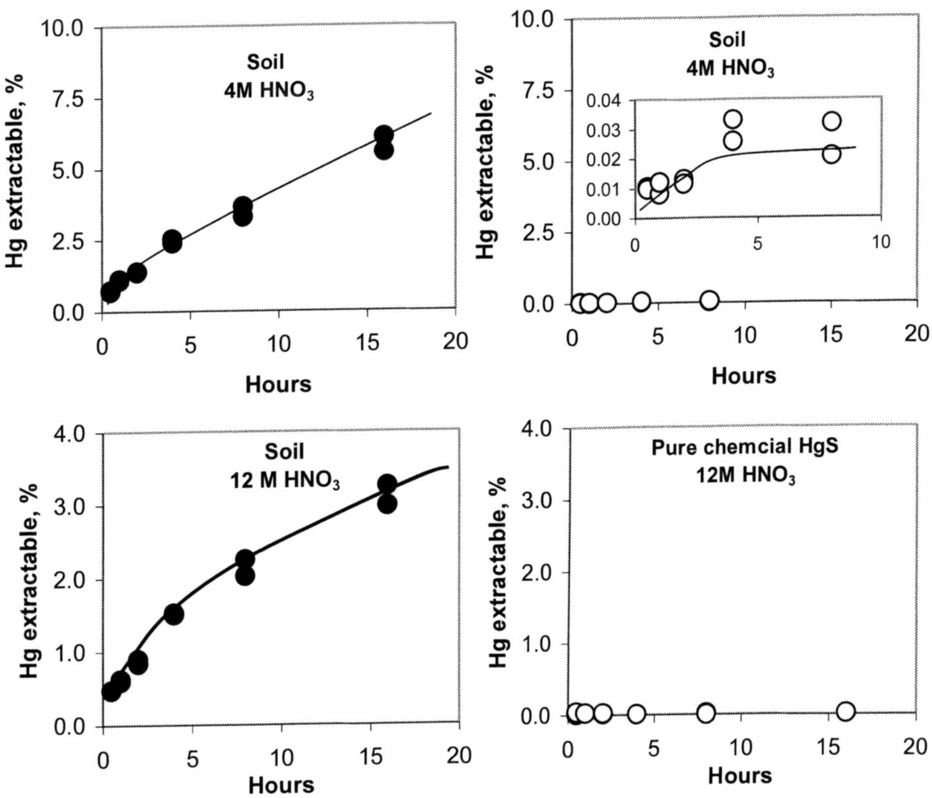

Figure 11. Kinetics of dissolution of HgS by 4M HNO_3 and 12 HNO_3 from cinnabar-contaminated Oak Ridge soils. (Han et al., 2008).

Table 5. Extractability of chemical HgS and Oak Ridge floodplain soils spiked with cinnabar by various chemical reagents (% of the HgS). (Han et al., 2008)

Reagents	Chemical HgS	HgS Contaminated Soils
Na_2S (saturated)		99-100%
$HCl:HNO_3:H_2O$, (v/v/v, 1:6:7)		98-100%
12 M HNO_3	0.001-0.086%	0.47-3.25%
4 M HNO_3	0.008-0.033%	0.69-6.08%
EDTA (0.05 M)		0.01%
H_2O_2		0.13-5.13%
NH_4OAc	0	0
$NH_2OH.HCl$, pH 2	0	0

However, the extractability of Hg from HgS-spiked Oak Ridge floodplain soil significantly increased (Han et al., 2008). Four M HNO_3 extracted 0.6-6.1% of mercury from cinnabar-contaminated soils, while 12 M extracted 3-4% of mercury from the soil. Extractability of HgS-contaminated soils after multiple seasons of planting was higher than that of pure chemical HgS. However, neutral salts, such as NH_4OAc, and other weak extractants (e.g., $NH_2OH \cdot HCl$, pH 2) did not extract HgS from pure chemical or from HgS-contaminated soils (Table 5, Figure 11).

Saturated solution of Na_2S extracted 99%-100% of HgS from cinnabar-contaminated Oak Ridge soils with 1000-2000 mg/kg Hg (Table 5). USEPA method 3200 (1:6:7 $HCl:HNO_3:H_2O$), which is recommended for extracting immobilized mercury from solid waste, extracted 98.9-100% of HgS from the current soil with 1000 mg/kg Hg (HgS). Revis *et al.* (1989) reported that a saturated solution of Na_2S extracted 98% of added HgS from soils and sediments while concentrated HNO_3 only extracted 1% of HgS. They suggested that saturated Na_2S is suitable for extracting HgS from soils and sediments.

The extractability by 4 M HNO_3 and 12 M HNO_3 of HgS from contaminated soils after multiple seasons of planting increased (Figure 11). The gradual increases in dissolution of mercury from cinnabar-contaminated soil by HNO_3 indicate that mercury has been transformed from the HgS form into other more soluble and HNO_3-extractable forms. It is noted that dissolution of Hg from pure chemical grade HgS also initially increased during the first four hours and thereafter its dissolution reached a plateau. The initial dissolution may be from a surface impure mercury form other than HgS. However, dissolution of HgS by 4 M and 12 M HNO_3 from contaminated soils continued to increase with time even after 16 hours. Various mechanisms including effects of dissolved organic matter and microorganisms may be involved with the stability of HgS; these will be discussed below.

A 4 M HNO_3 extraction is customarily used for extraction of micronutrients from soils (Lund *et al.*, 1976; Mullins and Sommers, 1986; Sposito *et al.*, 1982; Han and Banin, 1999). The average extractability of 4 M HNO_3 for other metals (Cu, Zn, Fe and Mn) from Mississippi Delta soils was between 76 to 85% of those by HF (Han *et al.*, 2007). This is consistent with reports by Han and Banin (1996) that 4 M HNO_3 extracted 76-100% of the total Mn and other metals extracted by HF from arid-zone soils. From HgS-contaminated Oak Ridge soil, 4 M HNO_3 should extract all non-cinnabar mercury (Han *et al.*, 2006).

6. Bioavailability of Mercury in Cinnabar-Spiked Floodplain Soil to Plants

Mercury in Oak Ridge floodplain soils spiked with insoluble HgS (1000 and 2000 mg/kg mercury) is to some extent bioavailable to plants (Table 6) (Han et al., 2008). Higher mercury concentrations were observed in both shoots and roots of fern and Indian mustard grown in HgS-contaminated soils than in control soils.

Mercury concentrations in fern and Indian mustard grown in control soils after a growing season were mostly higher than those found in plants from uncontaminated soils. This may be due to direct foliage adsorption of mercury vapor from air by plants. Direct foliage absorption of mercury vapor from the air was observed in trees and mosses (Lindberg *et al.*, 1991). Direct uptake of elemental Hg via stomata is controlled by stomatal and mesophyll resistances (Lindberg *et al.*, 1991). Therefore mercury in ferns/Indian mustard grown in the control soils may have contributions from direct absorption of airborne mercury vapor, which originated from volatilization of mercury from soils as well as from fern foliages of fern in soils with soluble mercury.

Table 6. Mercury concentrations in plants grown in Oak Ridge floodplain soils spiked with HgS (Han et al., 2008)

Plant (variety)	Year	Treatment HgS		Shoots (mg/kg)	Roots (mg/kg)
Pteris vittata	Spring, 2004	CK	Avg	4.1	7.4
(48 Days growing)			Stdev	1.6	2.9
		1000	Avg	14.3	36.9
			Stdev	5	5.8
		2000	Avg	10.6	45.5
			Stdev	1.6	8.7
Pteris vittata	Spring, 2005	CK	Avg	0.79	
(60 Days growing)			Stdev	1.76	
		1000	Avg	13.35	132.38
			Stdev	14.36	75.58
		2000	Avg	24.17	296.80
			Stdev	24.71	228.89
Indian mustard	Spring, 2006	CK	Avg	0.20	
(Long standing)			Stdev	0.32	
(53 Days growing)		1000	Avg	26.78	84.01
			Stdev	25.24	39.35
		2000	Avg	15.18	205.39
			Stdev	23.24	45.52
Indian mustard	Spring, 2006	CK	Avg	0.95	
(Florida broadleaf)			Stdev	1.07	
(53 Days growing)		1000	Avg	34.65	17.14
			Stdev	29.16	11.12
		2000	Avg	78.77	86.96
			Stdev	50.77	42.68

However, mercury concentrations in both shoots and roots of ferns and Indian mustard grown in HgS-treated soils were significantly higher than those in control soils, indicating some bioavailability of mercury in soils treated with HgS and that the major contribution was through root uptake. A dual mechanism of mercury uptake by alfalfa has been observed: roots accumulated Hg in proportion to the soil levels and aerial plant shoots absorbed Hg vapor directly from the atmosphere (Lindberg *et al.*, 1979).

There are a number of mechanisms triggering solubility of mercury in HgS-contaminated soil. Soil dissolved organic carbon increases dissolution of mercury from HgS (Ravichandran *et al.*, 1998; Ravichandran, 2004), thus increasing mercury bioavailability in soils. Hydrophobic acid (a mixture of both humic and fulvic acids) dissolved more mercury than hydrophilic acids and other nonacid fractions of dissolved organic matter (Ravichandran, 2004). The possible mechanisms of dissolution of HgS were suggested to include surface complexation of mercury and oxidation of surface sulfur species by the organic matter (Ravichandran, 2004). On the other hand, mercury in Oak Ridge soils contaminated with insoluble HgS (1000 and 2000 mg/kg mercury) was mainly distributed in cinnabar form (HgS) (86-97%) (Han *et al.*, 2006). However, H_2O_2 still extracted 0.13 - 5% of mercury from HgS-contaminated Oak Ridge soils. The H_2O_2 extractable Hg was mostly from soil organically bound Hg present/or transferred in soils. The contribution from oxidization of HgS during the extraction of the organically bound fraction is minimum (Han *et al.*, 2006). Also, as the present study indicated, EDTA increases solubility of mercury from HgS-contaminated soils. Other inorganic (chloride and sulfate) and organic ligands (salicylic acid, acetic acid, cysteine) were not found to enhance the dissolution of mercury from the mineral cinnabar (Ravichandran, 2004). This indicates the positive effects of the rhizosphere on bioavailability and stability of HgS in soils. Plant roots can alter their microenvironment and mobilize otherwise stable Hg compounds.

CONCLUSION

The present study clearly shows that the total mercury in floodplain soils of EFPC significantly decreased after a series of remediation, especially after excavation of the highly contaminated soils/sediments along the creek. The major mercury in the current floodplain soils of EFPC decades after remediation is mainly in non-cinnabar mercury bound in clay minerals. This form is not so stable as cinnabar form, but less bioavailable than other solid-phase components. The results also show linear relationships between mercury concentrations in earthworms (both mature and immature groups) and non-cinnabar mercury form, which is more bioavailable than cinnabar mercury to earthworms. Native earthworms may be used as a potential mercury ecological bio-indicator (bio-marker) for demonstrating mercury bioavailability and ecotoxicity. Long-term monitoring mercury bioavailability and speciation in floodplain soils is of importance since these soils act as both a medium-term sink and a long-term source to downstream mercury. The long-term stability, mobility and bioavailability of mercury contaminants in these floodplains still need to be monitored continuously and closely.

Our study also shows bioavailability of mercury in cinnabar spiked Oak Ridge floodplain soils to some extent after multiple seasons of planting. Extractability of HgS by 4M HNO_3

and 12 M HNO_3 in cinnabar-spiked Oak Ridge floodplain soils was significantly higher than that of pure HgS. The extractability of HgS by HNO_3 increased with extraction time. Planting seems to increase the extractability and bioavailability of mercury from HgS-spiked soil. In addition, saturated Na_2S and $HCl:HNO_3:H_2O$ (v/v/v 1:6:7) solutions are suitable for extraction of HgS (cinnabar form) from contaminated soils and wastes. The current study may help our understanding of the recent increase of mercury levels in water of Lower East Fork Popular Creek (LEFPC) of Oak Ridge. Further studies on effects of rhizosphere chemistry and soil minerals on solubility and bioavailability of floodplain Oak Ridge soils are needed.

REFERENCES

Barnett, M.O., Harris, L.A., Turner, R.R., Henson, T.J., Melton, R.E., Stevenson, R.J., (1995). Characterization of mercury species in contaminated floodplain soils. *Water Air and Soil Pollution* 80, 1105-1108.

Biester, H., Scholz, C., (1997). Determination of mercury binding forms in contaminated soils: Mercury pyrolysis versus sequential extractions. *Environmental Science and Technology* 31, 233-239.

Burton, D.T., Turley, S.D., Fisher, D.J., Green, D.J., Shedd, T.R., (2006). Bioaccumulation of total mercury and monomethylmercury in the earthworm *Eisenia Fetida*. Water Air and Soil Pollution 170, 37-54.

Han, F.X., Banin, A. (1996). Solid-phase manganese fractionation changes in saturated arid-zone soils: Pathways and kinetics. *Soil Science Society of America Journal 60*, 1072-1080.

Han, F.X., Banin, A. (1999). Long-term transformations and redistribution of potentially toxic heavy metals in arid-zone soils. II: Under the field capacity regime. *Water, Air and Soil Pollution 114*, 221-250.

Han, F.X., Banin, A., (1997). Long-term transformations and redistribution of potentially toxic heavy metals in arid-zone soils. I: Under saturated conditions. *Water Air and Soil Pollution* 94, 399-423.

Han, F.X., Banin, A., Kingery, W.L., Triplett, G.B., Zhou, L.X., Zheng, S.J., Ding, W.X., (2003). New approach to studies of redistribution of heavy metals in soils. *Advances in Environmental Research* 8(1), 113-120.

Han, F.X., Banin, A., Su, Y., Monts, D.L., Plodinec, M.J., Kingery, W.L., Triplett, G.B., (2002). Industrial age anthropogenic inputs of heavy metals into the pedosphere. *Naturwissenschaften* 89, 497-504.

Han, F.X., Kingery, W.L., Hargreaves, J.E., Walker, T.W. (2007). Effects of land uses on solid-phase distribution of micronutrients in selected vertisols of the Mississippi River Delta. *Geoderma 142*, 96-103.

Han, F.X., Patterson, W.D., Xia, Y., Sridhar, B.B.M., Su, Y., (2006b). Rapid determination of mercury in plant and soil samples using inductively coupled plasma atomic emission spectroscopy, a comparative study. *Water Air and Soil Pollution* 170, 161-171.

Han, F.X., Shiyab, S., Chen, J., Su, Y., Monts, D.L., Waggoner, C.A., Matta, F.B., (2008). Extractability and bioavailability of mercury from a mercury sulfide contaminated soil from Oak Ridge, Tennessee, USA. *Water Air and Soil Pollution* 194, 67-75.

Han, F.X., Su, Y., Monts, D.L., Waggoner, G.A., Plodinec, M.J., (2006a). Binding, Distribution, and Plant Uptake of Mercury in a Soil from Oak Ridge, Tennessee, USA. Science of Total Environment 368, 753-768.

Han, F.X., Y. Su, Z. Shi, Y. Xia, W. Tian, V. Philips, D.L. Monts, (2012). Mercury distribution and speciation in floodplain soils and uptake into native earthworms (Diplocardia spp.). *Geoderma* 170 (2012) 261–268.

Lindberg, S.E., Jackson, D.R., Huckabee, J.W., Janzen, S.A., Levin, M.J., Lund, J.R. (1979). Atmospheric emission and plant uptake of mercury from agricultural soils near the Almaden mercury mine. *Journal of Environmental Quality 8*, 572-578.

Lindberg, S.E., Turner, R.R., Meyers, T.P., Taylor Jr. G.E., & Schroeder, W.H. (1991). Atmospheric concentrations and deposition of Hg to a deciduous forest at Walker Branch Watershed, Tennessee, USA. *Water Air and Soil Pollution 56*, 577-594.

Liu, G., Cabrera, J., Allen, M., Cai, Y., (2006). Mercury characterization in a soil sample collected nearby the DOE Oak Ridge Reservation utilizing sequential extraction and thermal desorption method. *Science of Total Environment* 369, 384-392.

Loar, J.M., (2004). State of East Fork Poplar Creek: Status of ecological recovery. US Department of Energy Mercury Workshop, Oak Ridge, TN., November, 2004.

Lund, L.J., Page A.L., Nelson, C.O. (1976). Movement of heavy metals below sewage disposal ponds. *Journal of Environmental Quality 5*, 330-334.

Mullins, G.L., Sommers, L.E. (1986). Characterization of cadmium and zinc in four soils treated with sewage sludge. *Journal of Environmental Quality 15*, 382-387.

NCEDR, (1996). Workshop on Decision-Making Related to the Clean-up of Mercury Contamination at Lower East Fork Poplar Creek, Oak Ridge, TN, August 14, 1996.

Olsen, C.R., Larsen, I.L., Lowry, P.D., Moriones, C.R., Ford, C.J., Dearstone, K.C., (1992). Turner RR, Kimmel BL, Brandt CC. Transport and accumulation of cesium-137 and mercury in the Clinch River and Watts Bar Reservoir System. Oak Ridge National Laboratory, ESD Publication 3471, Oak Ridge, TN, 1992.

Pant, P., Allen, M., (2007). Interaction of soil and mercury as a function of soil organic carbon: Some field evidence. Bulletin of *Environmental Contamination and Toxicology* 78, 539-542.

Philips, E., (2004). Upper East Fork Poplar Creek Watershed. US Department of Energy Mercury Workshop, Oak Ridge, TN., November, 2004.

Ravichandran, M. (2004). Interactions between mercury and dissolved organic mater - a review. *Chemosphere 55*, 319-331

Ravichandran, M., Aiken, G.R., Redd, M.M., Ryan, J.N. (1998). Enhanced dissolution of cinnabar (mercuric sulfide) by dissolved organic matter isolated from the Florida Everglades. *Environmental Science and Technology 32*, 3305-3311.

Sposito, G., Lund, L.J., Chang, A.C., (1982). Trace metal chemistry in arid-zone field soils amended with sewage sludge: I. Fractionation of Ni, Cu, Zn, Cd, and Pb in solid phases. *Soil Science Society of America Journal* 46, 260-264.

Turner, R.R., Olsen, C.R., Wilcox, W.J. Jr., (1985). Environmental fate of Hg and 137 Cs discharged from Oak Ridge facilities, in: Hemphill, D.D., (Ed). Trace Substances in the Environment. Elsevier/North-Holland Biomed. Press, New York.

US Department of Health and Human Services, (2004). Public health assessment, Y-12 Uranium release, Oak Ridge Reservation (USDOE), Oak Ridge, Anderson County, Tennessee, http://www.atsdr.cdc.gov/hac/PHA/oakridgey12/oak_toc.html.

USDOE, (1998). Report on the Remedial Investigation of the Upper East Fork Poplar Creek Characterization Area at the Oak Ridge Y-12 Plant, Oak Ridge, Tennessee, Volume 1, DOE/OR/01-1641/V1&D2, pp. 3-10 to 3-106. http://newweb.ead.anl.gov/techcon/Projects/mercury/description/.

USEPA, (1989). National Priorities List for Uncontrolled Hazardous Waste Sites. Federal Register 54 (223), 48184 – 48189.

USEPA, (2008). National Listing of Fish Advisories. General Fact Sheet: 2008 National Listing. http://www.epa.gov/fishadvisories/advisories/fs2008.html#listing.

USGS, (2000). Mercury in the Environment, Fact Sheet 146-00. http://minerals.usgs.gov/mercury.

In: Floodplains
Editor: Enner Herenio Alcântara

ISBN: 978-1-62808-998-1

Chapter 2

FLOODPLAINS OF REGULATED RIVERS: PLANT-SOIL INTERACTIONS IN CHANGING ECOSYSTEMS

Takashi Asaeda [*,1,2] ***and M.H. Rashid*** [†,2,3]

[1]Institute for Environmental Science and Technology, Saitama University, Sakura-ku, Saitama, Japan

[2]Department of Environmental Science, Saitama University, Sakura-ku, Saitama, Japan

[3]Bangladesh Agricultural University, Mymensingh, Bangladesh

ABSTRACT

River floodplains are one of the most degraded ecosystems in the current world. Historically human civilization has been centered on rivers. Therefore, due to their close proximity with civilization, rivers receive the first pulse of pressure induced by human intervention. Naturally, a river is a highly dynamic ecosystem. The dynamics of this system depend on flooding disturbances. Frequent flooding keeps the river floodplain habitable for particular types of flora and fauna. The substrates are usually coarse and nutrient limited. Therefore, plant population is few and only plants having special adaptation can grow in this habitat. However, in the last few decades many rivers throughout the world have undergone flow regulation due to flood control, electricity generation, etc. Since the flow regulation has been started, the environments of those river floodplains have been changing. In many cases, successional patterns of herbs and trees have changed and exotic plant invasion has been reported. Though researchers and river administrators have been continuously seeking the reasons associated with these ecosystem changes, in many cases it has been explained by river water eutrophication, flow regulation, dams and weirs construction, etc. However, whatever the human role might be in river floodplain degradation, many researchers reported the direct link

* Corresponding author. Address: Institute for Environmental Science and Technology, Saitama University, 255 Shimo-okubo, Sakura-ku, Saitama, 338-8570, Japan. Tel.: +81-48-858-3563 Fax: +81-48-858-9574; Email: asaeda@mail.saitama-u.ac.jp.

† Email: mhrashid@bau.edu.bd.

between forestation and abrupt shift of stoichiometric balance of substrate. In this article, we will be discussing the patterns and associated factors related to plant-soil interactions in regulated rivers' floodplains with special emphasis on Japanese rivers.

Keywords: Riparian vegetation, dam, invasion, ecosystem

INTRODUCTION

The dynamic interaction between water and land is the principal process that produced river-floodplains, maintains them, and has affected the adaptations of biota that have evolved therein (Bayley, 1995). The ecosystems evolved on the land water interface of a floodplain, i.e. the riparian ecosystem are unusually complex, dynamic, and diverse and possess numerous economic, societal, and biological values (Sharitz *et al.*, 1992). These qualities make them key ecosystems for preserving biodiversity (Naiman *et al.*, 1993) and for understanding how environmental change may affect interactions between adjacent landscape elements (Decamps, 1993).

Apart from the biodiversity value, a healthy floodplain plant com-munity provides many other ecosystem services. Maintaining healthy riparian plant communities, particularly along rivers and streams, is critical for modulating nutrient inputs, moderating water temperature, regulating flow, erosion control and providing other ecological and aesthetic services (Palink *et al.*, 2000; Richardson, 2008). Riparian vegetation also serves as dispersal corridors and habitats for birds, amphibians, and other organisms (Naiman *et al.*, 1993; Richardson and Danehy, 2007; Welsh and Droege, 2001).

River floodplains are disturbance-dominated ecosystems characterized by high levels of habitat diversity and biota adapted to exploit the spatio-temporal heterogeneity (Décamps, 1996; Mitsch and Gosselink, 1993; Petts, 1989; Salo *et al.*, 1986; Ward and Stanford, 1995; Welcomme, 1985). The fluvial actions of flooding and channel migration create a shifting mosaic of habitat patches across the riverine landscape. Ecotones, connectivity and succession play major roles in structuring the spatio-temporal heterogeneity leading to the high biodiversity that characterizes river floodplains. Flood events alter the morphological condition of the river channel. At the same time, floods create large disturbances in riparian habitats (Nilsson, 1987) by washing away herbs and trees (Asaeda *et al.*, 2011b; Asaeda and Rajapakse, 2008; Gomes and Asaeda, 2009) and dispersing their propagules (Barsoum, 2001; Karrenberg *et al.*, 2002). Thus, flooding plays significant role in shaping and maintaining the complex mosaic of vegetation in riparian habitats (Tockner *et al.*, 1998). Since the riparian vegetation occurs in the ecotone between land and water, many bio-physical factors such as geomorphology, microclimate, soil water, nutrients, and disturbance regimes are at a transition, which influence the plant community (Lamb and Mallik, 2003; Stewart and Mallik, 2006). Consequently, the potential for affecting riparian habitats by changing natural discharge regimes with flow regulation will be high near stream banks and it will decrease as one moves away from the water's edge.

Fluvial Dynamics and Ecological Succession of Natural Rivers

The riparian plant communities along natural rivers are subjected to periodic flooding that varies in frequency and intensity (Mallik and Richardson, 2009). Flood regimes influence riparian ecosystems both directly and indirectly. When floods inundate floodplain soils, for example, the oxygen available to plant roots is depleted quickly. Different plant species have specific tolerances for oxygen stress (Kozlowski, 1984). The duration of oxygen stress associated with the duration of flooding or "hydroperiod" can thus directly influence the composition and productivity of riparian vegetation species and communities (Richter and Richter, 2000). Floods indirectly shape riparian ecosystems on meandering rivers through their influence on sediment erosion and deposition. Floods build and reshape floodplains by driving the lateral migration of river channels, effecting cutoffs of meander bends, and eroding and depositing sediments on the floodplain surface (Scott *et al.*, 1996; Shankman, 1993; Stromberg *et al.*, 1991). These geomorphic changes have significant implications for the successional dynamics of riparian ecosystems (Amoros *et al.*, 1987; Bravard *et al.*, 1986; Malanson, 1993; Medley, 1992). Therefore, periodic flooding destroys and creates habitats for plant colonization contributing to species adaptive traits and richness in the riparian area (Hibbs and Bower, 2001; Mallik *et al.*, 2001; Rood *et al.*, 1998; Rood *et al.*, 2003a; Rood *et al.*, 1995).

The spatio-temporal heterogeneity of riverine floodplains is attained through successional processes (Décamps, 1996; Salo *et al.*, 1986; Terborgh and Petren, 1991; Ward and Stanford, 1995). Native plants and animals that occupy floodplain environments typically have life histories that are coordinated with the seasonality of river flows, so the loss of the natural patters impedes growth and reproduction (Johnson, 2000; Karrenberg *et al.*, 2002; Lytle and Poff, 2004).

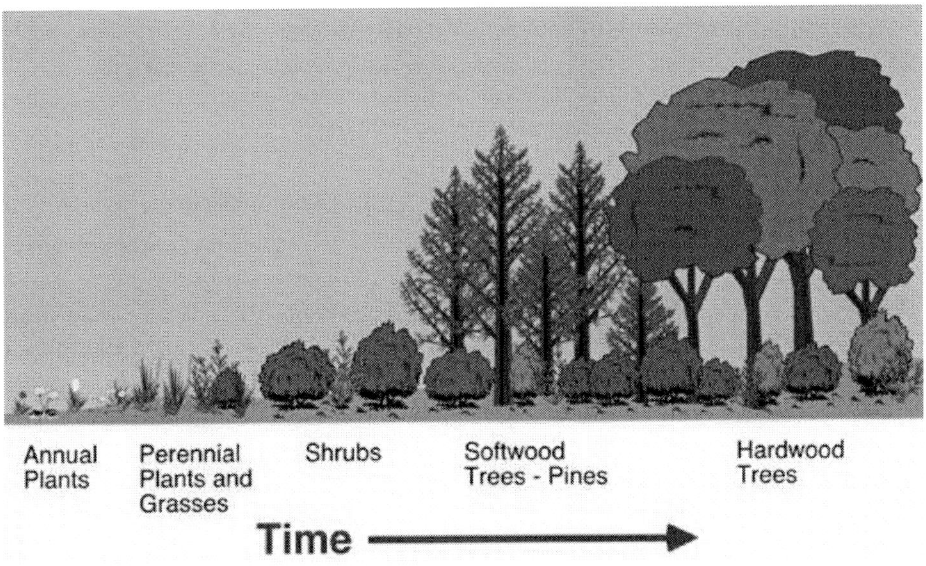

Figure 1. A simple scheme of succession of plant species (Pidwirny, 2006).

Fluvial dynamics and channel migration maintain a diversity of successional stages among the riparian vegetation. Salo *et al.* (1986) and Terborgh and Petren (1991) provide vivid descriptions of the role of natural disturbance by fluvial action in creating a mosaic of alluvial forest stands in different successional stages. Forests on the concave bends of laterally migrating rivers are undercut by erosion and primary succession is initiated on point bars of alluvium deposited as annual increments on convex bends. The mature forest stage on the floodplain of the Manu River in Upper Amazonia has high structural diversity (five vertical strata) and\200 species of trees per hectare (Terborgh and Petren, 1991). Therefore, human-induced changes in flooding regimes from damming and other water development can have large consequences for the ecological integrity of riparian ecosystems (Poff *et al.*, 1997; Richter *et al.*, 1997).

Suppression of Natural Disturbance

The regulation of streams and rivers has been integral to human population growth and technological innovation (Dudgeon, 2000; King *et al.*, 2003; Poff and Hart, 2002; Tharme, 2003). Humans have built millions of dams or that kind in 5000 years and most of them have been constructed in the last 100 years (WCD-World Commision on Dams, 2000). Various forms of river regulation, including dam construction, modification of natural flow dynamics, dredging, channel straightening, bank stabilization and construction of artificial levees, have altered successional trajectories and disrupted connectivity between ecological units of the floodplain aquifer complex in many of the world's rivers (Dudgeon, 2000; Manral *et al.*, 2012; Poff and Hart, 2002). Regulation has fragmented river systems (Dynesius and Nilsson, 1994). River regulation disrupts the natural disturbance regimes that maintain a diversity of successional stages and high levels of connectivity across the riverine landscape, resulting in a loss of habitat heterogeneity and biodiversity. Therefore, the construction of dams and weirs and other forms of flow regulations have a significant impact on the downstream environment of river.

Poff and Hart (2002) listed the impacts of dam construction on riverine ecology in the following way:

a) Dams alter the downstream flux of water and sediment, which modifies biogeochemical cycles as well as the structure and dynamics of aquatic and riparian habitat.
b) They change water temperatures, which influences organismal bioenergetics and vital rates.
c) Dams create barriers to upstream-downstream movement of organisms and nutrients, which hinders biotic exchange.

However, in this chapter, we will only be reviewing the responses of plant-soil system to the river regulation.

The environmental degradation as a result of flow regulation can be carried out in many ways such as alteration of sediment dynamics and changes of river morphology (Friedman *et al.*, 1998; Grant *et al.*, 2003; Ligon *et al.*, 1995) and the hydrological alterations. In case of hydrological alteration, the frequency and magnitude of floods decrease and there is an increase in water flow in times of drought (Johnson *et al.*, 1976; Naiman and Décamps, 1997; Shafroth *et al.*, 2002a; Walker, 1985). The impacts of river regulation are initially subtle, progressive change over years or decades often produce severe cumulative ecological impacts (Ligon *et al.*, 1995; Nilsson and Svedmark, 2002; Rood and Mahoney, 1990).

Figure 2. Urbanized Ara River, Tokyo.

Figure 3. Weirs on Tama River, Tokyo.

Succession, Competition and Composition of Plants under Regulated Flow Regime

The ecosystem of riparian landscapes is dynamic. Under natural conditions, large disturbances occur due to flooding (Brunet and Astin, 2000), which shapes the vegetative structure (Müller, 1995). In such conditions, poor colonists or long-lived species are excluded from the system and these species are dominated by small, short-lived species (Naiman *et al.*, 2005). However, at low levels of disturbance, as for example in a stream bar (Figure 4), sedimentation and erosion of sand and gravel by stream flow are repeated in the rain and typhoon season every year. These small disturbances prevent the establishment of seedlings while producing new safe sites for them (Sakio, 1997). Therefore, in cases of regulated rivers, where low levels of disturbances are most common, competitive interactions will result in lower diversity due to exclusion of species. Such communities often are dominated by large, relatively long-lived species (Naiman *et al.*, 2005; Ward and Stanford, 1983). Thereby, riparian landscapes are regarded as important corridors for plant dispersal (Johansson *et al.*, 1996; Naiman and Décamps, 1997) and play a central role in the processes of invasion and naturalization of exotic plants (Pyšek and Prach, 1993). Many successful invaders in natural vegetation were first observed in floodplains.

Figure 4. Gravel bars are a common feature of regulated rivers.

Today, human disturbances, such as dam construction, flow stabilization and gravel excavation, have altered natural disturbance regimes in most rivers. River regulation has varying effects on the downstream environments, particularly on the sediment regime, and therefore the reported changes of vegetation succession and competition are abundant. The importance of sediment quantities of the river channel (Asaeda and Rajapakse, 2008; Steiger and Gurnell, 2003; Steiger *et al.*, 2001; Steiger *et al.*, 2005) as well as characteristics of sediment (Braatne *et al.*, 1996; Scott *et al.*, 1996) in establishing riparian vegetation are well known. Sediment characteristics fundamentally affect the moisture and nutrient regimes that are so important for young plants, although too much silt accumulation or removal can adversely impact some plants after establishment (Langlades and Décamps, 1995).

Long term sediment cut-off makes incision in the downstream channel, coarser rive-bed, segregation of bed materials (Gurnell and Petts, 2002; Pizzuto, 2002; Steiger *et al.*, 2001), armoring of bed coat (Grant *et al.*, 2003), and the desiccation of riparian vegetation (Steiger *et al.*, 2001) in the channel. These processes eventually lower the channel-bed and increase the elevation difference between riparian floodplain and river flow. Low moisture, low nutrient levels and coarse sediment in the post-dam construction period are known therefore inhibit colonization of terrestrial flora in the channel (Steiger *et al.*, 2001; Stromberg *et al.*, 2007; van Oorschot *et al.*, 1998), while in the floodplain, low frequency of inundation promotes the thick colonization of terrestrial plants.

Figure 5. Temporal variation of vegetation in Tama River floodplain, Tokyo. The top photo was taken in 1974, whearas the bottom photo was taken in 2001 in the same bar.

Despite coarsening and incision of riverbed following upstream dams and weirs construction (Pizzuto, 2002), some works reported that segregated fine sediment can accumulate on floodplain habitats (Gurnell and Petts, 2002; Steiger *et al.*, 2001). Some researchers (Asaeda *et al.*, 2010; Asaeda *et al.*, 2009; Azami *et al.*, 2004; Naiman *et al.*, 1998) observed that downstream river channels became vegetated after dam construction.

In a field study, Asaeda *et al.* (2011d) investigated the competition and succession of two invasive lianas in relation to soil and plant characters, and flooding regime in the floodplain of a regulated river (Tama river) in Japan. The Tama River, originating from the mountains in the western district of Tokyo, exemplifies problems that are typical of Japanese rivers. The construction of the upstream Ogochi dam in 1957 (Steele, 2006) substantially reduced the flood volume. The frequency and depth of submergence of the floodplain was reduced. Several weirs constructed along the river to allow only fine sediments to the downstream. These regulations lead to an intense accumulation of fine sediments along the floodplain (Owens *et al.*, 1999; Richards *et al.*, 2002; Steiger *et al.*, 2001). As a result, the floodplain has gradually become vegetated since the 1960s. The most dominant liana species in the floodplain is *Sicyos angulatus*. It was first sighted in Japan in 1952 (NIAES, 2000) and in the Tama River floodplain in 1977 (Okuda *et al.*, 1979). In the recent past, many colonies of *S. angulatus* have been replaced by another liana, *P. lobata*. Today, these two species are apparent competitors, and dominance can switch suddenly from one species to the other. In their study Asaeda *et al.* (2011d) collected data on soil particle size, soil moisture content, organic matter content, total carbon (TC), total nitrogen (TN), total phosphorus (TP), K, Ca, Mg, Cu and Zn of soils of tree sites along Tama river. They also estimated plant parameters viz., aboveground biomass (AGB) and belowground biomass (BGB) of *P. lobata* and *S. angulatus*.

Figure 6. Invasive growth of Pueraria lobata in Tama river floodplain, Tokyo.

Figure 7. Seedling of Sicyos angulatus growing on fine soil.

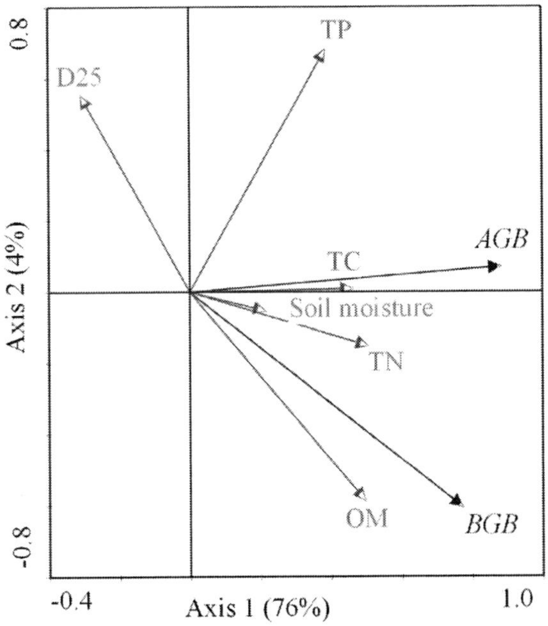

Figure 8. Redundancy analysis (RDA) bi-plot diagram for *Pueraria lobata*[1].

Figure 8 depicts the relationship between biomass and soil factor for *Pueraria*. From the figure, it can be seen that soil particle size (D_{25}) has no effect on the AGB of *P. lobata*.

[1] Solid-line arrows [total carbon (TC), total nitrogen (TN), total phosphorus (TP), soil particle size (D25), potassium (K), calcium (Ca), magnesium (Mg), copper (Cu), zinc (Zn), sodium (Na), soil organic matter (OM), soil moisture content (Moisture)]are explanatory variables. Dotted arrows [above-ground biomass (AGB) and below-ground biomass (BGB)] are response variables.

However, BGB has negative correlation with soil particle size, though the relationship is not significant. This species has very strong and deep root system and it has the capability to penetrate soil of any texture. Therefore, its biomass is not dependent on soil type. *S. angulatus*, on the other hand, has strong correlation with both AGB and BGB (Figure 9). Likewise, soil organic matter and soil moisture are positively correlated with the biomass of *S. angulatus*. Regarding *P. lobata*, soil moisture and organic matter have much correlation with BGB than that of AGB. Thus, higher soil OM, soil moisture and fine soil particle favors the growth of *P. lobata*, however, it can grow in any kind of soil. Total nitrogen (TN) of soil is equally correlated with AGB and BGB of *P. lobata*, though AGB is highly dependent on soil TP. From RDA, it is clear that soil TP is more important to *P. lobata* than TN, whereas, for *S. angulatus*, soil TN is limiting nutrient.

Phosphorus is a very important macro nutrient for symbiotic legumes, since it is directly involved in the nitrogen fixation process of the *Rhizobium* bacteria (Tsvetkova and Georgiev, 2003). In P limited soil, the numbers of nodules become reduced and ultimately the plant suffers from N deficiency if the soil is not enriched with inorganic N. Therefore, for *P. lobata*, P is more important than N particularly in N deficient soil.

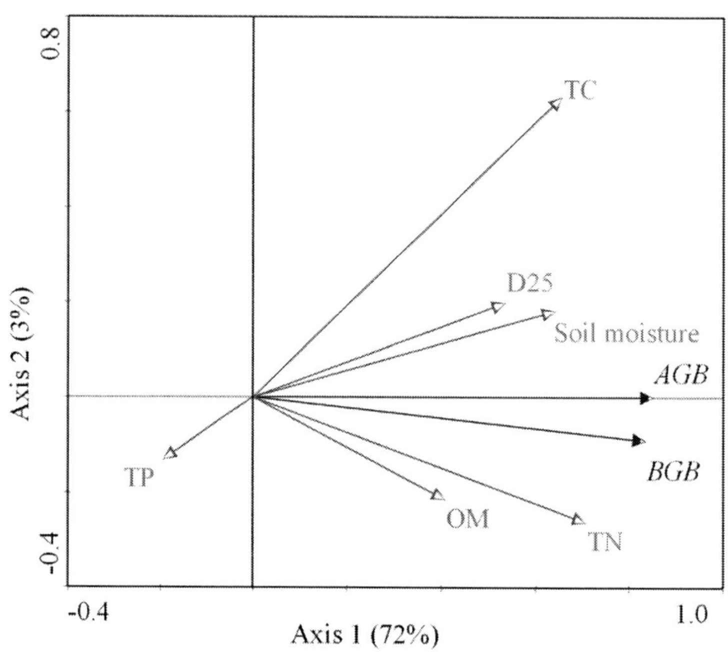

Figure 9. Redundancy analysis (RDA) bi-plot diagram for *Sicyos angulatus*[2].

The original invaders of the floodplain were *Miscanthus sacchariflorus* and *Phragmites japonica*, which tolerate coarse-sediment sites with sterile conditions (Asaeda and Rajapakse, 2008; Neukirchen *et al.*, 1999). *P. lobata* colonized underneath existing *Miscanthus* and

[2] Solid-line arrows [total carbon (TC), total nitrogen (TN), total phosphorus (TP), soil particle size (D25), potassium (K), calcium (Ca), magnesium (Mg), copper (Cu), zinc (Zn), sodium (Na), soil organic matter (OM), soil moisture content (Moisture)]are explanatory variables. Dotted arrows [above-ground biomass (AGB) and below-ground biomass (BGB)] are response variables.

Phragmites. Since the sediment was still too coarse for *P. lobata*, it did not dominate. However, colonization by *M. sacchariflorus* gradually alters sediments into finer matter by the accumulation of fine wash load during floods and the decomposition of organic matter (Asaeda *et al.*, 2009). Instead of *P. lobata*, which was a relatively slower invader, pioneering annuals such as *Humulus japonicus* and *Ambrosia trifida* seemed to colonize before *S. angulatus*. The large stolons and rhizomes of *P. lobata* mean that it takes a long period for this plant to develop a colony. After the sediment became sufficiently fine and fertile, *S. angulatus* invaded. With its small resource allocation to roots and a great ability to expand vines, it rapidly widened its territory. However, small roots are easily washed away in floods. In contrast, *P. lobata* sustained itself and spread after the flood. The accumulation of fine sediment is accelerated inside a colony during submergence in floods. The dead leaves of *P. lobata* accumulated annually on the sediment surface and were gradually decomposed by microbial activities under relatively wet conditions shaded by overlying canopies. In addition, symbiosis with *Rhizobium* may enhance the nitrogen supply to *P. lobata*, and its leaves and root produce allelopathic substances (Rashid *et al.*, 2010a; Rashid *et al.*, 2010b). These provide further advantages to *P. lobata* over *S. angulatus*. However, if the floodplain is deeply submerged, the surface layer is washed out with all vegetation on it.

Both soil factors and flooding regimes affected the succession and competition of liana species in the Tama River floodplain. Soil factors appeared to be the primary determinant of the pioneering species' colonization. Flooding events governed the deposition of wash load and changed habitat characteristics: after the pioneering species colonized, the soil was modified and other species invaded the study locations. The result of interspecific competition depended on the morphological characteristics of the interacting species. However, severe flooding could change the species composition at any time, after which the succession process restarts anew.

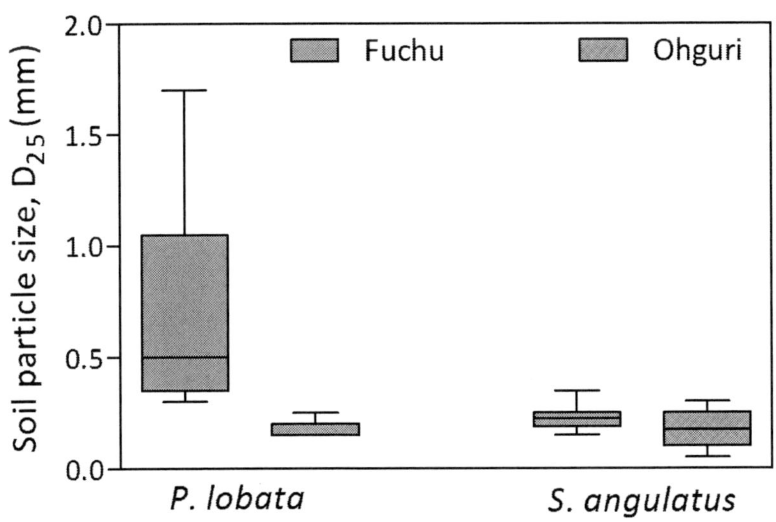

Figure 10. Preference of P. lobata and Sicyos angulatus for soil particle size.

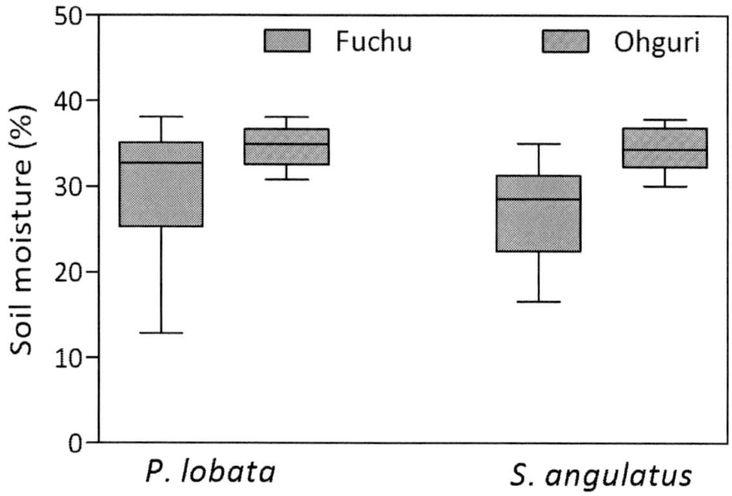

Figure 11. Preference of P. lobata and Sicyos angulatus for soil moisture.

Figure 12. Vine plants add s substantial amount of organic matter into the soil.

In a separate study based on volunteer monitoring data on the same river (Tama river) floodplain, Uchida *et al.* (2012) reported the association of *Sicyos angulatus* with other riparian species as controlled by major floods. The monitoring was carried out by volunteer participation of local residents. The authors determined the association of *S. angulatus* colonies as influenced by environmental variables, viz. soil depth and frequency of major floods by detrended correspondence analysis (DCA). The results revealed that *Miscanthus sacchariflorus* was the most frequent (60.4%) understory herbaceous species found in association with *S. angulatus* colonies. The DCA coordination triplot gave that the association of *S. angulatus* changed after a major flood and its understory layer was replaced by *Lolium* X *hybirdum*, *Phalaris arundinacea* and *Bromus catharticus*. *M. sacchariflorus* was almost absent in the understory group, and the soil depth of quadrats became deeper than it was in the pre-flood condition. They concluded that *S. angulatus* were strongly invading *M. sacchariflorus*, and that influence in combination with both the invasion of *S. angulatus* and the disturbance from flooding caused *M. sacchariflorus* colonies to change to pasture. Uchida *et al.* (2012) also suggested that all plants in this study, with a high frequency of appearance in *S. angulatus* colonies, exert an influence on emergent plant communities. It is also explainable that *S. angulatus* and other alien plants associated with native lianas, *P. lobata* and *H. japonica* influence native emergent plants strongly in riparian areas of the Tama River.

Okuda *et al.* (1995) reported that *S. angulatus* in the Tama River floodplain often forms a pure community (mono-stand) with almost no mixture with other plants. In fewer cases, however, other plants such as *P. lobata*, *H. japonica*, *C. japonica*, *M. sacchariflorus* and *P. australis* have been found to grow together with *S. angulatus*. Other invasive plants, viz. *A. trifida*, *S. altissima*, etc., sometimes grow in *S. angulatus* communities. Okuda *et al.* (1995) also reported that *S. angulatus* communities grew in eutrophic soil around the convergence point of tributaries of Tama River.

Phragmites japonica, a pioneering perennial species on floodplains in East Asia, has an extremely high ability to expand its territory by extending stolons for more than 10 m in a year with its high plasticity (Satake *et al.*, 2002). As floods accumulate washed coarse sediment on the surface of floodplain, the soil of flood plain is unsuitable for *P. japonica*, as moisture and nutrients in the surface sediment layer are very limited. Little moisture retained in the coarse sediment apparently restricts the growth of *P. japonica*, as the uptake of the same and nutrients is highly limited (Asaeda *et al.*, 2009; Lippert *et al.*, 1999; Moriuchi and Winn, 2005). Moisture content in fine sediment is higher than in coarse sediment as water vapour transfer is far smaller for fine beds compared with coarse beds (Asaeda and Ca, 1993). Therefore, sandy sediment sites are preferable for the growth of *P. japonica*, compared with stony sites. Asaeda *et al.* (2011a) developed an organ specific growth model to predict the annual growth of *P. japonica* in different habitats. The translocation rates were provided by functions of the environmental conditions such as habitat grain size. Using the observed environmental data, they simulated the growth patterns with a suitable agreement.

Aboveground biomass of *P. japonica* is easily removed subjected to floods and thereafter the photosynthesis is curtailed throughout the rest growth periods of the year, although the secondary shoots re-grow if the flood occurs early in the growing season (Asaeda and Rajapakse, 2008; Asaeda *et al.*, 2006). Subject to the annual floods, the annual maximum AGB and BGB gradually decrease year by year. In three years the annual maximum AGB and BGB reduced by 59% and 75% in sandy site and 68% and 76% in stony site, respectively by July floods. The AGB and BGB reduced by 18% and 39% in sandy site and 32% and 49% in stony site, respectively by September floods. There was no major reduction by December floods either in sandy or stony sites. The effect of floods was largest in case that they occur in July when *P. japonica* is in active growth stage, while floods in the senescence season do not affect the growth.

Subjected to biennial floods, long term biomass reduction rate was much smaller than the case of annual floods. Although BGB substantially reduces in the year when exposed to a flood, it recovers considerably in the next year as production continues throughout the year without interruption. In the third year, therefore, the upward translocation from the BGB in the second year and resulting AGB do not differ much from that of the first year. Year after year variation of the annual maximum aboveground biomass differs between the cases of July biennial and September biennial floods, although it gradually decreases in a long period.

When exposed to July flood, AGB of *P. japonica* is curtailed when the plant is still growing, therefore, the annual maximum AGB substantially decreases in the same year. However, if it undergoes flood after the growing period of the plant, the BGB is affected rather than the annual maximum AGB. Therefore, the flood affects the next year AGB rather than the annual maximum AGB of the year (Figure 15 and Figure 16).

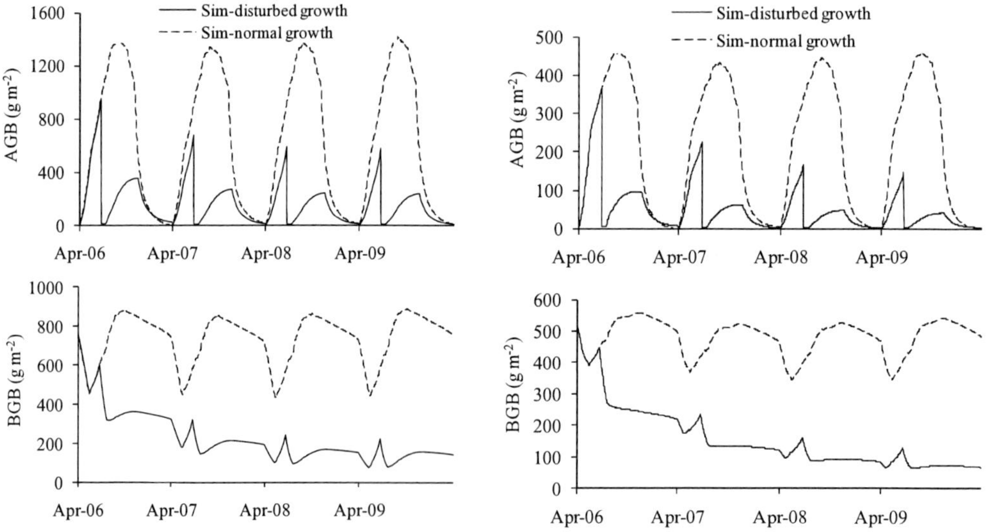

Figure 13. Comparison of simulation trends of annual flood disturbed growth with normal growth at (a) sandy site (b) stony site (Asaeda *et al.*, 2011a).

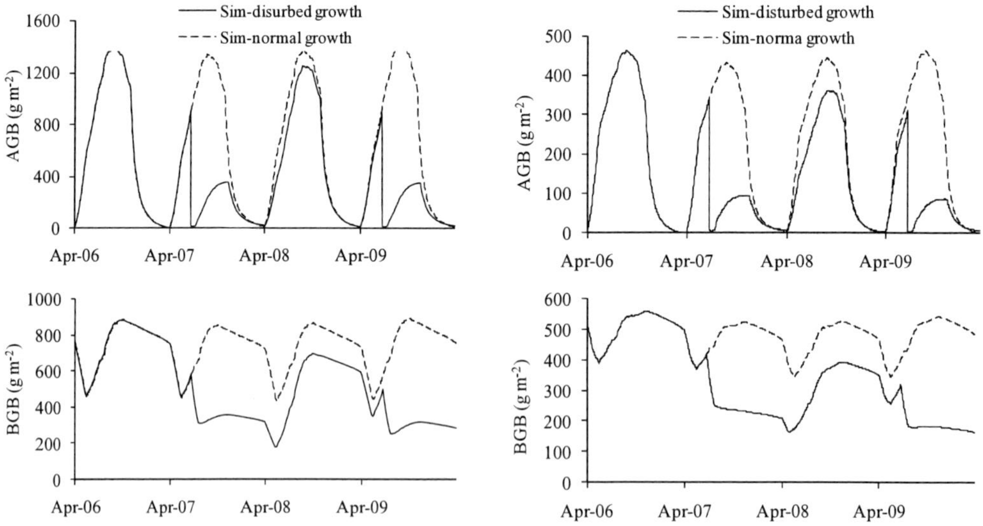

Figure 14. Comparison of simulation trends of biennial flood disturbed growth with normal growth at (a) sandy site (b) stony site (Asaeda *et al.*, 2011a).

Due to the reduced supply of photosynthesis products from the aboveground organs, the rhizome biomass gradually decreases by the losses of respiration and mortality (Asaeda *et al.*, 2006; Hai *et al.*, 2006). Then the aboveground biomass in the next year also decreases as the early growth is supported by the translocation of resources from the rhizome system (Fiala, 1973; Kvet, 1971). *P. japonica* stands subjected to annual floods, therefore, gradually shrinks year after year, particularly if floods occur when the plants are in active growth stage. Substrate texture changes due to flood flows. Thus, in case of low flow and shallow inundation, the substrate texture changes from coarse to fine, as fine sediment accumulates (Steiger *et al.*, 2001). In such cases, the growth of *P. japonica* is enhanced. Subjected to high flow and deep inundation, fine sediments wash away, remaining coarse sediments on the ground surface. The growth of *P. japonica* is thus disturbed and its stands shrink.

Mallik and Richardson (2009) studied tree growth up- and downstream of dams constructed on three temperate rivers of British Columbia and reported that tree canopy cover of western red cedar and red alder was significantly lower in downstream reaches of all the three rivers compared to the upstream reaches. Other studies (Hibbs and Bower, 2001; Shafroth *et al.*, 2002b; Swanson *et al.*, 1998) also supported their findings that the flow regulation in the downstream sections results in very few of the disturbed patches necessary for creating favorable seedbeds for tree colonization. Mallik and Richardson (2009) also concluded that the absence of tree canopy will have significant effects on downstream riparian species richness and diversity, as well as physical, chemical and biological characteristics of the rivers. In the same study, they reported a significant increase of herbaceous vegetation coverage in the downstream reaches.

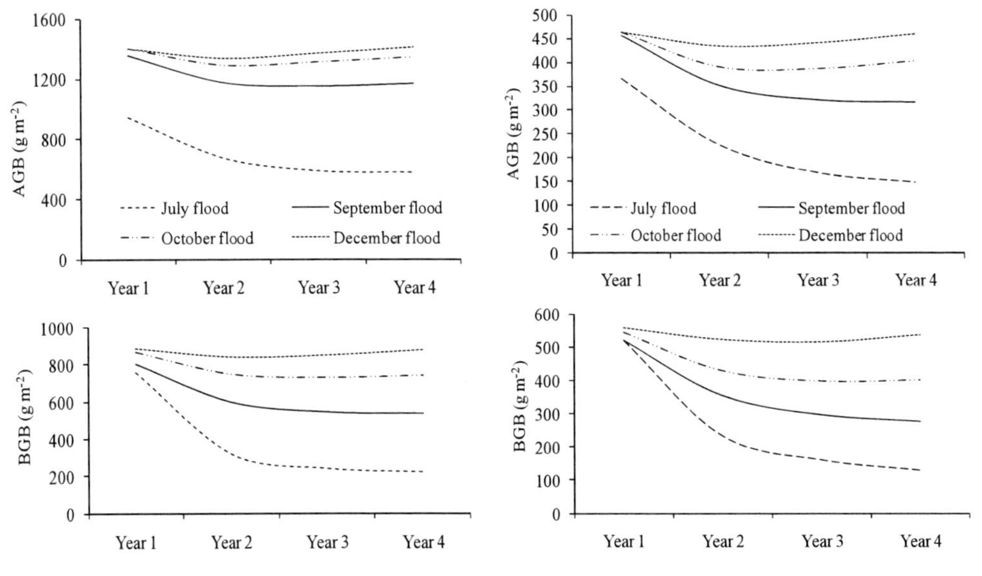

Figure 15. Simulated maximum biomass of P. japonica as affected by different annual flooding time at (a) sandy site (b) stony site (Asaeda et al., 2011a).

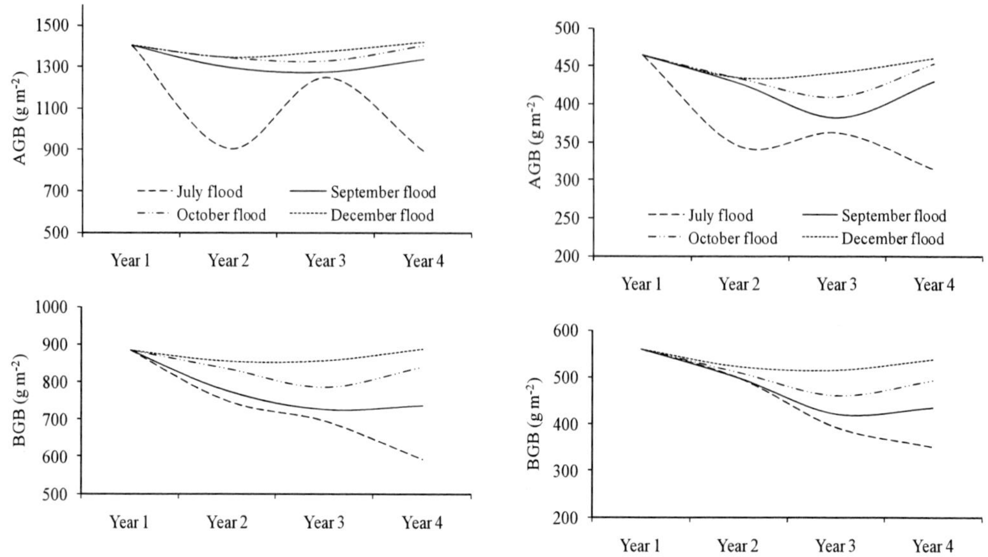

Figure 16. Simulated maximum biomass of P. japonica as affected by different biennial flooding time at (a) sandy site (b) stony site (Asaeda et al., 2011a).

SEDIMENT RELEASE AND FLOODPLAIN VEGETATION

Dams accumulate nearly all of the imported sediments other than the wash load (Klaver *et al.*, 2007; Tedoru and Wehrli, 2005; Walker, 1985) and the sediment storage decreases both the reservoir capacity and the operating efficiency of the dam. In contrast, a condition called "sediment-shadow" is created downstream where sediment-starved flows commonly erode channel boundaries and create long-term channel instabilities (Asaeda *et al.*, 2010; Wohl and Rathburn, 2003).

Therefore, sediments are deliberately released from the reservoir through a "sediment gate" or by other means such as bypassing, dredging, flushing, sluicing, and upstream sediment trapping (Liu *et al.*, 2004), in an attempt to either reduce the downstream channel instability or to increase reservoir capacity. Once sediment is released from the upstream dam, the nutrient stoichiometry of the riparian zone, especially that of N and P, shifts to a new equilibrium (Asaeda *et al.*, 2011c) and sediment is accumulated on the riverbed. As a result, one more plant species might out-colonize the others or new plant may invade the place being favored by habitat characteristics or soil nutrient condition (Asaeda *et al.*, 2011c). Following sediment release from the dam, the geomorphology of downstream channel changes (Acker *et al.*, 2008; Maingi and Marsh, 2002) and its adjustment to increase sediment influx depends on the magnitude, frequency, duration, and grain-size distribution of the sediment release, and on the downstream channel characteristics (Wohl and Rathburn, 2003). Sediment may be accumulated selectively at sites of locally reduced transport capacity, such as pools (Wilcock *et al.*, 1996) or edges of sand bars, and thus the inundation pattern of a channel during the wet season or a flash flood might change. Generalized sediment accumulation throughout a channel results in fining-in streambed grain-size distribution (Wilcock *et al.*, 1996), whereas the sediment particle size, moisture, and nutrient status of higher sites vary due to the lack of annual inundation (Asaeda *et al.*, 2011c).

Figure 17. Sediment Flushing of Reservoir Xiao Langdi, China (Photo source: The China Times).

Figure 18. Sediment deposited on bar of Kurobe River, Japan.

Figure 19. Progress of vegetation in downstream of Kurobe River, Japan after sediment release from upstream dams.

Asaeda and Rashid (2012) studied downstream sediment bars of Kurobe River, Japan to see the effects of sediment released from its upstream dams on herbaceous vegetation encroachment and to investigate the edaphic factors involved in the process. They reported that biomass of herbaceous plants on the downstream bars has substantially increased after starting sediment release and had a high correlation with the fraction of fine sediments rather than with the moisture content. More than 90% of the total nitrogen (TN) and total phosphorus (TP) in the substrate were in an organic form, which originated from the nutrients of the sediment accumulated in the reservoir. They also stated that the limiting factor for plant growth on the bar was TN rather than TP. With the accumulation of organic nitrogen from the reservoir, the nitrogen limitation had a less significant role in plant colonization through litter mineralization, which accelerates the colonization speed of vegetation. They concluded that sediment release from the dam enhances TN in the sediment and minimizes sediment size, which allows the intensive colonization of vegetation downstream.

CONCLUSION

Restoration and rehabilitation processes are long term undertakings (Buijse *et al.*, 2002) and the inherent complexity of floodplain ecosystems and the limited scientific understanding of the processes constraints the planning of restoration (Adams and Perrow, 1999; Brookes *et al.*, 1996; Buijse *et al.*, 2002).

Several authors reported adverse effects of regulated flow of dammed rivers on downstream riparian plant communities and their successful mitigation by using riparian Recruitment Box Model (Richter and Richter, 2000; Rood *et al.*, 1998; Rood *et al.*, 2003b; Rood *et al.*, 1995; Rood *et al.*, 2005). Recruitment Box Model is developed by combining the timing of tree seedling recruitment and establishment with flow pattern associated with river elevation and flow ramping (gradual flow decline after flood peak) that allow successful tree seedling establishment (Amlin and Rood, 2002; Kalischuk *et al.*, 2001).

Mallik and Richardson (2009) also attested to the suggestion of Rood *et al.* (2005) that for the existing dams, mitigation of downstream ecological impacts can be achieved by a better coordinated flow regulation using Recruitment Box Model and for the new dams well informed ecological impact assessment and mitigation measures should be incorporated at their design and construction phases followed by appropriate flow regulation and downstream biophysical monitoring. Using Indicators of Hydrologic Alteration (IHA) and Range Variability Approach (RVA), Hu *et al.* (2008) have shown that flow pattern of dammed rivers can be brought back to near normal by restoring the range of intra- and inter-annual variation in hydrologic regimes to protect native biodiversity and other aquatic values. Efforts to restore riparian habitats and other riverine ecosystem have also included the management of flow release downstream of dams to more closely mimic natural flow (Poff *et al.*, 1997).

REFERENCES

Acker SA, Beechie TJ, Shafroth PB. 2008. Effects of a Natural Dam-Break Flood on Geomorphology and Vegetation on the Elwha River, Washington, U.S.A. *Northwest Science* 82: 210-223. DOI. citeulike-article-id:9205986.

Adams WM, Perrow MR. 1999. Scientific and institutional constraints on the restoration of European floodplains. *Geological Society, London, Special Publications* 163: 89-97. DOI. 10.1144/gsl.sp.1999.163.01.07.

Amlin NM, Rood SB. 2002. Comparative tolerances of riparian willows and cottonwoods to water-table decline. *Wetlands* 22: 338-346. DOI. 10.1672/0277-5212(2002)022 [0338:CTORWA]2.0.CO;2.

Amoros C, Roux AL, Reygrobellet JL, Bravard JP, Pautou G. 1987. A method for applied ecological studies of fluvial hydrosystems. *Regulated Rivers: Research & Management* 1: 17-36. DOI. 10.1002/rrr.3450010104.

Asaeda T, Baniya MB, Rashid MH. 2011a. Effect of floods on the growth of *Phragmites japonica* on the sediment bar of regulated rivers: a modelling approach. *International Journal of River Basin Management* 9: 211-220. DOI. 10.1080/15715124.2011.613837.

Asaeda T, Ca V. 1993. The subsurface transport of heat and moisture and its effect on the environment: A numerical model. *Boundary-Layer Meteorology* 65: 159-179. DOI. citeulike-article-id:6897487.

Asaeda T, Gomes PIA, Sakamoto K, Rashid MH. 2011b. Tree colonization trends on a sediment bar after a major flood *River Research and Applications* DOI: 10.1002/rra.1372. DOI. citeulike-article-id:6893555.

Asaeda T, Gomes PIA, Takeda E. 2010. Spatial and temporal tree colonization in a midstream sediment bar and the mechanisms governing tree mortality during a flood event. *River Research and Applications* 26: 960-976. DOI. 10.1002/rra.1313.

Asaeda T, Rajapakse L. 2008. Effects of spates of different magnitudes on a *Phragmites japonica* population on a sandbar of a frequently disturbed river. *River Research and Applications* 24: 1310-1324. DOI. 10.1002/rra.1128.

Asaeda T, Rajapakse L, Manatunge J, Sahara N. 2006. The effect of summer harvesting of *Phragmites australis* on growth characteristics and rhizome resource storage. *Hydrobiologia* 553: 327-335. DOI. citeulike-article-id:8704658.

Asaeda T, Rashid MH. 2012. The impacts of sediment released from dams on downstream sediment bar vegetation. *Journal of Hydrology* 430-431: 25-38. DOI. 10.1016/ j.jhydrol.2012.01.040.

Asaeda T, Rashid MH, Kotagiri S, Uchida T. 2011c. The role of soil characteristics in the succession of two herbaceous lianas in a modified river floodplain. *River Research and Applications* DOI: 10.1002/rra.1374. DOI. citeulike-article-id:6893560.

Asaeda T, Rashid MH, Kotagiri S, Uchida T. 2011d. The role of soil characteristics in the succession of two herbaceous lianas in a modified river floodplain. *River Research and Applications* 27: 591-601. DOI. 10.1002/rra.1374.

Asaeda T, Siong K, Kawashima T, Sakamoto K. 2009. Growth of *Phragmites japonica* on a sandbar of regulated river: morphological adaptation of the plant to low water and nutrient availability in the substrate. *River Research and Applications* 25: 874-891. DOI. 10.1002/rra.1191.

Azami K, Suzuki H, Toki S. 2004. Changes in riparian vegetation communities below a large dam in a monsoonal region: Futase Dam, Japan. *River Research and Applications* 20: 549-563. DOI. 10.1002/rra.763.

Barsoum N. 2001. Relative contributions of sexual and asexual regeneration strategies in *Populus nigra* and *Salix alba* during the first years of establishment on a braided gravel bed river. *Evolutionary Ecology* 15: 255-279. DOI. 10.1023/A:1016028730129.

Bayley PB. 1995. Understanding Large River- Floodplain Ecosystems. *BioScience* 45: 153-158.

Braatne JH, Rood SB, Heilman PE. 1996. Life history, ecology, and conservation of riparian cottonwoods in North America. In *Biology of Populus and Its Implications for Management and Conservation,* Stettler RF, Bradshaw HD, Heilman PE, Hinckley TM (eds). NRC Research Press, National Research Council of Canada: Ottawa, Ontario.

Bravard J-P, Amoros C, Pautou G. 1986. Impact of Civil Engineering Works on the Successions of Communities in a Fluvial System: A Methodological and Predictive Approach Applied to a Section of the Upper Rhône River, France. *Oikos* 47: 92-111. DOI. 10.2307/3565924.

Brookes A, Baker J, Redmond C. 1996. Flooplain restoration and riparian zone management. In *River Channel Restoration: Guiding Principles for Sustainable Projects,* Brookes A, Shields Jr FD (eds). John Wiley & Sons: Chichester; 75-101.

Brunet RC, Astin B. 2000. A 12-month sediment and nutrient budget in a floodplain reach of the River Adour, southwest France. *Regulated Rivers: Research & Management* 16: 267-277. DOI. citeulike-article-id:6894043.

Buijse AD, Coops H, Staras M, Jans LH, Van Geest GJ, Grift RE, Ibelings BW, Oosterberg W, Roozen FCJM. 2002. Restoration strategies for river floodplains along large lowland rivers in Europe. *Freshwater Biology* 47: 889-907. DOI. 10.1046/j.1365-2427.2002.00915.x.

Decamps H. 1993. River Margins and Environmental Change. *Ecological Applications* 3: 441-445. DOI. 10.2307/1941913.

Décamps H. 1996. The renewal of floodplain forests along rivers: a landscape perspective. *Verhandlungen Internationale Vereinigung Limnologie* 26: 35-59.

Dudgeon D. 2000. Large-Scale Hydrological Changes in Tropical Asia: Prospects for Riverine Biodiversity. *BioScience* 50: 793-806. DOI. 10.1641/0006-3568(2000)050[0793:LSHCIT]2.0.CO;2.

Dynesius M, Nilsson C. 1994. Fragmentation and Flow Regulation of River Systems in the Northern Third of the World. *Science* 266: 753-762. DOI. 10.1126/science.266.5186.753.

Fiala K. 1973. Growth and production of underground organs of *Typha angustifolia* L., *Typha latifolia* L. and *Phragmites australis* Trin. *Polish Archives of Hydrobiologia* 20: 59-66. DOI. citeulike-article-id:8704667.

Friedman JM, Osterkamp WR, Scott ML, Auble GT. 1998. Downstream effects of dams on channel geometry and bottomland vegetation: Regional patterns in the great plains. *Wetlands* 18: 619-633. DOI. citeulike-article-id:10142020.

Gomes PIA, Asaeda T. 2009. Spatial and temporal heterogeneity of *Eragrostis curvula* in the downstream flood meadow of a regulated river. *Annales de Limnologie* 45: 181-193. DOI. 10.1051/limn/2009015.

Grant GE, Schmidt JC, Lewis SL. 2003. A Geological Framework for Interpreting Downstream Effects of Dams on Rivers. In *A Peculiar River: Geology, Geomorphology, and Hydrology of the Deschutes River,* O'Conner JE, Grant GE (eds). American Geophysical Union: Washington, DC, USA; 203-219.

Gurnell AM, Petts GE. 2002. Island-dominated landscapes of large floodplain rivers, a European perspective. *Freshwater Biology* 47: 581-600. DOI. 10.1046/j.1365-2427.2002.00923.x.

Hai D, Asaeda T, Manatunge J. 2006. Latitudinal effect on the growth dynamics of harvested stands of *Typha*: A modeling approach. *Estuarine, Coastal and Shelf Science* 70: 613-620. DOI. citeulike-article-id:5721867.

Hibbs DE, Bower AL. 2001. Riparian forests in the Oregon Coast Range. *Forest Ecology and Management* 154: 201-213. DOI. http://dx.doi.org/10.1016/S0378-1127(00)00623-X.

Hu W-w, Wang G-x, Deng W, Li S-n. 2008. The influence of dams on ecohydrological conditions in the Huaihe River basin, China. *Ecological Engineering* 33: 233-241. DOI. http://dx.doi.org/10.1016/j.ecoleng.2008.04.003.

Johansson M, Nilsson C, Nilsson E. 1996. Do Rivers Function as Corridors for Plant Dispersal? *Journal of Vegetation Science* 7: 593-598. DOI. citeulike-article-id:7231790.

Johnson WC. 2000. Tree recruitment and survival in rivers: influence of hydrological processes. *Hydrological Processes* 14: 3051-3074. DOI. 10.1002/1099-1085(200011/12)14:16/17<3051::AID-HYP134>3.0.CO;2-1.

Johnson WC, Burgess RL, Keammerer WR. 1976. Forest Overstory Vegetation and Environment on the Missouri River Floodplain in North Dakota. *Ecological Monographs* 46: 59-84. DOI. 10.2307/1942394.

Kalischuk AR, Rood SB, Mahoney JM. 2001. Environmental influences on seedling growth of cottonwood species following a major flood. *Forest Ecology and Management* 144: 75-89. DOI. http://dx.doi.org/10.1016/S0378-1127(00)00359-5.

Karrenberg S, Edwards PJ, Kollmann J. 2002. The life history of Salicaceae living in the active zone of floodplains. *Freshwater Biology* 47: 733-748. DOI. citeulike-article-id:9200528.

King J, Brown C, Sabet H. 2003. A scenario-based holistic approach to environmental flow assessments for rivers. *River Research and Applications* 19: 619-639. DOI. 10.1002/rra.709.

Klaver G, van Os B, Negrel P, Petelet-Giraud E. 2007. Influence of hydropower dams on the composition of the suspended and riverbank sediments in the Danube. *Environmental Pollution* 148: 718-728. DOI. 10.1016/j.envpol.2007.01.037.

Kozlowski TT. 1984. Flooding and plant growth. Academic Press: San Diego, California.

Kvet J. 1971. Growth analysis approach to the production ecology of reedswamp plant communities. *Hydrobiologia* 12: 15-40. DOI. citeulike-article-id:8704701.

Lamb EG, Mallik AU. 2003. Plant species traits across a riparian-zone/forest ecotone. *Journal of Vegetation Science* 14: 853-858. DOI. 10.1111/j.1654-1103.2003.tb02218.x.

Langlades LR, Décamps O. 1995. Silt accumulation and plant colonization along a gravel bar. *Comptes Rendus de l'Académie des Sciences* 318: 1073-1082.

Ligon F, Dietrich W, Trush W. 1995. Downstream Ecological Effects of Dams. *BioScience* 45: 183-192. DOI. citeulike-article-id:6959753.

Lippert I, Rolletschek H, Kühl H, Kohl J-G. 1999. Internal and external nutrient cycles in stands of *Phragmites australis* – a model for two ecotypes. *Hydrobiologia* 408-409: 343-348. DOI. citeulike-article-id:8704685.

Liu J, Minami S, Otsuki H, Liu B, Ashida K. 2004. Prediction of Concerted Sediment Flushing. *Journal of Hydraulic Engineering* 130: 1089-1096. DOI. citeulike-article-id:6897538.

Lytle DA, Poff NL. 2004. Adaptation to natural flow regimes. *Trends in ecology & evolution (Personal edition)* 19: 94-100.

Maingi J, Marsh S. 2002. Quantifying hydrologic impacts following dam construction along the Tana River, Kenya. *Journal of Arid Environments* 50: 53-79. DOI. citeulike-article-id:9205943.

Malanson GP. 1993. *Riparian Landscapes*. Cambridge University Press: Cambridge, UK.

Mallik A, Richardson J. 2009. Riparian vegetation change in upstream and downstream reaches of three temperate rivers dammed for hydroelectric generation in British Columbia, Canada. *Ecological Engineering* 35: 810-819. DOI. citeulike-article-id:4486532.

Mallik AU, Lamb EG, Rasid H. 2001. Vegetation zonation among the microhabitats of an artificial river channel: analysis and application of below-ground species trait patterns. *Ecological Engineering* 18: 135-146.

Manral U, Raha A, Solanki R, Hussain SA, Talukdar G, Mohan D, Veeraswamy GG. 2012. Hydrological characteristics and flood plain vegetation of human impacted wetlands: a case study from Okhla Bird Sanctuary, National Capital Region, New Delhi, India. *Asian Journal of Conservation Biology* 1: 110-119.

Medley KE. 1992. Patterns of forest diversity along the Tana River, Kenya. *Journal of Tropical Ecology* 8: 353-371. DOI. doi:10.1017/S0266467400006684.

Mitsch WJ, Gosselink JG. 1993. *Wetlands*: Van Nostrand Reinhold: New York.

Moriuchi K, Winn A. 2005. Relationships among Growth, Development and Plastic Response to Environment Quality in a Perennial Plant. *New Phytologist* 166. DOI. citeulike-article-id:8704695.

Müller N. 1995. River dynamics and floodplain vegetation and their alterations due to human impact. *Archiv für Hydrobiologie* 101: 477-512. DOI. citeulike-article-id:7231827.

Naiman R, Décamps H, McClain M. 2005. *Riparia: Ecology, Conservation, and Management of Streamside Communities*. Academic Press.

Naiman R, Decamps H, Pollock M. 1993. The Role of Riparian Corridors in Maintaining Regional Biodiversity. *Ecological Applications* 3: 209-212. DOI. citeulike-article-id:3015431.

Naiman RJ, Décamps H. 1997. The ecology of interfaces: Riparian Zones. *Annual Review of Ecology and Systematics* 28: 621-658.

Naiman RJ, Fetherston KL, McKay SJ, Chen J. 1998. Riparian forest. Naiman RJ, Bilby RE, Kantor S (eds). Springer Verlag: New York; 289-323.

Neukirchen D, Himken M, Lammel J, Czypionka-Krause U, Olfs HW. 1999. Spatial and temporal distribution of the root system and root nutrient content of an established *Miscanthus* crop. *European Journal of Agronomy* 11: 301-309. DOI. citeulike-article-id:7231828.

NIAES. 2000. Assessing the Ecological Impact of Alien Species and Managing the Risks. Priority Research Areas of NIAES (2006-2008), National Institute for Agro-Environmental Studies.

Nilsson C. 1987. Distribution of Stream-Edge Vegetation Along a Gradient of Current Velocity. *Journal of Ecology* 75: 513-522.

Nilsson C, Svedmark M. 2002. Basic Principles and Ecological Consequences of Changing Water Regimes: Riparian Plant Communities. *Environmental Management* 30: 468-480. DOI. citeulike-article-id:7208552.

Okuda S, Kobune A, Hatase Y. 1995. *The floodplain vegetation map of the Tama River*. The Tokyu Foundation for Better Environment, Tokyo.

Okuda S, Sone N, Fujima H, Fuji T. 1979. *The floodplain vegetation map of the Tama River*. The Tokyu Foundation for Better Environment, Tokyo.

Owens P, Walling D, Leeks G. 1999. Deposition and storage of fine-grained sediment within the main channel system of the River Tweed, Scotland. *Earth Surface Processes and Landforms* 24: 1061-1076. DOI. citeulike-article-id:7231839.

Palink BJ, Zassada JC, Hedman CW. 2000. Ecological principles for riparian silviculture. In *Riparian Management in Forests of the Continental Eastern Unite States,* Verry E, Hornbeck JW, Dolloff CA (eds). Lewis Pubs: Washington, DC; 233–254.

Petts GE. 1989. *Historical Change of Large Alluvial Rivers: Western Europe*. John Wiley and Sons: Chichester, United Kingdom.

Pidwirny M. 2006. Plant Succession: Fundamentals of Physical Geography. 2nd Edition.

Pizzuto JIM. 2002. Effects of Dam Removal on River Form and Process. *BioScience* 52: 683-691. DOI. 10.1641/0006-3568(2002)052[0683:eodror]2.0.co;2.

Poff NL, Allan JD, Bain MB, Karr JR, Prestegaard KL, Richter BD, Sparks RE, Stromberg JC. 1997. The natural flow regime: a paradigm for river conservation and restoration. *BioScience* 47: 769-784.

Poff NL, Hart DD. 2002. How Dams Vary and Why It Matters for the Emerging Science of Dam Removal. *BioScience* 52: 659-668. DOI. 10.1641/0006-3568(2002) 052[0659:hdvawi]2.0.co;2.

Pyšek P, Prach K. 1993. Plant Invasions and the Role of Riparian Habitats: A Comparison of Four Species Alien to Central Europe. *Journal of Biogeography* 20: 413-420. DOI. citeulike-article-id:469.

Rashid MH, Asaeda T, Uddin MN. 2010a. The Allelopathic Potential of Kudzu (*Pueraria montana* Lour.). *Weed Science* 58: 47-55. DOI. 10.1614/WS-09-106.1.

Rashid MH, Asaeda T, Uddin MN. 2010b. Litter-mediated allelopathic effects of kudzu (*Pueraria montana*) on *Bidens pilosa* and *Lolium perenne* and its persistence in soil. *Weed Biology and Management* 10: 48-56. DOI. 10.1111/j.1445-6664.2010.00366.x.

Richards K, Brasington J, Hughes F. 2002. Geomorphic dynamics of floodplains: ecological implications and a potential modelling strategy. *Freshwater Biology* 47: 559-579. DOI. 10.1046/j.1365-2427.2002.00920.x.

Richardson JS. 2008. Aquatic arthropods and forestry: effects of large-scale land use on aquatic systems in Nearctic temperate regions. *The Canadian Entomologist* 140: 495-509. DOI. 10.4039/n07-LS04.

Richardson JS, Danehy RJ. 2007. A Synthesis of the Ecology of Headwater Streams and their Riparian Zones in Temperate Forests. *Forest Science* 53: 131-147.

Richter BD, Braun DP, Mendelson MA, Master LL. 1997. Threats to Imperiled Freshwater Fauna. *Conservation Biology* 11: 1081-1093. DOI. 10.1046/j.1523-1739.1997.96236.x.

Richter BD, Richter HE. 2000. Prescribing Flood Regimes to Sustain Riparian Ecosystems along Meandering Rivers. *Conservation Biology* 14: 1467-1478. DOI. 10.2307/2641799.

Rood S, Kalischuk A, Mahoney J. 1998. Initial cottonwood seedling recruitment following the flood of the century of the Oldman River, Alberta, Canada. *Wetlands* 18: 557-570. DOI. 10.1007/BF03161672.

Rood S, Mahoney J. 1990. Collapse of riparian poplar forests downstream from dams in western prairies: Probable causes and prospects for mitigation. *Environmental Management* 14: 451-464. DOI. 10.1007/BF02394134.

Rood SB, Braatne JH, Hughes FM. 2003a. Ecophysiology of riparian cottonwoods: stream flow dependency, water relations and restoration. *Tree Physiol* 23: 1113-1124. DOI. 10.1093/treephys/23.16.1113.

Rood SB, Gourley CR, Ammon EM, Heki LG, Klotz JR, Morrison ML, Mosley DAN, Scoppettone GG, Swanson S, Wagner PL. 2003b. Flows for Floodplain Forests: A Successful Riparian Restoration. *BioScience* 53: 647-656. DOI. 10.1641/0006-3568(2003)053[0647:FFFFAS]2.0.CO;2.

Rood SB, Mahoney JM, Reid DE, Zilm L. 1995. Instream flows and the decline of riparian cottonwoods along the St. Mary River, Alberta. *Canadian Journal of Botany* 73: 1250-1260. DOI. 10.1139/b95-136.

Rood SB, Samuelson GM, Braatne JH, Gourley CR, Hughes FMR, Mahoney JM. 2005. Managing River Flows to Restore Floodplain Forests. *Frontiers in Ecology and the Environment* 3: 193-201. DOI. 10.2307/3868463.

Sakio H. 1997. Effects of natural disturbance on the regeneration of riparian forests in a Chichibu Mountains, central Japan. *Plant Ecology* 132: 181-195. DOI. 10.1023/A:1009775923208.

Salo J, Kalliola R, Hakkinen I, Makinen Y, Niemela P, Puhakka M, Coley PD. 1986. River dynamics and the diversity of Amazon lowland forest. *Nature* 322: 254-258.

Satake Y, Ohi J, Kitamura S, Tanri T, Tomishige T. 2002. *Wild Flowers of Japan, Herbaceous Plants*. Heibonsha: Tokyo.

Scott ML, Friedman JM, Auble GT. 1996. Fluvial process and the establishment of bottomland trees. *Geomorphology* 14: 327-339. DOI. 10.1016/0169-555X(95)00046-8.

Shafroth PB, Friedman JM, Auble GT, Scott ML, Braatne JH. 2002a. Potential Responses of Riparian Vegetation to Dam Removal. *BioScience* 52: 703-712. DOI. 10.1641/0006-3568(2002)052[0703:PRORVT]2.0.CO;2.

Shafroth PB, Stromberg JC, Patten DT. 2002b. Riparian Vegetation Response to Altered Disturbance and Stress Regimes. *Ecological Applications* 12: 107-123. DOI. 10.1890/1051-0761(2002)012[0107:RVRTAD]2.0.CO;2.

Shankman D. 1993. Channel Migration and Vegetation Patterns in the Southeastern Coastal Plain. *Conservation Biology* 7: 176-183. DOI. 10.1046/j.1523-1739.1993.07010176.x.

Sharitz RR, Boring LR, Lear DHV, Pinder JE. 1992. Integrating Ecological Concepts with Natural Resource Management of Southern Forests. *Ecological Applications* 2: 226-237. DOI. 10.2307/1941857.

Steele MW. 2006. The History of Tama River: Social Reconstruction. In *A History of Water: The World of Water, vol. 3,* Tvedt T, Jackobsson E (eds). I. B. Tauris & Co. Ltd.; 217-238.

Steiger J, Gurnell AM. 2003. Spatial hydrogeomorphological influences on sediment and nutrient deposition in riparian zones: observations from the Garonne River, France. *Geomorphology* 49: 1-23. DOI. 10.1016/S0169-555X(02)00144-7.

Steiger J, Gurnell AM, Petts GE. 2001. Sediment deposition along the channel margins of a reach of the middle River Severn, UK. *Regulated Rivers: Research & Management* 17: 443-460. DOI. 10.1002/rrr.644.

Steiger J, Tabacchi E, Dufour S, Corenblit D, Peiry JL. 2005. Hydrogeomorphic processes affecting riparian habitat within alluvial channel-floodplain river systems: a review for the temperate zone. *River Research and Applications* 21: 719-737. DOI. citeulike-article-id:6897559.

Stewart KJ, Mallik AU. 2006. Bryophytes responses to microclimatic edge effects across riparian buffers. *Ecological Applications* 16: 1474-1486. DOI. 10.1890/1051-0761(2006)016[1474:BRTMEE]2.0.CO;2.

Stromberg JC, Beauchamp VB, Dixon MD, Lite SJ, Paradzick C. 2007. Importance of low-flow and high-flow characteristics to restoration of riparian vegetation along rivers in arid south-western United States. *Freshwater Biology* 52: 651-679. DOI. 10.1111/j.1365-2427.2006.01713.x.

Stromberg JC, Patten DT, Richter BD. 1991. Flood flows and dynamics of Sonoran riparian forests. *Rivers* 2: 221-235.

Swanson FJ, Johnson SL, Gregory SV, Acker SA. 1998. Flood disturbance in a forested mountain landscape. *BioScience* 48: 681–689.

Tedoru C, Wehrli B. 2005. Retension of sediments and nutrients in the Iron Gate I Reservoir on the Danube River. *Biogeochemistry* 76: 539-565. DOI. citeulike-article-id:7231851.

Terborgh J, Petren K. 1991. Development of habitat structure through succes-sion in an Amazonian floodplain forest. In *The Physical Arrangement of Objects in Space,* Bell SS, McCoy ED, Mushinsky HR (eds). Chapman and Hall: London; 28-46.

Tharme RE. 2003. A global perspective on environmental flow assessment: emerging trends in the development and application of environmental flow methodologies for rivers. *River Research and Applications* 19: 397-441. DOI. 10.1002/rra.736.

Tockner K, Schiemer F, Ward JV. 1998. Conservation by restoration: the management concept for a river-floodplain system on the Danube River in Austria. *Aquatic Conservation: Marine and Freshwater Ecosystem* 8: 71-86. DOI. citeulike-article-id:2515534.

Tsvetkova GE, Georgiev GI. 2003. Effect of phosphorus nutrition on the nodulation, nitrogen fixation and nutrient use efficiency of *Bradyrhizobium japonicum* soybean (*Glycine max* Merr.) symbiosis. *Bulgarian Journal of Plant Physiology* special issue: 331-335. DOI. citeulike-article-id:7233151.

Uchida T, Nomura R, Asaeda T, Rashid MH. 2012. Co-existence of *Sicyos angulatus* and native plant species in the floodplain of Tama River, Japan. *International Journal of Biodiversity and Conservation* 4: 336-347. DOI. 10.5897/IJBC12.010

van Oorschot M, Hayes C, van Strien I. 1998. The influence of soil desiccation on plant production, nutrient uptake and plant nutrient availability in two French floodplain grasslands. *Regulated Rivers: Research & Management* 14: 313-327. DOI. 10.1002/(SICI)1099-1646(199807/08)14:4<313::AID-RRR506>3.0.CO;2-U.

Walker KF. 1985. A review of the ecological effects of river regulation in Australia. *Hydrobiologia* 125: 111-129. DOI. 10.1007/BF00045929.

Ward JV, Stanford JA. 1983. The serial discontinuity concept of lotic ecosystems. In *Dynamics of lotic ecosystems,* Fontaine TD, Bartell SM (eds). Ann Arbor Scientific Publishers: Ann Arbor, Michigan; 29-42.

Ward JV, Stanford JA. 1995. The serial discontinuity concept: Extending the model to floodplain rivers. *Regulated Rivers: Research & Management* 10: 159-168. DOI. 10.1002/rrr.3450100211.

WCD-World Commision on Dams. 2000. *Dams and Development: A New Framework for Decision-making*. Earthscan Publications: London, UK.

Welcomme RL. 1985. *River Fisheries. FAO Fisheries Technical Paper 262*. ood and Agriculture Organization (FAO) of the United Nations: Rome, Italy.

Welsh HH, Droege S. 2001. A Case for Using Plethodontid Salamanders for Monitoring Biodiversity and Ecosystem Integrity of North American Forests

Caso Sobre el Uso de Salamandras Plethodóntidas en el Monitoreo de la Biodiversidad y la Integridad del Ecosistema de Bosques de Norteamérica. *Conservation Biology* 15: 558-569. DOI. 10.1046/j.1523-1739.2001.015003558.x.

Wilcock P, Kondolf M, Matthews G, Barta A. 1996. Specification of Sediment Maintenance Flows for a Large Gravel-Bed River. *Water Resources Research* 32: 2911-2921. DOI. citeulike-article-id:8975973.

Wohl E, Rathburn S. 2003. Mitigation of sedimentation hazards downstream from reservoirs. *International Journal of Sediment Research* 18: 97-106.

In: Floodplains
Editor: Enner Herenio Alcântara
ISBN: 978-1-62808-998-1

Chapter 3

EARTHWORM COMMUNITIES AS INDICATORS FOR EVALUATING FLOODPLAIN RESTORATION SUCCESS

Renée-Claire Le Bayon[1*], Géraldine Bullinger-Weber[1,2], Jean-Michel Gobat[1] and Claire Guenat[2]
[1]Laboratory Soil & Vegetation, University of Neuchâtel, Neuchâtel, Switzerland
[2]Laboratory ECOS & WSL, École Polytechnique Fédérale de Lausanne (EPFL), School of Architecture, Civil and Environmental Engineering (ENAC), Lausanne, Switzerland

ABSTRACT

Floodplains are known to be areas of extraordinary biodiversity with a mosaic of shifting habitats with high interdependency. Such complex ecosystems fulfil a wide range of ecosystem services concerning economic, social and ecological functions. Moreover, functioning as complex ecotones at various spatio-temporal scales, floodplains usually provide a diversity of habitats and microhabitats for many organisms from river channels to uplands. However, these ecosystems have been largely subjected to human pressure through the embanking of rivers. Such damages have led to river restoration projects in order to re-establish a near-natural alluvial dynamics and to recover floodplain naturalness and biodiversity. In Switzerland, several projects have been implemented to maintain or recreate floodplain ecological functions, in particular for flood protection and biodiversity. In such context, this chapter highlights biological and pedological features as indicator tools to evaluate the success of floodplain restoration, focusing on earthworm communities that are assumed to vary in terms of species diversity, abundance and biomass along a gradient of naturalness from the embanked system to the near-natural reference. The first section of the chapter deals with floodplain restoration and potential consequences on soil functions. Then, the second section presents an overview of earthworm communities and activities as ecosystem engineers, their habitats and their

[*] Email: claire.lebayon@unine.ch, Phone: +41 (0) 32 718 23 65.

potential resilience to disturbance and stress. Concerning near-natural floodplains in Switzerland, the third section focuses on earthworm communities and environmental variables that may affect their distribution taking into account two spatial variables: 1) an altitudinal gradient from subalpine to hill levels, and, 2) a gradient perpendicular to the river, stratified by vegetation. The impact of fluvial dynamics is also discussed. The fourth section addresses the comparison of two floodplains that are partly embanked and restored (The Emme and Thur Rivers) and highlights the potential of earthworms as bioindicators of early stages of river restoration. Finally, the last section proposes future prospects to assess the success of ecological restoration, i.e. the self-sustainability of restored floodplains.

Keywords: Floodplains, earthworms, river restoration, alluvial soils

1. Floodplain Restoration

Floodplains are considered as complex ecosystems that provide a wide range of ecosystem services, which could be defined as the beneficial flows arising from natural capital stocks and fulfilling human needs (Dominati et al., 2010). These ecosystem services concern economic, social and ecological functions such as flood protection, recreation areas, biodiversity reservoir and nutrients cycling (Mitra et al., 2005). In a context of global warming, climate change may thus induce deep modifications in the stream flow primarily through variations in precipitation, hydrological conditions and river morphological characteristics which are of the most important issues impacting floodplains (Bertrand et al., 2012; Karamouz et al., 2011). Such modifications have consequences on biogeochemical cycles being concerned especially their central roles in alluvial systems that need investigation to assess likely impacts.

Moreover, within a framework of prevention from flood events, floodplain ecosystems have been deeply modified by dam construction and embanking (Tockner and Stanford, 2002). River infrastructures have however usually negative impacts on freshwater ecosystems as for instance loss of biodiversity as well as river channelling that accelerates water flow. Hence, current floodplain management is now being revised to recreate functional and diverse alluvial ecosystems (Wohl et al., 2005) and removal of in-stream structures is often seen as a viable option for sustainable watershed management (O'Hanley, 2011). River restoration projects are then off-the-moment (Palmer et al., 2005; Palmer and Bernhardt, 2006) notably in Switzerland where the modification of the Federal Law on the Protection of Waters (814.20, Article 38a) on January 1st 2011 has led regional authorities to define, schedule, and carry out restoration strategies for rivers (Palmer and Bernhardt, 2006; Wohl et al., 2005).

Assessing the outcome of river restoration projects is essential for adaptive management, evaluation of project efficiency, optimization of future programs, and gaining public acceptance (Woolsey et al., 2007). Potential indicators or criteria for evaluating river restoration success have been highlighted by several authors who particularly pointed out water quality (especially N and P contents), viability of target species (i.e. nesting possibilities for the Little Ring Plover (*Charadrius dubius*) and increased rates of ecosystems functions (carbon storage, biodiversity) (Jansson et al., 2005; Ruiz-Jaen et al., 2005; Woolsey et al., 2007; Henry et al., 2002). However, few research considered soils, humus forms or pedofauna as potential indicators of restoration success (Cui et al., 2009) albeit soils integrate information on ecosystem structure and could record past and present events of fluvial dynamics (Gerrard, 1987; Gerrard, 1992; Daniels, 2003; Bullinger-Weber and Gobat, 2006). Thus, pedological ecological functions such as flood regulation or carbon storage may be used as indicators of restoration success (Lorenz, 2003). Despite all these advantages, soils and their inhabitants remain poorly considered and a better knowledge is needed regarding consequences of river restoration on essential functions of soils (notably support for soil biota and organic matter storage).

2. Ecology of Earthworm Communities

2.1. Habitats

Earthworms constitute the largest terrestrial faunal biomass and occur worldwide preferring moist habitats of moderate temperature (Lee, 1985; Edwards and Bohlen, 1996) and being the most abundant in forests and grasslands (Coleman and Whitman, 2005). Their spatial distribution is widely heterogeneous depending on environmental factors, i.e. plant cover, and soil properties (texture, organic matter content), and internal population processes, i.e. reproduction rates and dispersal mode (Bullinger-Weber et al., 2012; Jimenez et al., 2011). Earthworms are generally split into three main ecological categories based on their behaviour and feeding ecology (Bouché, 1977): i) epigeic species often prefer substrates enriched with organic matter, and usually live in plant litter onto the soil surface; ii) endogeic species inhabit organo-mineral soil layers and, iii) anecic species take advantages of both the surface litter as a source of food and the mineral soil as a refuge in which they did burrows (Coleman and Whitman, 2005).

2.2. Ecosystem Engineering

Ecosystem engineers change biotic or abiotic materials in their environment thereby creating or modifying habitats and hence controlling availability of resources to other species (Jones et al., 1994, 1997; Lavelle, 1997; Lavelle et al., 1997; Decaëns, 2010). Ecosystem engineers are also well-known to provide ecosystem services (Fonte and Six, 2010).

Earthworms are thus usually classified as allogenic engineers (Lal, 1991; Berke, 2010) belonging to burrowing and excavating organisms (Shipitalo and Le Bayon, 2004; Edwards, 2004), and as chemical engineers through the addition of mucus enriched in carbon, nitrogen and phosphorus during the gut transit (McInerney et al., 2001; Schrader and Zhang, 1997; Le Bayon and Binet, 2006). Furthermore, being major bioturbators in terrestrial ecosystems, earthworms largely contribute to the formation of stable microaggregates within macroaggregates leading to a crumb soil structure that helps the protection of organic carbon (Bossuyt et al., 2005). Until now, most of the studies on earthworm communities have been conducted on mature soils and describe interactions between biota and structure in cultivated soils (Davidson and Grieve, 2006a, 2006b).

Few studies have been carried out in alluvial soils despite the fact that they are unique among terrestrial ecosystems to experiment recurrent primary succession (Bechtold and

Naiman, 2009), and constitute a relevant model to study initial stages of soil formation and in particular the role of earthworms in the topsoil structure formation. The presence of water stable macro-aggregates due to earthworm activities has been showed from pioneer stages of topsoil formation under willow forests (Guenat et al., 1999). Bullinger-Weber et al. (2007) show that epigeic earthworms and enchytraeids are the first engineers producing in a short-term soil structure and then, if texture is favourable, anecic and endogeic earthworms may colonize the different soil layers improving physical and nutrient conditions and creating long-term stable aggregates.

2.3. Resilience to Disturbance and Stress

Across various range of habitats, earthworms display a wide array of morphological, physiological, and behavioural adaptations to environmental conditions. For example, many species are able to enter a temporary dormant state (diapause or quiescent state) or produce resistant cocoons during unfavourable periods (Coleman and Whitman, 2005). Several studies focused on earthworm frost tolerance (Joergensen et al., 2008), their accommodation to heavy metals concentration such as copper (Fisker et al., 2012, 2013), atrazine and cadmium (Wang et al., 2012) and amendments (Lapied et al., 2009). Some others highlight impact of combined effects on earthworms (*Dendrobaena octaedra*) such as synergistic interactions between heavy metals and frost survival, and antagonistic ones between polycyclic aromatic hydrocarbons (PAHs) and frost survival (Bindesbol, 2008). In a floodplain context, Cornelis et al. (2009) observed high copper concentrations both in soil and in *Lumbricus rubellus* tissues while earthworm abundance and biomass were not affected. Thus, effects at the cellular level therefore did not result in a reduced functioning of earthworm communities (Cornelis et al., 2009). Moreover, flood events may greatly influence earthworm communities (see section 3.3 for more details).

3. Earthworm Communities in Near-Natural Floodplains

3.1. Earthworm Communities' Composition

In near-natural floodplains, succession and community-assembly changes occur more rapidly than in any other ecosystems due to high turnover of habitats and ecosystems (Milner and Tockner, 2010). According to Petts and Amoros (1996), successions of animal and plant communities pledged to alluvial systems are generally arranged along topographic gradients where pedological changes occur over time-scales from months to several hundred years. This variability of pedological stages leads hence to a particularly high soil heterogeneity and habitat diversity compared to other terrestrial ecosystems (Cierjacks et al., 2011). However, research projects on earthworm communities in floodplains are scared so far (Ivask et al., 2007; Plum and Filser, 2005; Zorn et al., 2005, 2008). Moreover, most of them were conducted at the hill level and focused on flooded meadows in northern Germany (Plum and Filser, 2005), flooded grasslands in river valley in Estonia (Ivask et al., 2007) or short grass and herbaceous vegetation in the Netherlands (Zorn et al., 2005, 2008). All these researchers recovered from 5 to a maximum of 8 species in their respective study sites. Gathering all results leads to the following diversity: epigeics (*Lumbricus rubellus*, *L. castaneus*, *Eiseniella tetraedra*, *Dendrodrilus rubidus*), endogeics (*Allolobophora chlorotica*, *Octolasion tyrtaeum*, *O. cyaneum*, *Aporrectodea rosea*, *A. caliginosa*) and one anecic species (*L. terrestris*).

Table 1. List of ecological categories and species of earthworms recovered in near-natural floodplains in Switzerland at several altitudinal levels. From Guenat et al. (1999), Bullinger-Weber et al. (2007), Salomé et al. (2011) and Bullinger-Weber et al. (2012)

Epigeic species
Bimastos eiseni (Levinsen, 1884)
Dendrobaena octaedra (Savigny, 1826)
Dendrobaena pygmea cognetti (Michaelsen, 1903)
Dendrobaena pygmea pygmea (Savigny, 1826)
Dendrodrilus rubidus rubidus (Savigny, 1826)
Dendrodrilus subrubicundus (Eisen, 1874)
Eiseniella tetraedra tetraedra (Savigny, 1826)
Eisenia andrei (Bouché, 1972)
Lumbricus castaneus (Savigny, 1826)
Lumbricus meliboeus (Rosa, 1884)
Lumbricus rubellus (Hoffmeister, 1843)
Octodrilus argoviensis (Bretscher, 1899)
Anecic species
Aporrectodea caliginosa nocturna (Evans, 1946)
Aporrectodea giardi giardi (Ribaucourt, 1901)
Aporrectodea longa longa (Ude, 1885)
Aporrectodea longa ripicola (Bouché, 1972)
Aporrectodea longa ripicola viridis (Bouché, 1972)
Lumbricus terrestris (Linnaeus, 1758)
Endogeic species
Allolobophora chlorotica chlorotica (Savigny, 1826)
Aporrectodea caliginosa alternitosa (Bouché 1972)
Aporrectodea caliginosa caliginosa (Savigny, 1826)
Aporrectodea handlirschi handlirschi (Rosa, 1905)
Aporrectodea icterica icterica (Savigny, 1826)
Aporrectodea rosea rosea (Savigny, 1826)
Octolasion cyaneum (Savigny, 1826)
Octolasion tyrtaeum lacteum (Oerley, 1885)
Octolasion tyrtaeum tyrtaeum (Savigny, 1826)

Other studies looked at earthworm communities in floodplains along an altitudinal gradient. Thus, overall in several Swiss near-natural floodplains, Guenat et al. (1999), Bullinger-Weber et al. (2007), Salomé et al. (2011) and Bullinger-Weber et al. (2012) found 27 species and subspecies from subalpine to hill levels (Table 1). This record is the highest compared to studies cited above and corresponds to two-thirds of all inventoried earthworm species and subspecies in Switzerland, confirming that floodplains are among the most diverse terrestrial ecosystems, as already shown for vascular plants, for example (Gallandat et al., 1993). All earthworm ecological categories are represented and their distribution was observed to be widely heterogeneous within the same floodplain (Bullinger-Weber et al., 2007) as well as in the same vegetation unit (Bullinger-Weber et al., 2012). Moreover, for the first time, the species *Lumbricus moliboeus* was recovered in carbonated soils while usually observed in acidic conditions (Bouché, 1972).

3.2. Environmental Variables Affecting Earthworm Communities

Several environmental variables, independently and/or interacting together, may influence the distribution and the composition of earthworm communities in near-natural floodplains, at different spatio-temporal scales. Hence, according to Emmeling (1995), main factors that may affect the soil macrofauna distribution in floodplains are soil organic matter content, soil moisture and flooding characteristics. Salomé (2011) enhanced these conclusions and highlighted a hierarchy of variables that govern the composition of earthworm communities. Different partitioning variance analyses demonstrated that the interaction between soil parameters and altitude explained the earthworm species diversity, total biomass, total abundance and species abundance, as well as biomass and abundance of earthworm ecological categories (data not shown).

The high temporal and spatial changes of these variables, mostly due to fluvial dynamics, create a broad mosaic of habitats. Consequences of this unpredictable environment are regularly observed on earthworm communities thus reflecting alluvial dynamics (Salomé et al., 2011; Bullinger-Weber et al., 2012). Going further into details, soil types and parameters influence earthworm communities (Guenat et al., 1999), especially depth and texture that drive the distribution of ecological categories. A recent study was conducted on this topic in the Rhine River floodplain (unpublished data; photo 1; Figure 1).

Photo 1. The near-natural site ("Rhäzuns") is located along the Rhine River (canton of Graubünden, Switzerland) is a site of national importance. The site lies at 600 m a.s.l, annual precipitation ranges between 800 and 1000 mm and mean annual temperature is 7.1°C .The mean annual flow is about 40 $m^3 s^{-1}$, with a minimum and a maximum annual discharge of 23 $m^3 s^{-1}$ and 60 $m^3 s^{-1}$, respectively. The alluvium deposits are mostly composed of calcareous pebbles and sand. The channel pattern corresponds to a braided river.

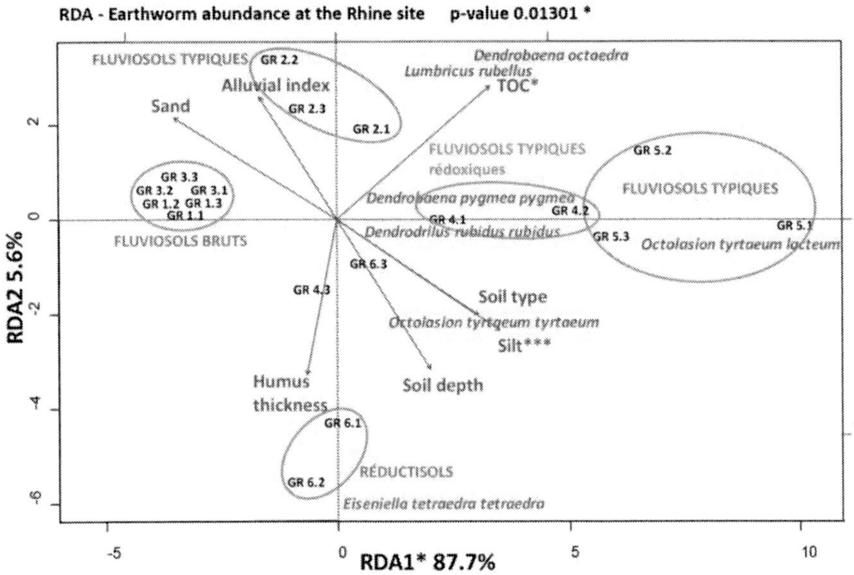

Figure 1. Redundancy analysis on earthworm abundance at the near-natural site "Rhäzuns" along the Rhine River. GR 1 to GR 6 indicated soil profiles with three replicates and their corresponding soil types (in green). Environmental variables are indicated in blue, TOC being the total organic carbon content and the alluvial index reflecting alluvial dynamics (calculated by dividing the total number of layers by the total depth of the profile). Earthworm species are specified in red. Levels of statistical significance are as followed: *P <0.05, **P <0.01, ***P <0.001.

Two main types of soils were identified: Fluvisols and Gleysols (IUSS Working Group WRB, 2006) which correspond to FLUVIOSOLS (GR 1- GR 5) and RÉDUCTISOLS (GR 6) according to the Référentiel Pédologique (Baize and Girard, 2009). This latter classification allows discriminating several stages of soil evolution, especially for fluviosols which are then divided in FLUVIOSOLS BRUTS, FLUVIOSOLS JUVÉNILES and FLUVIOSOLS TYPIQUES. Based on the degree of maturation of humiferous topsoil, this subdivision is essential to describe precisely and better understand the functioning of each soil type and to relate it to earthworm communities. This kind of soil succession results from fluvial dynamics that control sedimentation and erosion processes, and in turn the *in situ* pedogenesis duration (Daniels, 2003). In our study case, we thus observed a gradient from bare soils (GR 1, FLUVIOSOLS BRUTS) to developed soils (GR 5, FLUVIOSOLS TYPIQUES). The frequency of flood events and their intensity over the year are essential to explain pedogenesis. At the Rhine site, not the duration of floods but the period of the year at which they occur as well as the water discharge appear to be primordial. Hence, in June when flooding is the highest, floods are rapid and violent thus carrying rough sediment (pebbles, coarse sand), eroding or burying river banks then in turn destroying habitats. All these variables govern the distribution and the composition of earthworm as revealed by alluvial index, soil type and soil texture (Figure 1). Organic matter and especially total organic carbon (TOC) regulates also the presence of epigeic earthworms such as *Dendrobaena octaedra* and *Lumbricus rubellus* (Figure 1). In addition, hydromorphy features led to the identification of FLUVIOSOLS TYPIQUES rédoxiques and RÉDUCTISOLS, which, despite occasional anoxic conditions, seem to be in favour of earthworms as a general trend due to their soil thickness and their fine texture (*Octolasion tyrtaeum tyrtaeum*). In addition, the thickness of humus favours especially epigeics to live in (*Eiseniella tetraedrea tetraedrea*). Both the fine texture and the humus thickness contribute also to a better water retention that allows earthworms to resist to desiccation, at the opposite from FLUVIOSOLS BRUTS where the moisture and temperature amplitudes are high. Looking at the repartition of earthworms along the gradient of soils and vegetation (Figure 2), the stability of habitats governs widely earthworm communities' distribution which gathered in FLUVIOSOLS TYPIQUES and RÉDUCTISOLS. Despite the fact that no anecic species were observed, epigeic species dominated in terms of species diversity and abundance (*L. rubellus* and *D. octaedra*). The endogeic *Octolasion tyrtaeum lacteum* was recovered everywhere while *Octolasion tyrtaeum tyrtaeum* and some individuals of *Aporrectodea rosea rosea* were collected in deep and fine-textured soils. Other studies confirmed that soil types and parameters influence earthworm communities (Guenat et al., 1999), especially depth and texture that drive the distribution of ecological categories. In this case of deep soils and fine texture, the highest abundance and biomass of earthworms are observed (Guenat et al., 1999). Moreover, a fine soil texture leads to a higher diversity of earthworm categories and species. Thus, epigeics are usually associated with coarse sandy texture in contrast to anecics and endogeics which prefer silty soils from alluvial forests (Guenat et al., 1999; Bullinger-Weber et al., 2007; Salomé et al., 2011).

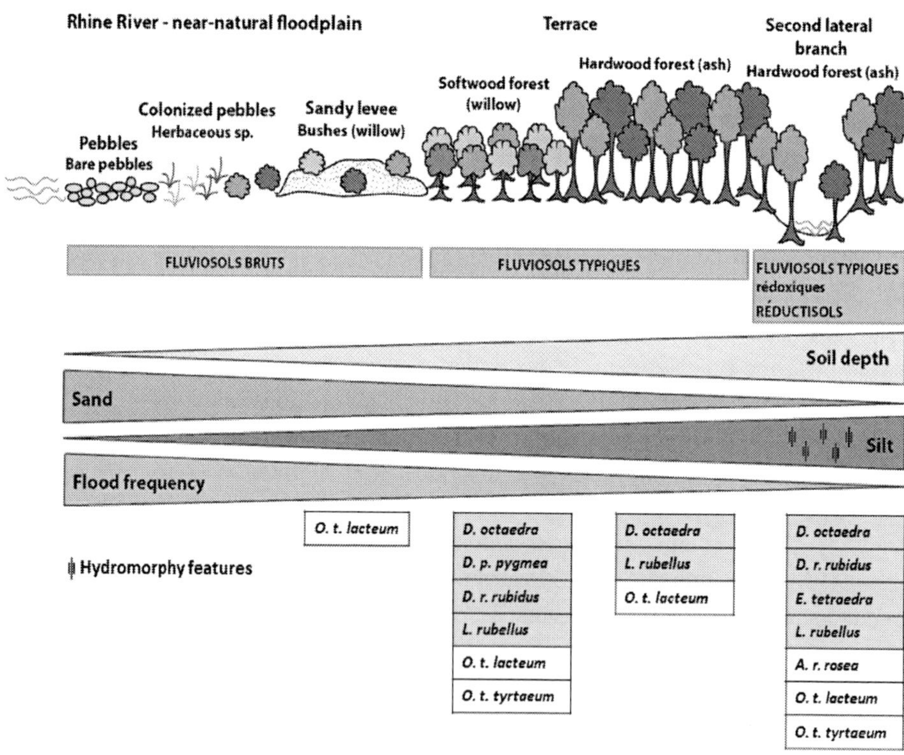

Figure 2. Schematic representation of the near-natural Rhine site. Vegetation succession and topography are represented from the riverbed (left) to the mature forest. All soils are calcareous-rich soils; soil depth, sand and silt contents as well as hydromorphy features and flood frequency are shown. Earthworm species are listed, epigeic and endogeic in grey and white boxes, respectively.

Focusing on humus forms is also an interesting approach regarding the aboveground-belowground ecological relationships. Humus forms hence reflect plant-soil interactions being the place where resonance between these communities takes place, both in functional and evolutionary sense (Ponge, 2013). In the floodplain context, mineral parental material plays a crucial role in the formation and evolution of humus forms. According to Jabiol et al. (2013) classification, most of the studied humus forms belong to the group of Fluvic Parahumus, especially in the vicinity of the river bed where coarse sediments predominate. As for soils, the variability of terraces' topography is correlated to a gradient of humus forms and a large panel of them (from Fluvic Parahumus to Mull, Moder and Amphi forms) could be observed in floodplains. However, research is currently in progress for a better understanding of these aboveground-belowground relationships between vegetation, humus forms soils and macrofauna communities.

3.3. Earthworm Communities and Fluvial Dynamics

As a general rule, earthworms can cope well with submergence short-term floods. However, their resistance differs according to ecological categories and species. Some of them are known to be well adapted to floods and have even been classified as riparian species (Roots, 1956; Sims and Gerard, 1999 in Schütz, 2008). Roots (1956) even reported that some species could survive for 31-50 weeks in submerged soils. Nevertheless, flooding may generally affect negatively earthworm populations by decreasing their numbers immediatly after flooding (Zorn et al., 2005). The duration of flooding seems to be the most limiting factor, as species' tolerance is related to high soil moisture and low aeration. The period of flooding is also crucial because effects will not be the same on earthworm communities if it occurs during the reproduction seasons when earthworms move a lot onto the soil surface (Spring and Autumn) or the estivation time when earthworms are deep in the soils (Summer and Winter). When flood occurs, Lumbricidae could use different strategies for flooding survival in wet grassland, *i.e* horizontal/vertical migration, physiological adaptation, and reproduction strategies (Plum, 2005). Zorn et al. (2005) showed that earthworm numbers and biomasses tend to decrease during flooding and that different earthworm species and ecological categories react differently towards these flooding dynamics. For instance, these authors showed that the anecic *Lumbricus terrestris* was found in high numbers only at the end of a flooding period while the endogeic *Allolobophora chlorotica* was hardly affected by flooding. Moreover, *Aporrectodea caliginosa* showed fluctuating numbers and biomasses during the sampling period that did not correlate with flooding frequency. Finally, the epigeic *Lumbricus rubellus* is a successful colonizer (Eijsackers, 2010) well adapted to flooded soils and that may escape to more favourable habitats when a flood event occurs (Simonsen and Klok, 2010; Zorn et al., 2008). As already mentioned above, not the duration of submersion but mostly the mechanical action of floods (erosion, sedimentation) regulates earthworm communities (abundance, diversity and dispersion) in our studied braided rivers.

4. Earthworm Communities in Restored Floodplains

4.1. Earthworm Communities' Composition

Some studies about earthworm communities have been reported in human transformed ecosystem (Thonon and Klok, 2007; Schütz et al., 2008). In Switzerland, Fournier et al. (2012) reported a total of 15 species and subspecies in a restored floodplain of the Thur River, whereas 9 species were found in the non-restored area corresponding to an embanked pasture.

Photo 2. The Emme River (canton of Bern) is one of the oldest river widening projects in Switzerland. The study sites are located between Aeflingen (restored stretch; photo at the top) and Oberburg (embanked stretch; photo at the bottom) in an agricultural floodplain and lies respectively at 560 m a.s.l. and 460 m a.s.l. Annual precipitation is about 1050 mm year^{-1}, as average annual temperature is 9.4 °C. Before river modification by embanking and dam construction, the channel pattern of Emme River was braided. The annual mean flow rate is about 19 m^3 s^{-1} with a minimum and maximum annual discharge of 9 m^3 s^{-1} and 28 m^3 s^{-1}, respectively. Alluvial deposits are mostly composed of calcareous pebbles and sand. Aiming at flood protection and water quality improvement, restoration of the site was conducted in two steps. At Aeflingen site, embankments were removed and thus the riverbed was widened (30 m width) in 1991/92 and secondly in 1998/99 on both sides of the river, along a 530 m stretch.

In addition, the average abundances were respectively 93 and 67 individuals per square meter. These authors confirmed the general trend previously observed that *Allolobophora chlorotica, Eiseniella tetraedra* and *Lumbricus rubellus* adapted to disturbed environment due to their r-strategy behaviour with fast maturation and high reproduction rates (Bouché, 1972; Bouché, 1977; Gerard, 1967; Satchell, 1967). These species may thus take advantage of the perturbation generated by the restoration process to increase in density and biomass. In thicker soils showing the finest texture, Fournier et al. (2012) found also anecic species such as *Aporrectodea longa*, *A. caliginosa nocturna*, and *L. terrestris, A. longa* being the most tolerant species to flooding.

A recent study was conducted comparing earthworm communities in embanked and restored stretch of the Emme floodplain (unpublished data; Photo 2; Table 2).

Table 2. Earthworm abundance (mean ± standard deviation) per square meter in the Emme River floodplain (embanked and restored stretches)

Ecological category	Earthworm species	N = 21 (ind m^{-2}) Restored	Embanked
Epigeic	*Dendrobaena octaedra*	14 ± 27.4	7 ± 6.0
	Dendrodrilus rubidus rubidus	2 ± 3.5	1 ± 1.5
	Lumbricus rubellus	1 ± 1.4	-
	Lumbricus friendi	-	0 ± 0.8
Endogeic	*Octolasion tyrtaeum lacteum*	2 ± 3.7	5 ± 6.3
	Octolasion tyrtaeum tyrtaeum	0 ± 0.9	-
	Aporrectodea rosea rosea	1 ± 3.5	5 ± 9.2
	Aporrectodea caliginosa caliginosa	1 ± 1.7	4 ± 5.0
Anecic	*Aporrectodea caliginosa nocturna*	4 ± 7.0	5 ± 4.8
	Lumbricus terrestris	1 ± 3.2	14 ± 10.4

All ecological categories were observed, anecics being more numerous in the embanked site (Table 2) while epigeic species dominated in the restored section of the floodplain. Despite the low number of individuals, 10 species and subspecies were observed, *Lumbricus rubellus* and *Octolasion tyrtaeum tyrtaeum* being only observed in the restored part.

Such comparison of earthworm communities in embanked, restored and near-natural floodplain could be helpful to evaluate river restoration success, and the main unresolved question is to know if earthworm communities could be potential indicators in such a context.

4.2. River Restoration: How Does It Change Earthworm Communities?

The Emme River

As for the near-natural site in the Rhine River floodplain, we looked at the repartition of earthworms along the gradient of soils and vegetation at the Emme site (Figure 3), taking into account soil types and parameters. It appears that the stability of habitats governs widely earthworm communities' distribution which gathered in FLUVIOSOLS TYPIQUES and

FLUVIOSOLS TYPIQUES polyphasés. No earthworm was recorded in FLUVIOSOLS BRUTS but all earthworm categories were observed elsewhere. Compared to the embanked stretch, earthworm species diversity is higher in the restored floodplain (9 species instead of 7). The steep slope behind the willow forest leads to a strong decrease of flood frequency thus maintening deep and stable soils favourable to earthworm colonization. As in the near-natural site, the endogeic *Octolasion tyrtaeum lacteum* was recovered everywhere while *Aporrectodea rosea rosea* were collected only in deep and fine-textured soils of the embanked floodplain. Moreover, anecic species found suitable conditions for installation in all sites.

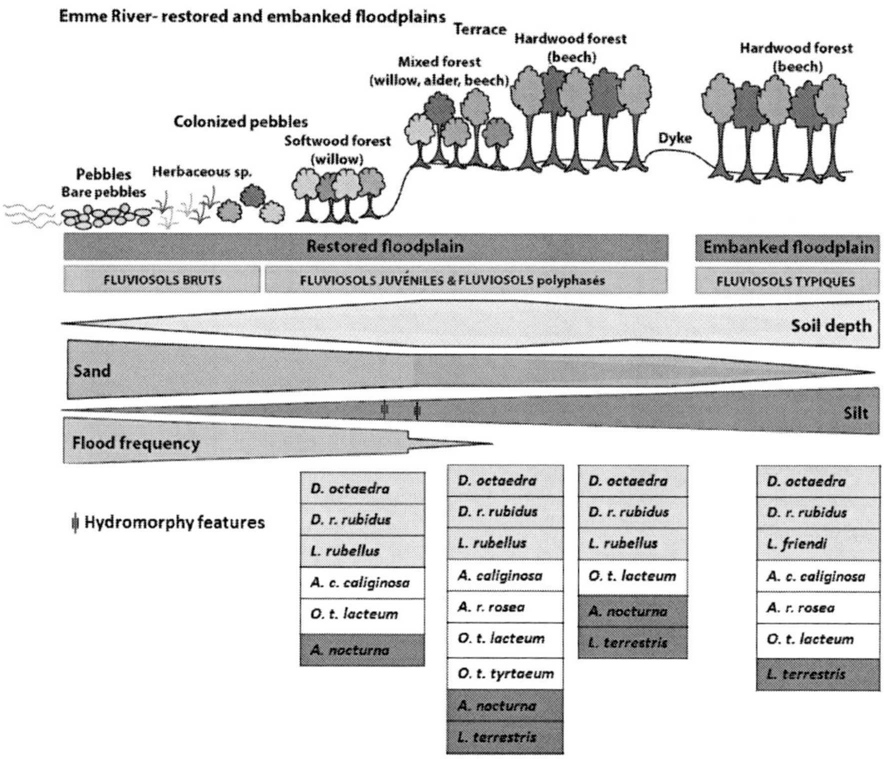

Figure 3. Schematic representation of the Emme site, with the restored and the embanked stretches. Vegetation succession and topography are represented from the riverbed (left) to the mature forest. All soils are calcareous-rich soils; soil depth, sand and silt contents as well as hydromorphy features and flood frequency are shown. Earthworm species are listed, epigeic, endogeic and anecic in grey, white and black boxes, respectively.

The Thur River

Another -study was conducted nearby the Thur River (Photo 3; Figure 4). In this case, earthworms were observed all along the gradient of soils, from FLUVIOSOLS BRUTS under herbaceous species to FLUVIOSOLS TYPIQUES and FLUVIOSOLS TYPIQUES rédoxiques.

With reference to the embanked stretch, i.e. pasture, earthworm species diversity is higher in the restored floodplain (11 species instead of 9). The species *Lumbricus rubellus* is present everywhere, as well as *Allolobophora chlorotica chlorotica*, both known to be successful colonizers as mentioned above.

Anecics were recovered in all soils, indicating suitable conditions such as a fine texture and a sufficient soil depth, *Aporrectodea longa ripicola* being a typically riparian species. As in the near-natural Rhine site, the endogeic *Aporrectodea rosea rosea* were collected only in deep and fine-textured soils.

Photo 3. The Thur River (cantons of Thurgau and Zürich) restoration is the biggest river widening project in Switzerland and includes post-restoration monitoring and evaluations of several stretches (Pasquale et al., 2011). The study site "Schäffäuli", is located near Frauenfeld in an agricultural floodplain and was before its channel rectification considered as a braided river. The average flow is 47 $m^3 s^{-1}$ with a minimum and a maximum annual discharge of 23 $m^3 s^{-1}$ and 76 $m^3 s^{-1}$, respectively. Alluvium deposits are mostly composed of calcareous pebbles. The site lies at 365 m a.s.l. and annual precipitation is about 1000 mm $year^{-1}$, as average annual temperature is 7.9 °C. Aiming at flood protection, restoration of the site was conducted in two steps. First, following a major flood in 1995, the riverbank protections were destroyed thus allowing river bank erosion. Secondly, in 2002, the riverbed was widened along a one-side 1.5 km stretch from 50 to 110 m, and the riverbanks were stabilized by plantation of willow bushes. Both reaches (restored, photo at the top, and embanked, photo at the bottom) are adjacent, and the embanked one, located upstream, is used as pasture. Flood events frequently occur in Autumn and in Spring and are of short duration (1 or 2 days).

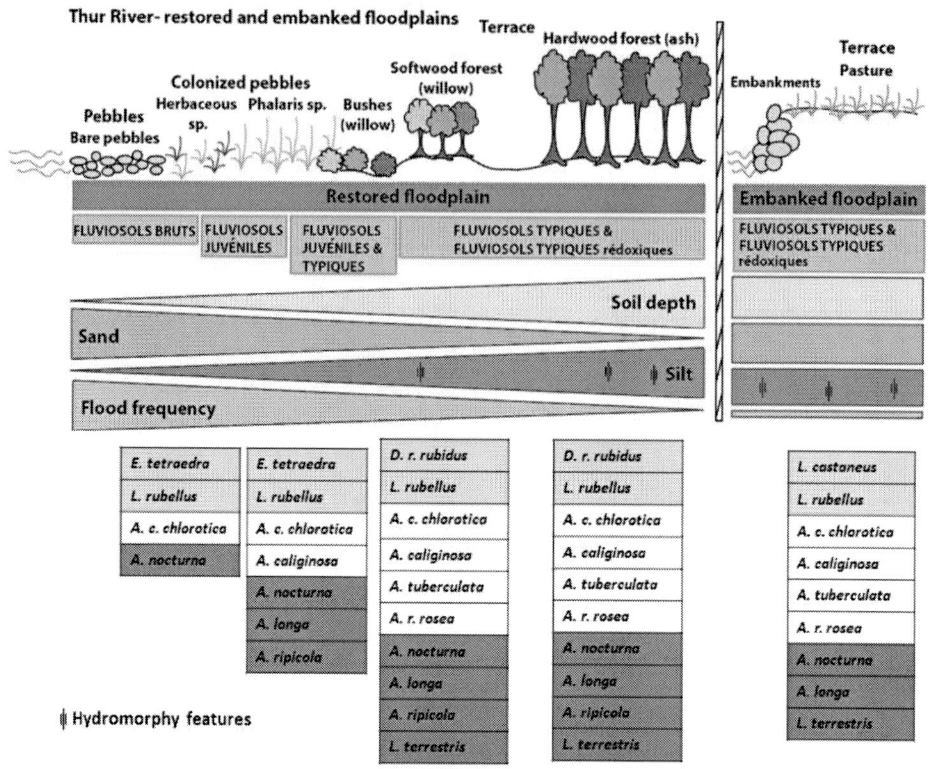

Figure 4. Schematic representation of the Thur site, with the restored and the embanked stretches. Vegetation succession and topography are represented from the riverbed (left) to the mature forest. All soils are calcareous-rich soils; soil depth, sand and silt contents as well as hydromorphy features and flood frequency are shown. Earthworm species are listed, epigeic, endogeic and anecic in grey, white and black boxes, respectively.

The Emme River versus *the Rhine River*

Regarding the comparison of earthworm communities in a near-natural floodplain (Rhine River), a restored floodplain and its respective embanked section (Emme River), it appears that the abundance of earthworm species is explained by soil parameters in relation with fluvial dynamics (Figure 5).

River restoration re-creates and/or maintains soil habitats that fit more or less to ecological needs of the different species. Indeed, river restoration in the studied braided rivers led to the modification of earthworm communities by enhancing some species and eliminating others. For example, the presence of species such as *Lumbricus rubellus* coincides with the presence of a well-developed structure (i.e. high proportion of water-stable aggregates and high organic matter content; Figure 5).

Some other species, such as *Lumbricus terrestris*, *L. friendi*, *Aporrectodea caliginosa nocturna*, *A. caliginosa caliginosa* related to silty soils, were present in both restored and embanked floodplains, but absent in the near-natural one. The presence of these species could be considered as relics of the prior embanked floodplain. By contrast, some species such as *Eiseniella tetraedra tetraedra* were recorded only in soils with hydromorphy features (RÉDUCTISOLS) typical of lateral branches of the Rhine River. These soils were not re-created in the Emme restored floodplain, where no lateral branch was formed, even if 20 years have elapsed since the restoration project has started. Finally, no species was found in all the bare soils (FLUVIOSOLS BRUTS) constituted of coarse alluvium deposits and subjected to frequent floods. These soils are characteristic of pioneer stages in near-natural floodplain, and are re-created through the river widening.

Thus, comparisons of restored and non-restored stretches along the Emme River confirmed that, as in near-natural systems such as the Rhine floodplain, pedological characteristics reflecting fluvial dynamics govern the abundance of earthworm species concomitantly to fluvial dynamics as already demonstrated by Fournier et al. (2012). These authors showed that high abundance of small epigeic species and low abundance of large anecic species characterized poorly developed gravel bare soils recently created by the river widening and most exposed to flooding.

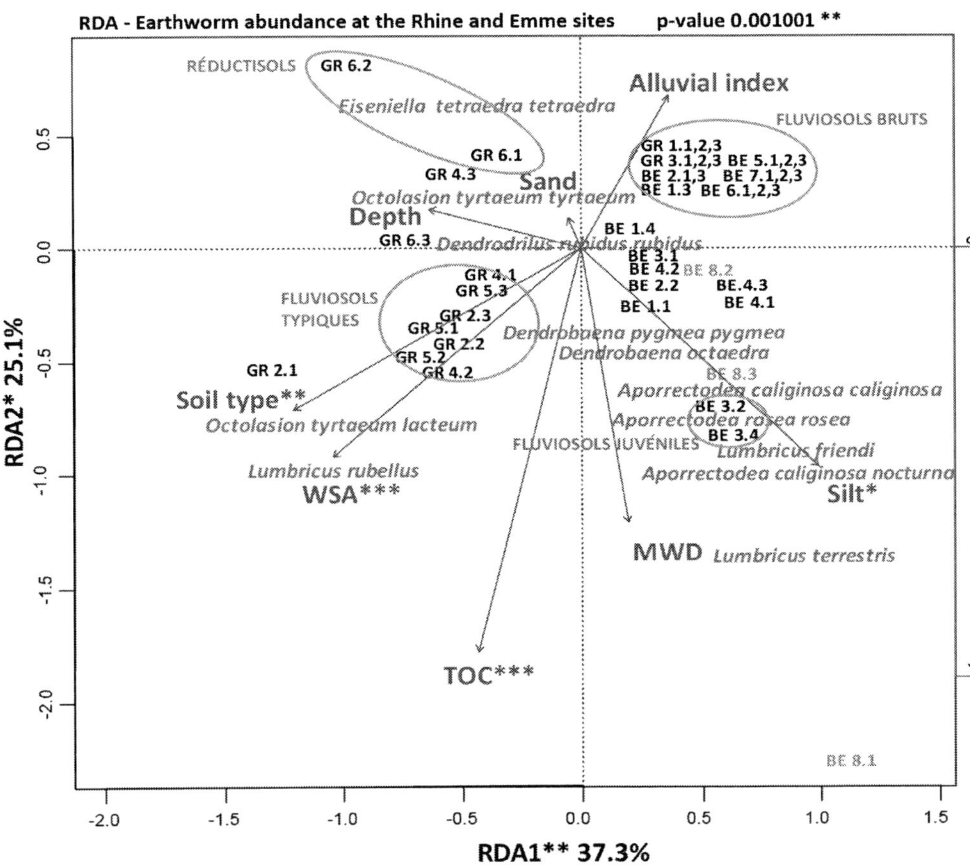

Figure 5. Redundancy Analysis on earthworm abundance at the near-natural site "Rhäzuns" along the Rhine River (GR), and at the Emme site (BE). GR 1 to GR 6, BE 1 to BE 8 indicated soil profiles with three replicates and their corresponding soil types (in green), BE 8.1, BE 8.2 and BE 8.3 in light blue representing the embanked site. Environmental variables are indicated in dark blue, TOC being the total organic carbon content, WSA the percentage of water-stable aggregates, MWD the mean water-stable aggregates diameter, and the alluvial index reflecting alluvial dynamics. Earthworm species are specified in red. Levels of statistical significance are as followed: *P <0.05, **P <0.01, ***P <0.001.

Near-Natural, Restored and Embanked Floodplains: What for Comparison?

As a general trend, we highlighted that our study sites (Emme, Rhine and Thur floodplains) respond similarly to flood disturbance regarding soil types, physico-chemical parameters and earthworm communities whatever they are restored or not. Flood intensity and seasonality, soil depth and stability, soil texture, and soil organic carbon content appear to be the main variables that govern and control earthworm distribution, abundance and species diversity.

River restoration has indeed a positive impact on earthworm communities through the creation or maintenance of diverse habitats. Except for the bare sediments, all these habitats are suitable for some species of earthworms and are then very quickly colonized by earthworms, over only several years after river restoration.

Comparing the Emme and the Thur flooplains, we observed that the stability of habitats obviously manages earthworm communities' distribution. We are also able to affirm that, regarding our results, river restoration enhance earthworm diversity and abundance especially when conducted along a wide and long reach. Thus, the Thur River widening was engaged in 1995 then 2002 while in 1991-92 then 1998-99 for the Emme River. So, it appears that restoration success regarding earthworm communities rehabilitation depends less on the time elapsed since restoration project has started, than on the used engineering technics and the size of the restored reach. A second challenge that needs further research is to assess if the Rhine River is a relevant near-natural reference. Indeed, earthworm diversity was lower in that place than at both the Emme and Thur sites perhaps due to upstream dams and embankements that may cause changes of the natural behaviour of the Rhine River. However, current researches are still in progress about this research of a near-natural floodplain reference.

By contrast to our studies, in some cases, some river rehabilitation may have a considerable negative impact on earthworm population and even may lead to extinction of some species (*Lumbricus rubellus*) in large areas of the restored floodplain. In the case of a lower River Rhine floodplain in Netherlands, the creation of a secondary channel and partial removal of embankment will, i) reduce the part of the floodplain area where the populations can sustain themselves, ii) modify the fluvial dynamics leading to more frequent flooding, especially during late spring and summer and hence a lower viability of the earthworm population, and iii) an increase in pollutants availability, mainly heavy metals, resulting from both more deposition of contaminated sediments and an increased exposure to stored contaminated sediment, may further decrease the area of the floodplain where earthworm population can sustain (Thonon and Klok, 2007).

4.3. Earthworm Communities As Bioindicators of Fluvial Dynamics and River Restoration Success

According to Henry et al. (2002), ecological restoration can be defined as returning an ecosystem to its condition prior to disturbance or to a state as similar as possible to that which prevailed prior disturbance. To assess the restoration success, we have compared the earthworm communities in a near-natural floodplain (considering as the state prior disturbance), a restored floodplain and an embanked floodplain. In the light of above results, earthworm communities vary depending on the gradient of naturalness from embanked to near-natural sites. Indeed, as the different earthworm ecological categories and species are more or less resistant to flood duration and frequency, earthworm communities can reveal the presence of different pedological habitats resulting from fluvial dynamics (flood frequency and duration, water-table fluctuations, texture of alluvial deposits, woody debris, etc.). Because earthworms are influenced by environment changes, they may therefore be particularly efficient for the purpose of soil bioindication in the context river restoration. Their potential role as bioindicators has already been highlighted by Bullinger-Weber et al. (2012) at the subalpine level. These authors showed that abundance of epigeics can to be used as bioindicator of fluvial dynamics. At the hill level, Fournier et al. (2012) confirmed that changes in flooding frequency have a predominant influence on earthworm community. Thus, the ratio of the relative abundances of epigeic and anecic species, and the differences in species composition within earthworm categories could be used as indicators of soil development and functioning in floodplains. Moreover, the use of ecological traits is one of the key that may improve the potential of earthworms as bioindicators (Fournier et al., 2012). As a complement to this study, we showed that, as earthworm communities rapidly colonize habitats created by river restoration, they can thus be used as bioindicators of early stages of river restoration.

Conclusion: Limits and Perspectives

In the framework of this chapter, we focused on study cases in Switzerland, in several braided rivers that carry calcareous-rich alluvial deposits. Regarding restoration impact on earthworm communities, it appears difficult to transpose and extrapolate our results to other fluvial systems. For instance, short-duration floods associated with large and rapid fluctuations of water-table and the coarse sandy texture of soil and sediment, avoid long-term submersion of soils, thus preventing earthworms from asphyxia as observed by Thonon and Klok (2007). In floodplains characterized by long-term anoxic conditions due to semi-permanent water table and loamy texture, earthworm communities may be more severely affected. In our study sites, damages to earthworm communities are mainly due to mechanical impact of floods that have drastic consequences on earthworm habitats through erosion and sediment deposits. Earthworms are infrequent in pioneer stages of soil formation while they may be maintained in areas less affected by flooding events.

The primarily colonisation by earthworms of bare soils that are typical of near-natural floodplains or re-created by river widening remains poorly studied. As highlighted by Eijackers (2010) no study has been found that factually combines primary colonisation by earthworms of bare soils with the impacts of earthworms on soil structure, texture and characteristics. In this context, manipulative mesoscosms could be used to better understand involved processes in soil structure formation and in particular the role of earthworm in the early stages of structure formation and stabilisation.

Moreover, in this chapter, we didn't take into account the chemical composition of soil, sediment and water despite it could play an important role on earthworm populations. Pollution by heavy metals or by pesticides tends to increase along the river coarse and often reachs severe values of contaminants in alluvial urban and agricultural lowland floodplains.

In addition, a better knowledge of the ecological needs (such as organic matter supply) and resilience of earthworm communities to disturbances that occur in floodplains (floods, pollution, sedimentation and erosion processes, hydric and thermic stress) could help to use earthworm communities as bioindicators of environmental conditions, of ecosystems functioning and in turn as bio-indicators of river restoration success. In this context, the floodplain characteristics (altitude, hydrological regime, previous landuses and management), the engineering technics applied to restore floodplains (length and width of the widening, partial or total embankment removal) as well as the time elapsed since the restoration should be considered. In addition, according to the aim of the floodplain restoration, the performance of the earthworm as bio-indicators should be compared with other bioindicators more frequently used such as plant communities or macro-invertebrates to assess the restoration success.

Long-term surveys would be also crucial to discriminate the effect of natural successional dynamics from fluctuations, and from human impacts on this ecosystem (Henry et al., 2002). In addition, according to these authors, such surveys also required to assess the success of ecological restoration, i.e. the self-sustainability (i.e. requiring minimal maintenance or management) of restored ecosystems.

ACKNOWLEDGMENTS

This research was supported by grants n° PMPDP2_123031 (MHV program) and n° 3100A0-116825 (FLOOD STRUBIO project) from the Swiss National Science Foundation. The authors would like to thank the CCES RECORD Catchment project, as well as Dr Clémence Salomé, Nathalie Moreira, Coraline Sahin and Didier Burkhalter for their help during the study.

REFERENCES

Baize, & Girard. (2009). Référentiel pédologique, Quae edition.

Bechtold, & Naiman. (2009). A quantitative model of soil organic matter accumulation during floodplain primary succession, *Ecosystems.*

Berke. (2010). Ecosystem engineering in the Marine Realm, *Integrative and Comparative Biology.*

Bertrand, Goldscheider, Gobat, & Hunkeler. (2012). From multi-scale conceptualization to a classification system for inland groundwater-dependent ecosystems. *Hydrogeology Journal.*

Bindesbol.. (2008). Interactions between climatic and toxic stress - Studies with the freeze tolerant earthworm Dendrobaena octaedra. PhD Thesis, 96 pp.

Bossuyt, Six, & Hendrix. (2005). Protection of soil carbon by microaggregates within earthworm casts, *Soil Biology and Biochemistry.*

Bouché. (1972). Lombriciens de France : écologie et systématique. INRA. Paris.

Bouché. (1977). Stratégies lombriciennes. *Ecological Bulletins.*

Bullinger-Weber, & Gobat. (2006). Identification of facies models in alluvial soil formation: The case of a Swiss alpine floodplain. *Geomorphology.*

Bullinger-Weber, Le Bayon, Guenat, & Gobat. (2007). Influence of some physicochemical and biological parameters on soil structure formation in alluvial soils. *European Journal of Soil Biology.*

Bullinger-Weber, Guenat, Gobat, & Le Bayon. (2012). Impact of flood deposits on earthworm communities in alder forests from a subalpine floodplain (Kandersteg, Switzerland). *European Journal of Soil Biology.*

Cierjacks, Kleinschmit, Kowarik, Graf, & Lang. (2011). Organic matter distribution in floodplain can be predicted using spatial and vegetation structure data. *River Research and Applications.*

Coleman, & Whitman. (2005). Linking species richness, biodiversity and ecosystem function in soil systems. *Pedobiologia.*

Cornelis, Koolhaas, Hamers, van Hoppe, van Roovert, Korsman, & Reinecke. (2009). Effects of metal pollution on earthworm communities in a contaminated floodplain area: Linking biomarker, community and functional responses. *Environmental Pollution.*

Cui, Yang, Yang, & Zhang. (2009). Evaluating the ecological performance of wetland restoration in the Yellow River Delta, China. *Ecological Engineering.*

Daniels. (2003). Floodplain aggradation and pedogenesis in a semiarid environment. *Geomorphology.*

Davidson, & Grieve. (2006a). Relationships between biodiversity and soil structure and function: Evidence from laboratory and field experiments. *Applied Soil Ecology.*

Davidson, & Grieve. (2006b). The influence of soil fauna on soil structural attributes under a limed and untreated upland grassland. *Land Degradation and Development.*

Decaëns. (2010). Macroecological patterns in soil communities. *Global Ecology and Biogeography.*

Dominati, Patterson, & Mackay. (2010). A framework for classifying and quantifying the natural capital and ecosystem services of soils. *Ecological Economics.* 69(9).

Edwards. (2004). Earthworm Ecology, 2nd ed., CRC Press LLC, St. Lucie Press, Boca Raton.

Edwards, & Bohlen. (1996). The Biology and Ecology of Earthworms. (3rd Ed.). Chapman & Hall, London.

Eijsackers. (2010). Earthworms as colonisers: Primary colonisation of contaminated land, and sediment and soil waste deposits. *Science of the Total Environment.*

Emmerling. (1995). Long-term effects of inundation dynamics and agricultural land-use on the distribution of soil macrofauna in fluviosols. *Biology and Fertility of Soils.*

Fisker, Sorensen, & Holmstrup. (2012). Costs of adaptation and expression of metallothionein in earthworm populations adapted to copper polluted soils. *Comparative Biochemistry and Physiology a-Molecular and Integrative Physiology.*

Fisker, Holmstrup, & Sorensen. (2013), Variation in metallothionein gene expression is associated with adaptation to copper in the earthworm Dendrobaena octaedra, *Comp. Biochem. Physiol. C-Toxicol. Pharmacol.*

Fonte, & Six. (2010). Earthworms and litter management contributions to ecosystem services in a tropical agroforestry system. *Ecological Applications.*

Fournier, Samaritani, Shrestha, Mitchell, & Le Bayon. (2012). Patterns of earthworm communities and species traits in relation to the perturbation gradient of a restored floodplain. *Applied Soil Ecology.*

Gallandat, Gobat, & Roulier. (1993). Cartographie des zones alluviales d'importance nationale, Cahier de l'environnement, OFEV, Berne.

Gerard. (1967). Factors affecting earthworms in pastures. *Journal of animal ecology.*

Gerrard. (1987). Alluvial soils, Hutchinson Ross, New York.

Gerrard. (1992). Soil geomorphology : an integration of pedology and geomorphology. Chapman & Hall, London.

Guenat, Bureau, Weber, & Toutain. (1999). Initial stages of soil formation in a riparian zone: Importance of biological agents and lithogenic inheritance in the development of the soil structure. *European Journal of Soil Biology.*

Henry, Amoros, & Roset. (2002). Restoration ecology of riverine wetlands: A 5-year post-operation survey on the Rhône River, France, *Ecological Engineering.*

IUSS Working Group WRB. (2006). World reference base for soil resources. FAO, Rome.

Ivask, Truu, Kuu, Truu, & Leito. (2007). Earthworm communities of flooded grasslands in Matsalu, Estonia. *European Journal of Soil Biology.*

Jabiol, Zanella, Ponge, Sartori, Englisch, van Delft, de Waal, & Le Bayon. (2013). A proposal for including humus forms in the *World Reference Base for Soil Resources.* (WRB-FAO). Geoderma.

Jansson, Backx, Boulton, Dixon, Dudgeon, Hughes, Nakamura, Stanley, & Tockner. (2005). Stating mechanisms and refining criteria for ecologically successful river restoration: a comment on Palmer et al. (2005). *Journal of Applied Ecology.*

Jimenez, Decaëns, Amezquita, Rao, Thomas, & Lavelle. (2011). Short-range spatial variability of soil physico-chemical variables related to earthworm clustering in a neotropical gallery forest, *Soil Biology and Biochemistry.*

Joergensen, Overgaard, Holmstrup, & Westh. (2008). Cryoprotectants are metabolic fuels during long term frost exposure in the earthworm Dendrobaena octaedra. *Comparative Biochemistry and Physiology a-Molecular and Integrative Physiology.*

Jones, Lawton, & Shachak. (1994). Organisms as ecosystem engineers. *Oikos.*

Jones, Lawton, & Shachak. (1997). Ecosystem engineering by organisms: Why semantics matters. *Trends in Ecology and Evolution.*

Karamouz, Noori, Moridi, & Ahmadi. (2011). Evaluation of floodplain variability considering impacts of climate change. *Hydrological Processes.*

Lal. (1991). Soil conservation and biodiversity. Biodiversity of microorganisms and invertebrates: its role in sustainable agriculture. Proceedings of the First Workshop on the Ecological Foundations of Sustainable Agriculture (WEFSA 1), London.

Lapied, Nahmani, & Rousseau. (2009). Influence of texture and amendments on soil properties and earthworm communities. *Applied Soil Ecology.*

Lavelle. (1997). Faunal activites and soil processes: Adaptive strategies that determine ecosystem function. *Advances in Ecological Research.*

Lavelle, Bignell, Lepage, Wolters, Roger, Ineson, Heal, & Dhillion. (1997). Soil function in a changing world: the role of invertebrate ecosystem engineers. *European Journal of Soil Biology.*

Le Bayon, & Binet. (2006). Earthworms change the distribution and availability of phosphorous in organic substrates. *Soil Biology and Biochemistry.*

Lee. (1985). Earthworms: Their ecology and relationships with soil and land use. Academic Press. Australia.

Lorenz. (2003). Bioindicators for ecosystem management, with special reference to freshwater systems. in Bioindicators and Biomonitors : Principles, concepts and applications. Markert, Breure, Zechmeister, Elsevier.

McInerney, Little, & Bolger. (2001). Effect of earthworm cast formation on the stabilization of organic matter in fine soil fractions. *European Journal of Soil Biology*.

Milner, & Tockner. (2010). River science - What has it contributed to general ecological theory?. *River Research and Applications*.

Mitra, Wassmann, & Vlek. (2005). An appraisal of global wetland area and its organic carbon stock. *Current Science*.

O'Hanley. (2011). Open rivers: Barrier removal planning and the restoration of free-flowing rivers. *Journal of Environmental Management*.

Palmer, & Bernhardt. (2006). Hydroecology and river restoration: Ripe for research and synthesis. *Water Resources Research*.

Palmer, Bernhardt, Allan, Lake, Alexander, Brooks, Carr, Clayton, Dah, Shah, Galat, Loss, Goodwin, Hart, Hassett, Jenkinson, Kondolf, Lave, Meyer, O'Donnell, Pagano, & Sudduth (2005). Standards for ecologically successful river restoration. *Journal of Applied Ecology*.

Pasquale, Perona, Schneider, Shrestha, Wombacher, & Burlando. (2011). Modern comprehensive approach to monitor the morphodynamic evolution of a restored river corridor. *Hydrology and Earth System Sciences*.

Petts, & Amoros. (1996). Fluvial Hydrosystems. Chapman & Hall. London.

Plum. (2005). Terrestrial invertebrates in flooded grassland: A literature review. Wetlands.

Plum, & Filser. (2005). Floods and drought: Response of earthworms and potworms. (Oligochaeta : Lumbricidae, Enchytraeidae) to hydrological extremes in wet grassland. *Pedobiologia*.

Ponge. (2013). Plant-soil feedbacks mediated by humus forms: A review. *Soil Biologyand Biochemistry*.

Roots. (1956). The water relations of earthworms. Resistance to dessication and immersion, and behaviour when submerged and when allowed a choice of environment. *Journal of Experimental Biology*.

Ruiz-Jaen, & Aide. (2005). Restoration success: How is it being measured? *Restoration Ecology*.

Salomé. (2011). La structure des sols en zones alluviales : importance des communautés lombriciennes et du matériel parental dans les premières étapes de la pédogénèse. *Doctorat es Sciences thesis*. Neuchâtel.

Salomé, Guenat, Bullinger Weber, Gobat, & Le Bayon. (2011). Earthworm communities in alluvial forests: Influence of altitude, vegetation stages and soil parameters. *Pedobiologia*.

Satchell. (1967). Lumbricidae, Soil biology (A. Burges and F. Raw, eds). Academic Press, London.

Schrader, & Zhang. (1997). Earthworm casting: stabilization or destabilization of soil structure ? *Soil Biology and Biochemistry*.

Schütz, Nagel, Dill, & Scheu. (2008). Structure and functioning of earthworm communities in woodland flooding systems used for drinking water production. *Applied Soil Ecology*.

Shipitalo, & Le Bayon. (2004). Quantifying the effects of earthworms on soil aggregation and porosity. *Earthworm Ecology* . Edwards, CRC Press, Boca Raton.

Simonsen, & Klok. (2010). Genetic and ecological impacts of heavy metal and flooding stress on the earthworm Lumbricus rubellus in floodplains of the Rhine river. *Soil Biology and Biochemistry.*

Sims, & Gerard. (1999). Earthworms: Note for the identification of British species. *Field Studies Council, Shrewsbury.*

Thonon, & Klok. (2007). Impact of a changed inundation regime caused by climate change and floodplain rehabilitation on population viability of earthworms in a lower River Rhine floodplain. *The Science of the Total Environment.*

Tockner, & Stanford. (2002). Riverine flood plains: present state and future trends. *Environmental Conservation.*

Wang, Zhu, Meng, Wang, Xie, & Zhang. (2012). The combined stress effects of atrazine and cadmium on the earthworm Eisenia fetida. *Environmental Toxicology and Chemistry.*

Wohl, Angermeier, Bledsoe, Kondolf, MacDonnell, Merritt, Palmer, Poff, & Tarboton. (2005). *River restoration Water Resources Research.*

Woolsey, Capelli, Gonser, Hoehn, Hostmann, Junker, Paetzold, Roulier, Schweizer, Tiegs, Tockner, Weber, & Peter. (2007). A strategy to assess river restoration success. *Freshwater Biology.*

Zorn, Van Gestel, & Eijsackers. (2005). Species-specific earthworm population responses in relation to flooding dynamics in a Dutch floodplain soil. *Pedobiologia.*

Zorn, Van Gestel, Morrien, Wagenaar, & Eijsackers. (2008). Flooding responses of three earthworm species, Allolobophora chlorotica, Aporrectodea caliginosa and Lumbricus rubellus, in a laboratory-controlled environment. *Soil Biology and Biochemistry.*

In: Floodplains
Editor: Enner Herenio Alcântara
ISBN: 978-1-62808-998-1

Chapter 4

THREE MAIN STAGES OF FLOODPLAIN EVOLUTION IN NORTHERN EURASIA AND THEIR ECOLOGICAL SIGNIFICANCE

Aleksey Sidorchuk*[*1], Olga Borisova[2], Aleksey Chernov[1] and Andrey Panin[1]
[1]Geographical Faculty, Moscow State University,
Moscow, Russia
[2]Institute of Geography, Russian Academy of Sciences,
Moscow, Russia

ABSTRACT

Three main stages of floodplain formation were identified in northern Eurasia: (1) floodplain formation by the rivers larger than modern ones, (2) floodplain formation by the rivers smaller than modern ones, and (3) floodplain formation by the rivers of the modern morphological type.

Large paleochannels of the first stage are common on river floodplains and low terraces on the East European and West Siberian plains. They are up to 10-15 times greater than modern river channels. The large paleochannels are dated to 11-15 thousand radiocarbon years B.P. (the Late Glacial time). The reconstructed surface runoff was then 1.4 times greater than the modern one on the northern mega-slope of the East European Plain, 2.3 times greater on its southern megaslope, and two times greater in West Siberia.

Small oxbows of the second stage are wide spread on the floodplains of lowland rivers in Northern Eurasia, mostly in the forest zone. They were formed during the Atlantic period of the Holocene. During the Holocene climatic optimum, water runoff from the northern mega-slope of the East European Plain was ~180 $km^3 a^{-1}$, which is 30% less than the present-day runoff from the same drainage area. The annual runoff in the Volga River basin was ~134 km^3, which is about one half of the present value. The runoff in the Don and Dnepr River basins during the Holocene optimum was 40% less, and that in the Ob' and Irtysh River basin was 30% less than the modern one.

* Corresponding author: Aleksey Sidorchuk: fluvial05@gmail.com.

The rivers of the present-day sizes and morphological types were formed during the climate cooling and wetting during the Subboreal and Subatlantic time of the Holocene, the general process being an increase in the river channel size and decrease in the width of meandering belt on the floodplain. The floodplain topography, formed during the previous stages, have significant influence on the modern flow hydraulics and the regime of sedimentation, being some kind of past events 'memory'.

Since the Holocene climatic optimum can be considered as a paleoanalogue of the global anthropogenic warming of the mid-21st century, information on the river floodplain evolution during the Late Glacial – Holocene can be used in the ecological scenarios of the global warming.

The transformation of large lateglacial paleochannels into smaller ones due to climate warming at the beginning of the Holocene represents a likely scenario of the ecological changes within the present-day permafrost zone. The revers transformation of the modern rivers into smaller ones is a likely scenario of the ecological changes within the present-day forest zone. A general decrease in surface runoff may cause considerable landscape transformation in this vast territory.

1. Introduction

An extensive literature has been focused on the studies of floodplains (or floodplain-channel complexes, according to Chernov (2009) of the modern rivers of the Northern Eurasia (Yelenevsky, 1936; Schanzer, 1951; Popov, 1969; Lewin, 1978; Chernov, 1983, 2006; Baryshnikov, 1984; Kondratyev et al., 1982; Chalov & Chernov, 1985; Nansen & Croke, 1992).

Considerably less attention has been paid to the history of the formation of the floodplain relief on this large territory (Chernov, 1983; Volkov, 1963; Makkaveev & Chalov, 1970; Rotnicki, 1991; Sidorchuk et al., 2000; Chalov et al., 2004), although the need in such investigations was acknowledged, for example, in the course of the studies on the riparian forest successions (Braslavskaya, 2004). Paleohydrological research based largely on the studies of the oxbows morphology show that within the periglacial belt of the Northern hemisphere the floodplains represent heterogeneous formations, where often only a small fraction of the area is formed by the erosion-accumulation work of the channel of the modern morphological type.

On the river floodplains and low terraces Dury (1954) and Volkov (1963) discovered widespread oxbows with larger sizes than those of the modern meanders of the respective river channels. Makkaveev (1969) showed that on the floodplains of meandering rivers of the Northern Eurasia there are oxbows both larger and smaller than the modern oxbow lakes.

Reconstruction of the main stages of the river floodplains formation, studies of the paleochannels morphology and age, as well as their role in the floodplain relief development, forms the basis for the qualitative and quantitative analysis of the evolution of the valley landscapes and for assessment of changes of the river discharge over time. Such reconstructions can be used to predict similar changes in the future due to the climate change and increasing human influence on the landscape.

Evolution of the floodplains on the lowlands of Northern Eurasia was investigated by the authors from the beginning of 1970s. This article mainly summarizes our own results with only a few references to the publications of other researchers on the different territories.

2. Methods

Floodplain Evolution Reconstruction

Floodplain is the part of the river valley, which is covered by riparian vegetation and is periodically inundated by the flood waters. Being in the vicinity to the river channel floodplain surface and sediment are transformed by erosional and sedimentary activity of the river. Therefore in the most cases a floodplain topography and sedimentary body are formed by the river channel lateral migration. As the result even the neighboring parts of a floodplain can be of different age, texture and height. Simultaneously, floodplain surface is covered with fine deposits, formed during the flood inundation. This sedimentation process is not even – floodplain height near the main channel increases much faster than at the distal parts. Some parts of floodplain surface, usually the lowermost ones, can be eroded during the flood. Therefore deposits of the same age can be situated at the different floodplain levels. Floodplain soils, vegetation species and succession stages are controlled by floodplain height and inundation frequency. At the same time vegetation species and succession stages control the floodplain surface accretion rate and soil formation. Fluvial processes are the main in floodplain evolution. Some other processes, for example, aeolian, can be also significant on sand floodplain deposits. The reconstruction of floodplain evolution must consider all the variability of these heterogeneous and uneven-aged sedimentary bodies with the complicated surface relief.

The most natural way to reconsrtruct the spatial and temporal evolution of some floodplain is to detect similar floodplain segments and to find the relative temporal correspondence of different uniform units (Panin et al., 1999). It is based on the investigation of large-scale maps and images of floodplain surface, on the coring of floodplain deposits and the analyses of their composition, structure and age. This work is often results in the geomorphological map of the floodplain, showing the main floodplain uniform units and the sequence of their formation (Figure 1). These uniform units are characterised by the age, surface morphology (qualitative and quantitative), soil and vegetation, deposit composition, texture and structure, as well as paleofloristic characteristic.

Morphological Analysis of the Paleochannels

The morphology of floodplain uniform unit gives information about the river channel pattern at the time of floodplain segment formation, the main geometrical characteristics of these channels (paleochannels). The most common river channel pattern on the plains of the Northern Eurasia is the meandering, only the largest rivers have several branches, divided by islands and braid bars.

Alluvial meandering rivers usually have one channel or distinct main channel, which is bending within vegetated floodplain. Usually their morphology is characterised at the bankfull stage, when the water level at meander crosses is close (in average) to the level of the vegetated floodplain surface. The main morphometric characteristics are channel mean width, meander mean wavelength and amplitude, and channel mean depth.

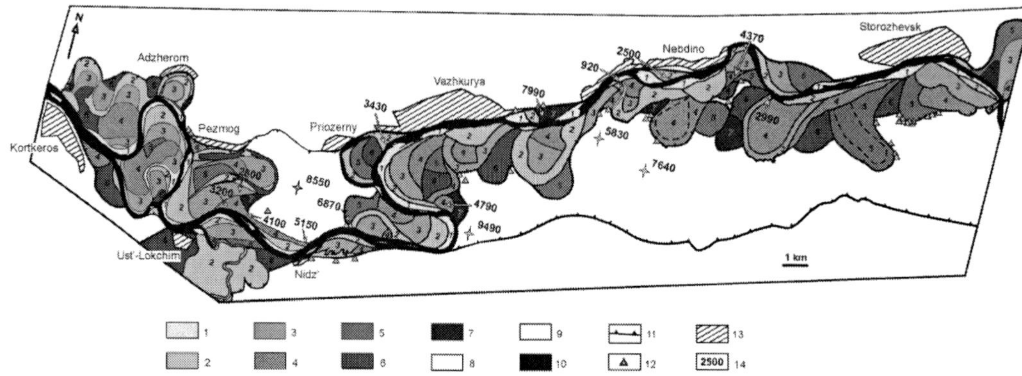

Keys: floodplain segments of different age: (1) modern (late Subatlantic, <1.0 C^{14} kyr BP), (2) middle Subatlantic (1.8-2.2 kyr BP), (3) early Subatlantic (2.2-2.5 kyr BP), (4) Subboreal (2.5-5.0 kyr BP), (5) late and middle Atlantic (5.0-7.0 kyr BP), (6) early Atlantic (7.0-8.0 kyr BP), (7) Boreal and Preboreal (8.0-10.0 kyr BP); fluvial terraces: (8) 1st terrace (8-10 m above low water level, 10.0-14.0 kyr BP), (9) 2nd terrace (30-35 m); (10) modern river channel and oxbow lakes on the floodplain; (11) bluffs of the terraces; (12) Neolithic Man sites; (13) modern settlements; (14) radiocarbon dates.

Figure 1. The geomorphic map of the Vychegda River floodplain near the Lokchim River mouth.

At the bankfull stage central bars and islands are usually submerged, therefore the channel bankfull width for this stage can be measured as a minimal distance between two opposite banks at a given point. The error of an individual measurement depends on the map or image scale and on the accuracy of floodplain edge estimate. As this error can be both positive and negative, this error is reduced by multiple measurements. Multiple measurements need to be done for estimating the mean width of any channel reach without large tributaries. Bankfull width varies along a channel meander, sometimes significantly. To reduce this variability, the width is measured at the crosses of meandering channel, i.e. at the points where the bend curvature sign changes to the opposite.

For the modern rivers the available number of crosses is 30-100 for different reaches without large tributaries. The error of an individual measurement depends on the river image scale and on the accuracy of floodplain edge estimate, and for large scale images is about ±5 % of a channel width. The relative error of the mean value estimate E_{rm} equals to

$$E_{rm} = \frac{\sigma}{M\sqrt{N}} \tag{1}$$

where σ is the estimate of standard deviation, M is the estimate of the mean value, and N is the number of measurements. For the modern rivers this error varies between 2-5%.

The available number of crosses at a single uniform reach of a paleochannel is usually less than for the recent rivers. The error of an individual measurement due to the scale is the same as for the modern rivers. The error of an individual measurement due to the accuracy of the floodplain edge estimate is much more uncertain. Usually the width of the paleochannel infill surface is visible and measurable, although it can be both less and more than its former bankfull width. In some cases oxbow lakes exist in paleochannels, and their sizes give some estimates of the former channel width. Individual estimates of the paleochannel width at channel crosses made by different experts vary by ±10%. Such sites are most favorable for estimations of the bankfull width, as channel banks are steep there, so that any width measured above the flat channel bottom would be close to the bankfull width. Sections at pools give much higher relative error of the bankfull width estimate due to a triangle cross-section shape and rapid width increase with water level.

The error of an individual measurement is rather large for paleochannels, but the relative error of the mean value estimate can be even larger. For example, the latter varies between 2 and 16% for the paleorivers in the Volga River basin. Therefore the mean cumulative error of the bankfull width estimate there can be ±15%, which is rather high.

Meander wavelength L (the minimal distance between two successive meander tops along the same bank) or meander step λ (the minimal distance between two successive points of axis line curvature sign alteration, which is half of wavelength on average) usually can be measured for a larger number of preserved fragments of a paleochannel than its width. The absolute error of an individual meander step measurement is not more than that of a width measurement; therefore its relative error is about 1-2% of meander wavelength. Relative error of the mean value estimate for the meander wavelength is rather large due to a high variability of this parameter. As it decreases with the number of measurements, according to (1), the number of measured meander wavelengths must be 3-10 times more than the number of measured widths to get the same accuracy of the mean estimate. In general, for the rivers on the plains of the Northern Eurasia the cumulative error of meander wavelength estimate is less than the error of paleochannel width estimate, its mean value being ±10%.

The correlation exists between the river channel bankfull width and meander wavelength (Leopold and Wolman, 1960):

$$L = k_w W \tag{2}$$

Coefficient k_w varies from 6 to 11 for different sets of data (see Leopold and Wolman, 1960), it is 11.2 for the rivers of the Northern Eurasia (Figure 2). So, it is possible to estimate paleochannel effective width as

$$W^* = \frac{W_{measured} + L / k_w}{2} \tag{3}$$

The use of formula (3) decreases the error of the mean paleochannel width estimate: its mean value for the paleochannels is ±12%.

The most indefinite measure to obtain is the paleochannel depth. In the most cases it requires coring of floodplain, sometimes it is possible to measure the depth of ox-bow lakes. In the cores it is necessary to estimate the position of the former channel bed, which is covered with infill deposits. Usually the stratigraphic contact between the fine floodplain deposits and coarser channel alluvium is taken for the former channel bed at the time of paleochannel avulsion. The maximum depth of a paleochannel during the powerful floods can be estimated by the position of so called "base" strata, which consists of the most coarse particles. These two measures show the range of paleochannel depth variation for a given section, and often this range is significant.

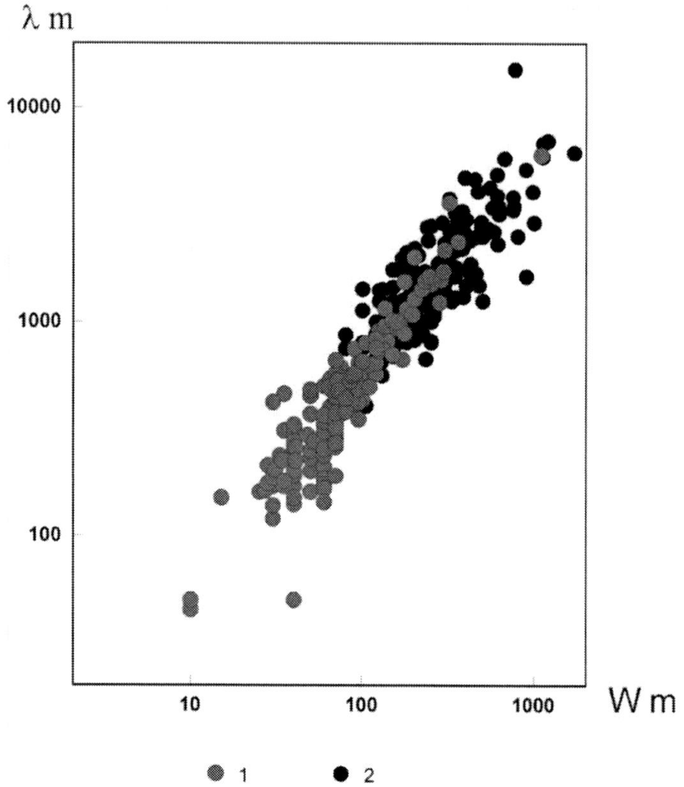

Figure 2. Relationship between the meander half-wavelength λ and the channel width W_m (1) for present-day and (2) ancient rivers.

Paleohydrological Reconstructions

The measures of the paleochannels are the main information used for paleohydrological reconstructions. There are two main ways of paleohydrological reconstruction, both first used by Dury (1964, 1965): 1) calculation of flow velocity and discharge in paleoriver for a given stage based on cross-section geometry; 2) calculation of flow discharge of some return period based on surface plan geometry of paleochannel and regime equations.

The first method was broadly discussed and used by Rotnicki (1983, 1991). The method has good theoretical basement on the equations of 1D and 2D hydraulics. It has significant limitations: 1) cross-section geometry of a paleochannel must be known in detail; 2) the water surface slope and hydraulic resistance to flow must be known for different water levels; 3) hydrologic regimen of the river-analogue must be known to estimate such important characteristics as former annual or mean maximum discharge. There are some suggestions on the methodology of such estimate (Sidorchuk & Borisova, 2000), which need further investigations.

The second method (so-called regime equations approach) is based on the relationships between channel morphology and flow hydrology. It is much more simple and gives discharge values of a given return period, such as annual or mean maximum discharge. Its advantages and limitations were discussed in (Sidorchuk et al., 2008, 2009). The method requires two main procedures: 1) measurement of paleochannel plan geometry characteristics, such as channel width W and meander wavelength L, estimation of their mean values and errors of estimates; 2) determination of transfer functions between paleochannel plan geometry and former flow discharge based on empirical information on such relationships for the modern rivers.

The relationships between river channel plan geometry and flow discharge are well known for the rivers since the work of Ferguson (1863, after Leliavskiy, 1955). From a great number of regime equations, we can use in paleohydrology only the simplest ones, such as the downstream relation between channel width W and flow discharge Q.

$$W = kQ^b \tag{4}$$

Formula (4) is an empirical formula. As any empirical formula, it shows some tendencies only within the field of measured values and can be wrong out of that field. The variability of coefficient k for different sets of empirical data makes it necessary to use a broad range of river sizes and river basin landscapes to work out the empirical formula, which suits the purposes of paleohydrological reconstructions.

About 450 sections of rivers in the Northern Eurasia with mean annual discharge Q_a from 1 to 13000 m^3 s^{-1} and channel bankfull width W_b from 15 up to 3000 m, with the drainage basins situated in a variety of landscapes from steppe to tundra, were used to determine formula (4) with the variable coefficient k (Figure 3). Based on these data, the relationship in question takes the form

$$Q_a = 0.012y^{0.73}W_b^{1.36} \tag{5}$$

where

$$y = 100\frac{Q_a}{Q_{\max}} \tag{6}$$

Coefficient y is annual/mean maximum discharge ratio, i.e. within-year discharge variability. Flow variability depends on the basin area F (km^2):

$$y=aF^{0.125} \tag{7}$$

Parameter a in formula (7) reflects the geographical distribution of climatic flow variability independent from the size of the river basin. It can be calculated from measured mean annual discharge Q_a, mean maximum discharge $Q_{\max}$ and basin area F. Parameter a typically varies between 1.5 for the river basins with high seasonal flow variability and over 4 for the river basins with low variability. For paleolandscapes for each paleochannel parameter a is estimated using recent fluvial analogues. Then it is possible to calculate y for a given paleolandscape and basin area F with formula (7), the mean annual discharge Q_a from the paleochannel width with formula (5), and the mean maximum discharge Q_{max} with formula (6).

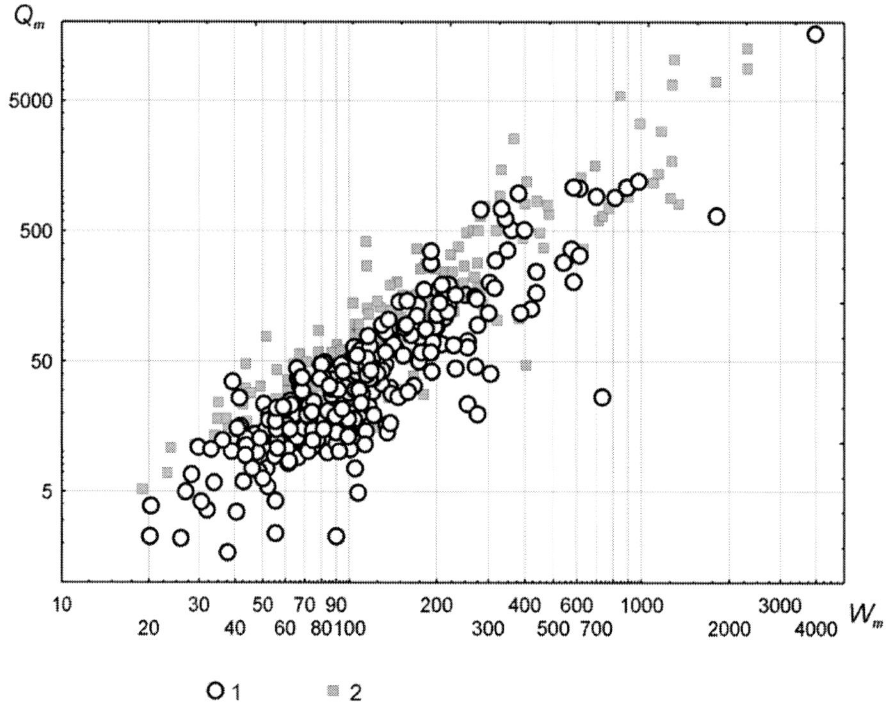

Figure 3. Relationship between the mean water discharge Q_m and channel width W_m for different discharge variability y: (1) $y \geq 12$ and (2) $y<12$.

Estimation of coefficient a in formula (7) for the past requires the knowledge of this relationship for the past landscape or its modern hydro-climatic analogue. Geographical influences on river flow bring about similarity of hydrological regimes for rivers in similar landscapes (Evstigneev, 1990). Geographical controls over river flow and their applications to paleohydrology lead to the principle of paleogeographical analogy (Sidorchuk and Borisova, 2000): a hydrological regime of a paleoriver within a paleolandscape must have been similar to that of a present-day river within the same type of landscape. Therefore, hydrological regimes of modern rivers in a certain type of landscape can be used for estimations of the paleohydrological regime in the same type of paleolandscape.

Paleofloristic Method of Paleolandscape and Hydro-Climate Reconstructions

Usually, methods of climate reconstruction rely either on the comparison of fossil and modern pollen assemblages and their associated modern climate, or on selected indicator plant species with certain climatic requirements. Paleobotanical data used for such reconstructions are of two main kinds: plant macrofossils (seeds, fruits, leaves, etc.) and pollen and spores. Macrofossils have the advantage that they can usually be identified to the species level. On the other hand, the occurrence of macrofossils is relatively restricted, and they usually belong mainly to aquatic and sub-aquatic plants. Pollen diagrams provide more detailed information on the history of specific plant taxa as well as vegetation on the whole. Because plants sensitive to climatic conditions are mainly medium to low pollen producers, sufficiently detail pollen data are essential. In the process of pollen identification, the highest possible taxonomic resolution should be achieved to obtain more complete resulting lists of fossil floras. If well-preserved, pollen of arboreal plants, as well as pollen and spores of certain groups of herbaceous plants (e.g. *Thalictrum*, *Lycopodium*, *Equisetum*) can be identified to genus or even species levels.

Iversen (1944) was first to use pollen occurrences of certain plant species to estimate paleotemperatures. This author established the relationship between present occurrences of *Ilex* and *Hedera* and summer as well as winter temperatures. Such relationships can be established by comparing present boundaries of the plant geographical range (its "area") and climatic data, e.g. the coincidence of “area” limits and certain isotherms. For example, Iversen showed that present-day ranges of *Ilex* and *Hedera* are limited by the 0°C and -2°C January isotherms respectively. A well-known example of this kind is the coincidence of the tree-line with the 10°C July isotherm in many lowland and mountain areas. This approach has subsequently been extended by increasing the number of indicator species and that of climatic indexes considered (e.g. Hintikka, 1963; Zagwijn, 1996). Grichuk (1969) further developed Iversen's approach into the mutual climatic range method. In this method as many species as possible are used to obtain paleoclimatic estimates for a specific site. Their geographical distributions are first converted to climatic ranges, expressed as climatic range diagrams – climatograms, and after that their mutual climatic field is determined. Method of climatograms has the advantage that even plant species without present-day geographical overlap can be used for paleoclimatic reconstructions. Moreover, as this approach is based on the requirements of individual species, finding modern analogues for the fossil pollen spectra or plant assemblages is not necessary.

Grichuk (1969) suggested yet another method of climatic reconstruction based on paleofloristic data, elaborating on the idea of Szafer (1946). This method consists of identifying a region where all the species of a fossil flora grow at the present time. The mutual area of all individual species of a certain fossil flora can be found by overlapping their present-day ranges. As many present day ranges of plants as possible should be used. Geographical analysis of the modern spatial distribution of all the plants of a certain fossil flora (compilation of a so-called arealogram) allows finding the location of the closest modern floristic analogue to the past vegetation at the site. By identifying the region where the majority of plant species found in a fossil flora are growing at the present time, it is possible to determine the closest modern landscape and hydro-climatic analogue of the past environment under consideration.

An accuracy of these reconstructions depends on the accuracy of paleofloristic definitions based on detailed pollen analysis, on the richness of the resulting fossil floras, on the accuracy of the data on present-day geographical ranges of plants – components of the fossil floras, on the sizes of the region-analogues, and the variability of hydro-climatic characteristics within them. The accuracy of palynological analysis and the richness of a paleoflora depend not only on palynologist’s personal experience, but also on the type of sediment, vegetation type, and the degree of pollen preservation. The lateglacial floras are usually relatively poor, so that not all analyzed samples provide sufficient data for paleofloristic reconstruction.

Usually the conditions suitable for all the species of a given fossil flora can be found within a comparatively small area. The present-day features of plant communities and the main hydro-climatic indices of such a region-analogue would be close, in most cases, if not identical, to those that existed at the sampled site in the past.

3. The Main Stages of the Floodplain Formation on the Plains of Northern Eurasia

The relief of floodplains was formed relatively recently: the oldest dated fluvial relief with an age of about 100,000 years is found on the floodplain of the Murrumbidgee River in Australia (Page et al., 1996). The river floodplains within the lowlands of Northern Eurasia, which are analyzed in this study, are usually no more than 17,000 years old, being formed after the beginning of deglaciation. Therefore the fluvial topography there is well preserved. On the floodplains of meandering rivers, the systems of natural levees marking wandering ancient channels are usually well defined. Hence the morphometric characteristics of the ancient channels, such as channel width, meanders amplitude and wavelength, can be measured fairly reliably. Field studies and analyses of the maps, aerial photographs and space images (Figure 4) show that the development of floodplains of meandering rivers on the plains of Northern Eurasia occurred in three main stages (Panin et al., 2011).

During the first stage the floodplains were formed by very large rivers. The channel width, meander amplitude and wavelength of these rivers were up to 13-15 times the size of the respective modern rivers. At this stage, very wide floodplains, corresponding to the widths of meandering belts of the large rivers, were formed. During the subsequent stages of development at some territories these floodplains partly came out of flood influence and became low terraces, and at other territories are still completely inundated by flood waters.

In the second stage, floodplains were formed by rivers 1.5-2 times smaller than modern rivers with steeper channel bends. Sometimes the floodplains of this stage occupy relatively small portions of floodplains created by large meandering channels of the previous stage, though often small meandering channels rework major portions of the first generation floodplains, so that the large oxbows are not preserved on such floodplain segments.

The third, current stage of the floodplains formation is associated with the increase in the river sizes accompanied by the decrease in curvature and amplitude of the river channel meanders. As a result, the channel meandering belts became narrower, which helped to preserve the fluvial relief formed during the previous stages in many areas of the river floodplains. Only on the largest rivers the inherited floodplain relief has been completely reworked. However, channel meandering in general leads to destruction of the ancient floodplain topography. Therefore to distinguish the typical stages of the floodplains formation it is necessary to study fairly long stretches of river valleys.

Sometimes the rivers which were meandering in the first stage of their evolution later formed several branches and/or anastomosing channel patterns. In that case the second and third stages of the floodplain formation cannot be distinguished.

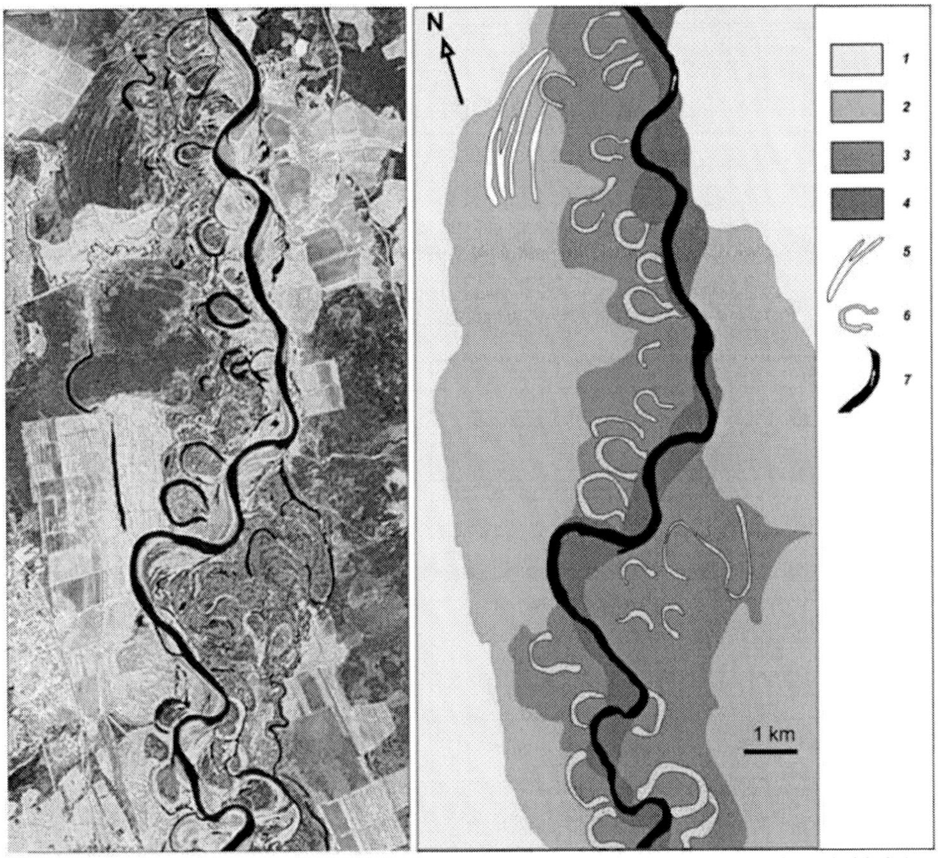

Keys: (1) Slopes of the valley and terraces, (2–4) floodplains of the first, second, and third (current) stages of formation, respectively, (5) natural levees with large radius of curvature on the 1st-stage floodplain, (6) oxbows of a channel with small radius of curvature on the 2nd-stage floodplain, (7) current channel with gentle meanders.

Figure 4. Floodplain morphology of the Sysola River upstream of Syktyvkar.

The First Stage: Floodplain Formation by the Rivers Larger Than Modern Ones

On the Russian Plain, in Western Siberia and in northern Kazakhstan the floodplains of the first stage are found in all landscape zones from the tundra to the dry steppe. These floodplains are clearly distinguished by the presence of extremely large natural levees, and of very large paleochannels, filled in by loam and peat deposits (Panin et al., 1992; Sidorchuk et al., 2001a; Panin & Sidorchuk, 2006). Currently about 200 well preserved large paleochannel fragments were found on floodplains in these regions (Table 1). Studies of the paleochannel fragments at the key-sites in the Seim, Khoper, Moskva, and Protva river basins showed that major parts of the modern floodplains were formed at this stage.

Table 1. Parameters of meanders and macromeanders of the rivers on the Northern Eurasia lowlands

Index	River	Latitude N	Longitude E	F km^2	λ_r m	W_r m	Q_r m^3/s	λ_p m	W_p m	Q_p m^3/s
33	Oma	66.68	46.40	4100	610	75	47.2	1750	150	66
26	Vym'	64.17	51.53	2700	300	40	38	1450	150	54
25	Chisva	63.55	51.33	1000	150	15	11.5	1000	120	33
17	Veslyana	62.83	51.33	3940	500	60	42.9	2800	250	118
19	Ched'yev	62.62	49.33	180	80		1.8	500		18
18	Yarenga	62.58	49.47	2450	270	40	11.1	1300	200	57
21	Paden'ga	61.87	42.57	1040	260	40	12.5	1250	125	40
22	Sheren'ga	61.77	42.67	300	50	10	1.8	310		9
23	Vamsheren'ga	61.77	42.75	300	45	10	1.8	430		13
24	Bolshaya Churga	61.48	42.37	560	190	40	6.7	750	80	21
16	Lokchim	61.60	51.60	6040	640	100	52.8	2900	1000	286
65	Vychegda	61.27	46.73	121000	6000	1100	1100	7000	1200	900
20	Luza	60.58	47.25	18100	1500	250	137.7	4630	450	266
27	Yug	59.88	45.50	4600	430	100	38	2000	380	124
7	Nemnyuga	65.42	43.77	800	230	35	7.2	750	100	24
6	Sula	64.88	48.45	1040	310	35	11.4	1130	100	33
4	Khar'yaga	67.15	56.58	970	230	50	10.7	810	150	32
5	Adz'va	67.08	60.12	8700	1540	175	100	1810	180	80
3	Soz'va	66.58	52.88	1400	480	50	15.4	1080	150	41
14	Laya	66.42	56.37	9530	2000	200	104.8	2200		139
1	Ersa	66.15	53.35	1800	250	50	19.8	1000	125	36
2	Slepaya	66.18	53.35	230	85		2.5	470	70	13
10	Tobysh	66.07	51.17	3250	530	100	35.8	1700	200	71

Index	River	Latitude N	Longitude E	F km^2	λ_r m	W_r m	Q_r m^3/s	λ_p m	W_p m	Q_p m^3/s
8	Lyzha	65.73	56.32	6000	810	140	66	3080		197
12	Pizhma	65.38	52.02	5470	580	100	52.5	1500	200	69
13	Neritsa	65.22	52.62	2900	380	70	31.6	1250	150	49
9	Lem'yu	63.97	56.63	3850	570	120	42.4	4800		319
11	Ayyuva	63.75	54.17	1970	500	60	19.9	2100	180	76
15	Bolshaya Lyaga	62.62	56.77	1330	450	50	16.7	1400	125	45
32	Pil'va	60.72	55.95	890	170	35	8	1200		36.7
31	Veslyana	60.43	52.7	3900	180	35	33.2	1700		67.5
67	Yaz'va	60.17	57.02	3470	651	100	63.1	1420	200	56.8
68	Vyatka	59.4	51.78	5340	339	70	52.2	1992	300	98.4
28	Ya'va	59.15	57.03	5230	830	120	62.8	2800	500	177.9
69	Kama	58.93	53.25	4080	499	70	30.9	1320	200	55.1
70	Kostroma	58.9	41.53	3560	318	70	27.3	2041	210	75.7
71	Vetluga	58.23	45.48	27500	751	120	191	2900	290	144.5
72	Nemda	57.82	43.28	4650	306	60	25.1	2303	610	189.6
73	Rutka	57.62	46.63	1610	114		11.4	1094	170	40
74	Mera	57.48	42.3	2380	260	70	12.9	1291	230	57.7
75	Kotorosl'	57.25	39.62	3860	184	60	23.3	1977	170	65.7
76	Medveditsa	57.18	36.43	3925	375	70	23.9	1491	210	61.4
77	Osuga	57.08	34.67	1740	238	60	10.5	1060	260	56.3
30	Usta	56.92	45.48	6030	340	60	42.2	2200	200	81.7
78	Linda	56.57	44.1	1010	107		5.4	1420	100	33.4
79	Teza	56.55	41.88	3450	446	70	18.8	2327	300	104.4

Table 1. (Continued)

Index	River	Latitude N	Longitude E	F km^2	λ_r m	W_r m	Q_r m^3/s	λ_p m	W_p m	Q_p m^3/s
80	Koloksha	56.48	39.68	510	144		2.9	668	233	37.6
81	Bolshaya Kokshaga	56.48	47.4	3670	204		18.6	1221	180	48.1
82	Bolshoi Kundysh	56.48	47.4	1590	151		8	1704	170	54
83	Uvod'	56.43	41.4	3770	173	60	21.4	2396	240	93.1
62	Kerzhets	56.42	44.87	4600	370	50	24.6	2196	200	79.6
84	Lukh	56.23	42.43	4450	165	60	24.2	1746	280	84.8
85	Buy	56.22	54.23	3985	365	60	17.5	1100	140	37.9
86	Suvorosh	56.18	42.87	1390	110		7.6	3747	320	142.3
87	Izh	56.17	53	5480	260		24	814	155	35.1
88	Klyaz'ma	55.97	37.83	492	144		3.5	750	150	26.2
89	Vorya (Ugra River basin)	55.93	38.23	1220	137		6.9	1044	140	32.8
90	Peksha	55.9	39.68	1010	168		5.7	820	200	37.6
91	Sherna	55.87	38.58	1890	126		10.7	1268	230	55.9
92	Mesha	55.77	49.8	2570	183		13.4	2034	200	71.1
93	Serezha	55.67	42.88	2135			7.8	855	130	28.8
94	P'yana	55.58	45.05	5960	190	70	28.6	1270	225	61
52	P’yana	55.52	44.32	7930	270	70	38.1	1300	300	80.8
95	Moskva	55.53	35.98	1410	313		9.1	769		20.9
96	Moskva	55.73	37.13	5000	572	90	37	1632	230	71
66	Moskva	55.65	37.78	9000	700	150	67.2	2500	500	175.8
97	Tesha	55.48	42.52	4700	143	60	23.5	1059	155	40.3

Index	River	Latitude N	Longitude E	F km^2	λ_r m	W_r m	Q_r m^3/s	λ_p m	W_p m	Q_p m^3/s
98	Sura	55.43	46.38	52600	1003	158	243.6	2975	357	178.1
99	Sviyaga	55.32	48.6	12420	660	70	31	2640	320	133.7
100	Protva	55.22	36.5	1800		60	9.6	1005	140	33.2
60	Protva	55.2	36.52	2170	380	80	11.6	800	180	36.3
101	Vorya (Klyaz'ma River basin)	55.12	35.05	1220	172		6.5	559	130	21.8
102	Shanya	55.12	35.6	306	140		1.7	404	105	13.6
103	Gus'	55.12	40.95	1560	138	30	5.7	759	120	24.6
104	Ugra	54.9	34.55	15700	527	85	87.4	2759	237	117.4
90	Alatyr’	54.87	46.43	11200	437	80	43.4	1416	130	47.7
45	Ikk	54.78	53.57	7660	140	40	31.9	1440	290	82.3
105	Malyi Cheremshan	54.73	50.1	9908	187	30	36	550		15.8
44	Dyema	54.68	55.32	12610	254	60	42.6	950	135	37.2
106	Bol'shoi Cheremshan	54.48	50.37	4942	258	41	18	723	123	27
107	Bol'shoi Cheremshan	54.2	49.67	9908	496	64	36	802	141	34
64	Urshak	54.48	55.87	3130	50	40	230		200	23.7
59	Zhizdra	53.85	35.12	1970	150	40	9.9	1000	250	53.5
108	Moksha	53.63	44.05	4840	420	30	17.6	1540	250	72.7
109	Ashkadar	53.53	55.92	3080	239		10.4	868	80	21.9
110	Kondurcha	53.65	50.25	3990			13.8	390	87	14.3
111	Kondurcha	53.52	50.37	4484	237	33	15.5	541	83	16.7
112	Sok	53.75	50.73	5534	162	27	19.2	644	128	26.4
114	Sok	53.42	50.28	11709	530	72	40.6	974	166	43.9

Table 1. (Continued)

Index	River	Latitude N	Longitude E	F km^2	λ_r m	W_r m	Q_r m^3/s	λ_p m	W_p m	Q_p m^3/s
46	Bol'shaya Kinel'	53.37	51.27	5970	160	25	19.4	920	200	46.7
115	Bol'shaya Kinel'	53.5	51.52	8453	224	41	27.5	805	163	37.8
116	Bol'shaya Kinel'	53.32	51.17	11659	295	48	33	909	149	38.7
117	Bol'shaya Kinel'	53.27	50.58	14799	475	73	41.9	993	140	39.9
118	Samara	53.03	51.3	28512	562	87	57.9	1265	173	57.6
119	Samara	53.23	50.53	30436	562	87	61.8	1324	187	63.1
120	Tok	52.9	53.63	5893	225	34	13.4	962	152	38.2
121	Bol'shoi Irgiz	52.33	50.62	2731	213	28	3.5	453	95	16.1
122	Bol'shoi Irgiz	52.42	50.05	4602	274	40	5.8	721	137	29.2
123	Bol'shoi Irgiz	52.25	49.68	6098	203	31	7.7	691	121	26.5
124	Bol'shoi Irgiz	52.02	48.87	20471	796	105	25.9	1673	179	69.4
125	Kamelik	52.08	49.48	8910	276	53	11.3	992		35
35	Ubort'	51.92	28.50	5260	175	30	10.7	880	120	43
1bs	Aidar	49.01	38.96	7015				845	174	48
2bs	Berezina	54.02	28.87	9557				991	260	74
53	Lesnoy Voronezh	53.02	40.63	1740	160	50	6.5	690	150	35
58	Bityug	51.70	40.52	7330	200	40	18.2	1300	350	115
3bs	Bitiug	51.03	40.06	7695				1177	211	69
4bs	Buzuluk (Khoper)	50.64	42.79	6390				1206	182	62
42	Buzuluk	50.53	42.57	6830	460	35	9.7	1380	300	85
5bs	Desna	51.13	31.03	84183				2145	385	194

Index	River	Latitude N	Longitude E	F km^2	λ_r m	W_r m	Q_r m^3/s	λ_p m	W_p m	Q_p m^3/s
6bs	Desna	51.44	31.95	71145				1806	278	136
7bs	Desna	52.45	33.56	24989				1513	297	115
8bs	Dnepr	54.97	32.98	7340				1997		136
9bs	Egorlyk	45.99	41.3	11220				938		50
10bs	Egorlyk	45.87	41.4	8707				845	171	48
11bs	Iput'	52.47	31.39	9400				809		41
12bs	Kalitva	48.42	40.96	10140				805	133	40
13bs	Khoper	51.27	42.33	19329				1906	328	140
41	Khopyer	51.32	42.37	19100	360	60	67	2500	800	453
57	Vorona	52.07	42.25	9540	430	70	31.3	1700	420	147
56	Savala	51.13	41.45	7720	250	50	20.4	800	250	77
54	Tersa	50.80	44.40	7320	250	40	14.5	1000	250	77
14bs	Khoper	52.46	43.67	8981				1358	184	71
15bs	Medveditsa (Don)	50.19	43.73	29612				1810	213	107
55	Medveditsa	51.52	44.63	7610	320	40	18.4	1200	250	77
16bs	Medveditsa (Don)	51.44	44.86	7582				973	184	55
17bs	Nerussa	52.44	33.9	5346				890	209	56
40	Orel'	48.82	34.40	9400	200	50	21.6	1790	350	113
18bs	Orel'	48.97	34.49	9158				1265	187	68
19bs	Orel'	48.93	34.59	9158				973	203	60
20bs	Orel'	49.16	34.93	7476				912	152	46
21bs	Orel'	49.13	35.12	5653				792	153	41
22bs	Psel	49.23	33.68	22158				1203	152	62
23bs	Psel	49.74	33.78	14692				841	162	49

Table 1. (Continued)

Index	River	Latitude N	Longitude E	F km^2	λ_r m	W_r m	Q_r m^3/s	λ_p m	W_p m	Q_p m^3/s
24bs	Psel	50.18	33.97	11735				712	122	35
38	Psyel	49.63	33.77	11300	330	40	30.5	1250	330	92
25bs	Psel	50.56	34.44	9417				671	121	33
26bs	Psel	51.04	35.22	6587				670		30
27bs	Ros'	49.49	31.48	11021				3701	458	267
28bs	Sal	47.33	41.34	20523				1392	288	105
29bs	Samara (Dnepr)	48.64	35.37	19903				1936	286	130
30bs	Seim	51.39	33.41	27070				1938	234	118
31bs	Seim	51.28	33.88	22257				1602	220	97
32bs	Seim	51.48	34.78	19678				1673	312	124
33bs	Seim	51.69	35.23	11418				1924	349	140
34	Seym	51.65	35.33	10700	170	40	37.1	3000	400	141
49	Svapa	51.65	35.33	6310	120	30	29.1	1400	250	75
34bs	Severskiy Donets	49.33	36.84	18706				742	152	44
35bs	Snov	51.66	31.64	8281				2311		167
36bs	Sozh	52.35	30.95	39336				1693	380	154
37bs	Sozh	53.86	31.8	6630					225	77
38bs	Styr'	50.77	25.34	7201				1762		114
36	Sula	50.25	33.35	14200	170	30	28.5	2500	400	146
39bs	Sula	49.65	32.71	18009				1595	263	106
40bs	Sula	49.62	32.88	18009				1662	272	112
41bs	Sula	49.81	32.9	15295				1727	190	91

Index	River	Latitude N	Longitude E	F km^2	λ_r m	W_r m	Q_r m^3/s	λ_p m	W_p m	Q_p m^3/s
42bs	Sula	50.22	33.32	6098				1260	238	77
43bs	Tersa	50.87	44.11	6589				963	171	51
37	Uday	50.30	32.53	6120			11.4	1500	300	83
44bs	Uday	50.21	32.76	6579				1204	210	69
45bs	Volchia	48.11	36.07	9627				771	139	40
46bs	Vorona	51.78	42.37	12746				1434	252	93
47bs	Vorskla	49	34.16	14433				1293	256	89
48bs	Vorskla	49.04	34.33	14433				1216	238	81
49bs	Vorskla	49.49	34.58	11124				838	115	38
50bs	Vorskla	49.73	34.63	9129				1259	207	72
51bs	Vorskla	50.14	34.74	6220				759	133	37
43	Ilovlya	49.78	44.50	8730			8	850	150	44
1ws	Yuribey	70.65	76.97	11700	1630	245	125.2	3820	750	520
2ws	Yuribey	70.6	76.32	8300	1420	220	88.8	3180	620	379
3ws	Solyenaya	69.48	84.07		745	90	0	1480	475	166
4ws	Taz	64.95	81.4	60410	2360	360	646.4	4050	980	830
5ws	Chayselka	64.93	81.07	12100	330	70	129.5	1630	900	399
6ws	Taz	64.18	81.85	55940	1570	285	598.6	2920	490	384
7ws	Ratta	63.38	83.98	3470	420	95	37.1	760		39
8ws	Taz	63.70	84.42	6850	630	110	73.3	1720	230	116
9ws	Yuredey-yakha	66.92	80.77	2680	640	120	28.7	1860	425	169
10ws	Bol'shaya Khye-yakha	66.90	79.92	2020	270		21.6	810	250	63

Table 1. (Continued)

Index	River	Latitude N	Longitude E	F km^2	λ_r m	W_r m	Q_r m^3/s	λ_p m	W_p m	Q_p m^3/s
11ws	Bol'shaya Khodyr-yakha	65.93	78.33	5040	910	125	50.2	2470	390	211
12ws	Bol'shaya Khodyr-yakha	65.73	78.87	2700	300	70	26.9	1440	260	97
13ws	Ventokoy_yakha	65.72	79.00	1880	420	75	18.7	1310	205	73
14ws	Kharam-Pur	64.17	78.35	3610	250	60	36	1100	260	84
15ws	Erkal-Nadey-Pur	63.00	78.98	3650	470	80	36.4	865	225	64
16ws	Yarudey	65.83	71.67	5685	965	165	52.8	3455	760	447
17ws	Yarudey	65.77	70.18	1080	270	60	10	910	185	48
18ws	Kheygee-yakha	65.32	72.75	7230	880	175	67.2	2630	580	310
19ws	Levaya Khetta	65.15	73.33	10850	1230	280	100.8	4230	550	450
20ws	Poluy	66.28	67.78	18540	1145	180	162.1	3880	605	478
21ws	Poluy	66.05	68.48	14375	955	135	125.7	3415	570	403
22ws	Poluy	65.55	69.08	8830	925	160	77.2	2630	390	238
23ws	Glubokiy Poluy	65.33	69.58	4440	470	85	38.8	2400	420	214
24ws	Sobtuegan	65.93	67.27	3090	340	70	27	1030	240	74
25ws	Kunovat	65.07	66.32	8500	495	110	74.3	1475	400	162
26ws	Kunovat	64.88	67.13	6660	540	90	58.2	1880	315	153
27ws	Kunovat	64.80	65.52	5730	430	70	50.1	2660	370	219
28ws	Kazym	63.55	69.12	7530	1015	150	81.3	2490	460	251

Index	River	Latitude N	Longitude E	F km^2	λ_r m	W_r m	Q_r m^3/s	λ_p m	W_p m	Q_p m^3/s
29ws	Bol'shoy Salym	60.98	71.25	15200	485	95	97.3	1680	455	211
30ws	Konda	59.43	66.87	55170	1080	195	247.2	6790	1130	1306
31ws	Kuma	59.18	66.88	5450	330	60	13.1	880	200	62
32ws	Tobol	57.68	66.43	333000	1550	230	388.9	5785	670	1037
33ws	Tobol	56.95	66.63	249000	640	95	109.5	5145	895	1094
34ws	Tobol	56.33	66.32	182100	485	85	48.1	4065	470	572
35ws	Iset'	56.55	66.25	58900	370	75	69.4	3110	360	334
36ws	Miass	55.98	64.45	21400	245	60	16.2	1875	275	162
37ws	Tobol	55.45	65.30	159000	540	65	42	2730	370	346
38ws	Tobol	54.57	64.88	143000	260		27.9	2760	490	416
39ws	Tobol	53.85	63.77	49800	120		0	1310	385	185
40ws	Vagay	57.78	69.22	15600	485	70	22	2020	320	182
41ws	Ishim	57.60	71.20	177000	730	125	78.4	4870	605	774
42ws	Ishim	57.52	71.00	171000	570	70	75.8	4715	395	596
43ws	Ishim	56.12	69.87	140000	485	65	44.1	3220	345	377
44ws	Ishim	54.83	69.05	110000	410	90	0	2595	510	397
45ws	Ishim	51.68	68.40	39000	280	70	0	1401	270	139
46ws	Ishim	51.32	70.65	7400			0	1200	160	69
47ws	Irtysh	55.35	73.13	337000	3590	320	965.2	15025	770	2915
48ws	Tromegan	61.42	74.57	23400	665	170	190.7	3455	615	457

Table 1. (Continued)

Index	River	Latitude N	Longitude E	F km^2	λ_r m	W_r m	Q_r m^3/s	λ_p m	W_p m	Q_p m^3/s
49ws	Kulegan	60.78	75.78	6400	470	90	45.7	1590	250	114
50ws	Vakh	61.13	77.33	73360	2160	300	682.9	6170	1710	1770
51ws	Vakh	61.10	78.30	59515	1720	295	554	5915	1120	1185
52ws	Tym	59.85	81.50	28330	925	165	219.7	3325	750	532
53ws	Vasyugan	59.10	78.97	35400	750	130	183.2	3275	380	338
54ws	Ket'	58.58	89.30	21780	350	95	140.7	2150	345	210
55ws	Ket'	58.17	90.87	9750	260	70	63	1485	280	124
56ws	Chaya	57.82	82.63	17110	545	75	54.1	1050	190	78
57ws	Chulym	57.33	90.30	50600	1275	200	83.4	2710	540	386
58ws	Inya	53.05	85.12	10340	275	70	33.8	1100	200	78
59ws	Tarsma	55.00	85.45	1800			3.7	450	85	17
60ws	Kargat	54.80	79.10	1870	240		2.5	890	180	50
61ws	Chumysh	53.55	85.35	11000	1155	135	88.3	1630	345	158
62ws	Kulunda	53.35	80.98	6500	120		3	570	125	33
63ws	Charysh	52.23	83.72	21500	1010	160	201.5	1750	275	153
64ws	Aley	52.45	82.78	18700	285	60	35	915	210	78
65ws	Peschanaya	52.05	85.05	4700	400	65	32.3	1100	155	59

* λ - meander step length is measured as a straight distance between the adjacent points of curvature change; W – channel bankfull width at crosses; Q – mean annual discharge; F – basin area. (index "r" refers to recent river channels, "p" refers to paleochannels).

A. The Protva River (the Volga River basin)

The Protva River basin is covered by mixed broad-leaved – coniferous forest, annual precipitation is 600 mm and runoff depth is app. 170 mm. About 75% of annual runoff is drained during the flood, when the water level rise near Borovsk (basin area 1800 km^2) is 4-5 m. Mean maximum flood discharge is 310 $m^3 s^{-1}$ here. The length of the Protva River is 282 km, but the remnants of large meandering paleochannels are found only at two reaches of the valley, near Borovsk (index 100) and Obninsk (index 60 in Table 1), 7-10 km long each. There the modern river channel forms large incised meanders with the mean half-wavelength of 1000 and 1300 m, respectively.

The necks of these meanders are at the level of the first river terrace. These large meanders do not fit the modern hydraulic conditions, as their wavelength/width ratio is too high: mean λ/W is more than 30. Several large meanders (one near Borovsk and four near Obninsk) were cut from the main channel due to meander neck narrowing and now form large ox-bows.

The bottoms of these ox-bows, filled with sediments, can be inundated only during the highest floods (Sidorchuk et al., 2009). One of these large ox-bows was cored at a former riffle (Figure 5). Coring shows U-shaped 250 m wide cross-section of the paleochannel, typical for riffles.

Level of this former riffle surface is only slightly higher than the levels of the riffles in the modern Protva River channel. Level of the first terrace surface (the former floodplain) is 8-9 m above the former riffle. Alluvium at the former riffle is very similar to the recent bed-load of the Protva River near Borovsk. It consists of coarse sand with gravel and small pebbles. The first terrace is composed of fine sand, very similar to the recent floodplain sediments.

The paleochannel is filled with a complex of alluvial sediment bodies: the lower composed of fine sand, the upper of silt and silt clay. Infill alluvium surface is 5 m above the former riffle surface, which is close to the level of high recent flood. Alluvium is covered by the layer of peat up to 3 m thick at the paleochannel axis. Several ^{14}C dates show the sequence of paleochannel infill.

The oldest date of 13200±120 ^{14}C yr B.P. (Ki-7316) was obtained from a sample collected right at the riffle surface. This date shows an approximate time when the large meander neck was cut. Infilling by sediments due to the flooding had different rates at different parts of the ox-bow. At the beginning of the Boreal period (about 9Kyr B.P.) peat accumulation started at the deepest part of the ox-bow, but its filling in with alluvial sediments continued at least till the Subboreal, according to the radiocarbon dates.

Protva River is the typical example of the complicated relations between the modern and ancient floodplain. The old lateglacial floodplain is now partly the low terrace of Protva River, and the modern floodplain is narrow at these sections. Partly the old floodplain is still flooded in spring and therefore the modern inherited floodplain width increase significantly. The river channel still follows the ancient macromeanders and meander wavelength do not correspond to the recent river hydrology.

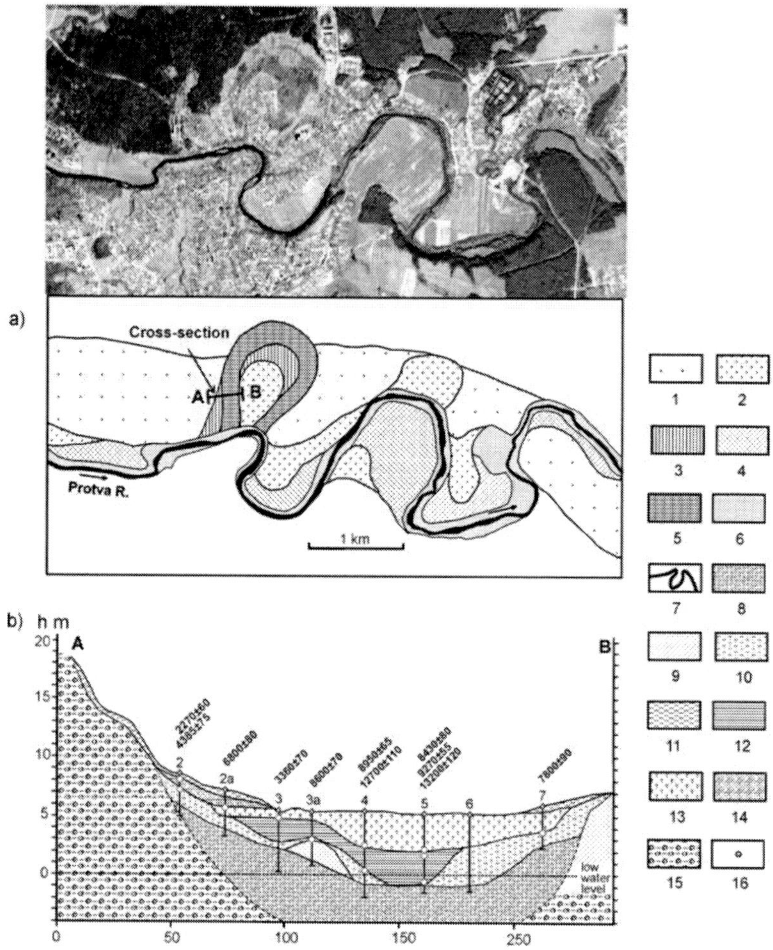

Figure 5. Paleochannel of the Protva River near Borovsk (A) and cross-section at the former channel riffle (B): (1) high terraces, (2) the first terrace, (3) point bars in lateglacial palaeoriver, (4) floodplain formed by activity of the large lateglacial paleoriver, (5) lateglacial palaeochannel, (6) floodplain formed by activity of the Holocene river, (7) the recent river channel, (8) gravel, (9) sand, (10) sandy loam, (11) loam, (12) clay, (13) peat, (14) travertine; (15) till; (16) ^{14}C samples.

B. The Moskva River (the Volga River basin)

In the Moskva River basin macromeanders do not occur along the most part of the river length which amounts to 473 km, but are very distinct at three reaches near Mozhaisk (index 95), near Zvenigorod (index 96), and within Moscow city (index 66 in Table 1). At the latter reach 18 sequential macrobends with the mean half-wavelength of 2200 m can be distinguished (Figure 6 shows the eastern part of the system). These macrobends are mainly inherited by the recent river and partly transformed by river flow. A few macrobends are abandoned by the recent channel (Sidorchuk et al., 2009). The ratio between the macromeander half-wavelength and the recent channel width λ/W is about 17 within this reach. It is much more than the ratio between the recent small meanders wavelength and channel width (λ/W~5). This discrepancy indicates an unfitness of the channel morphology to the recent flow. The majority of the large paleochannel fragments are included into the recent floodplain, thus significantly increasing its width. Several inherited and abandoned macrobends are incised into the first and even higher terraces. On the high bank at the apex of one of such macrobends the initial town of Moscow was founded more than 850 years ago. Now the most part of the floodplain within the Moscow is urbanized and only on the old maps the former floodplain configuration is clearly seen (Figure 6).

The abandoned paleochannel of the Moskva River (the last bend in a sequence, Figure 6) was cored at former pool section. It has a classical triangle shape with the maximum depth near the concave bank at the apex of paleomeander. The altitude of the pool bottom (105 m a.s.l.) is lower than the altitudes of the recent pools in adjacent part of the Moskva River channel. The paleopool bed was covered with coarse sand with gravel and small pebbles. Being cut from the paleochannel, the meander was filled with clay, silt and fine sand. Two radiocarbon dates, 15300 and 14100 ^{14}C yr B.P., were obtained from the infilling sediment close to the former pool bottom. These dates indicate an approximate time when the large meander was cut off the main channel.

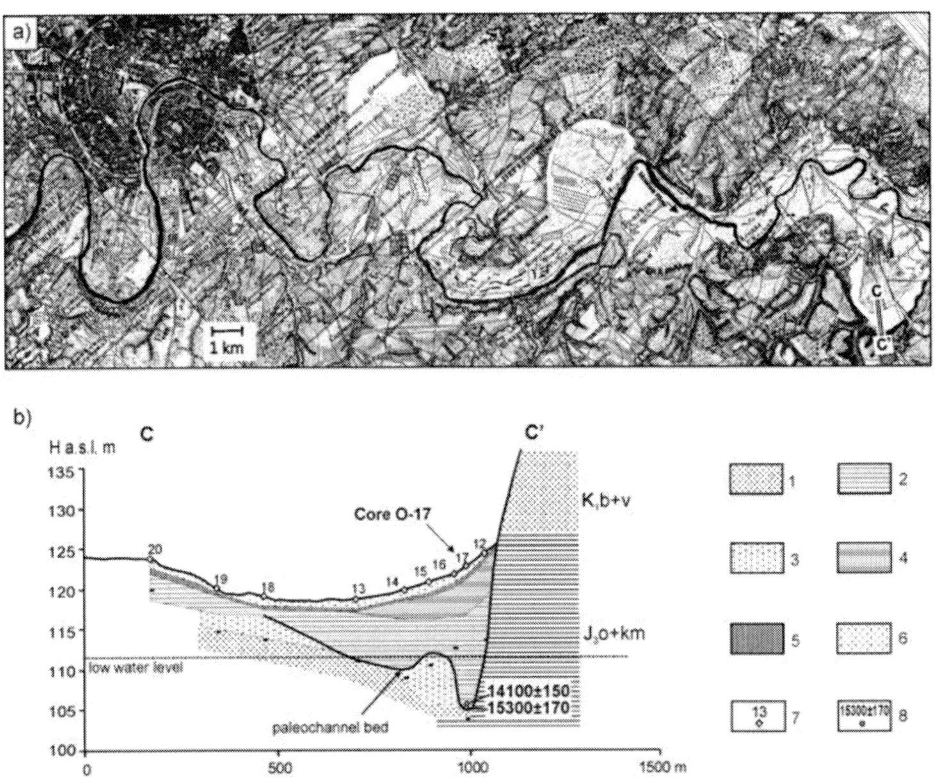

Figure 6. The sequence of macromeanders of the Moskva River as seen on the map of the mid-19th century (a) and cross-section C-C' at one of the pools of the paleochannel (b). Keys: (1) sand with gravel; (2) silted sand; (3) silt; (4) clay; (5) loam with peat; (6) peat; (7) cores locations and numbers; (8) ^{14}C samples and dates.

C. Paleochannels of the Seim and Svapa Rivers (the Dnepr River basin)

The floodplain of the Seim River and its main tributaries is characterized by a sequence of arcs with the radii of curvature exceeding that of the modern river channel bends by an order of amplitude (Figure 7). Systems of natural levees and large abandoned oxbows are well defined on the air photographs. Systems of levees and hollows with the relative relief of 0.5-1.5 m indicate steeply curved meandering paleochannels with the half-wavelength of 1.7-3.3 km and the width of 350-700 m. This floodplain with the remains of large paleochannels lies largely at 2.0-2.5 m above low water level (LWL) and usually are inundated during floods. Only the tops of the highest levees are as high as 3-3.5 m and usually stay above the flood level.

The large paleochannel of the Seim River forms highly curved meanders with the mean half-wavelength of 3 km and the mean length along the channel 7 km (Borisova et al., 2006). The texture of infilling of one of these large paleochannels was investigated by coring along the profile A'-A" in the upper part of the bend near the Kudintsevo village (Figure 8). The paleochannel cross-section has an asymmetrical triangular shape with the deepest part (app. 10 m below LWL) near the steep concave bank of the paleomeander. The paleochannel fill includes three main units. The lowermost unit is fine silty sand 5-7 m thick, which belongs to the initial stage of the paleochannel infill. The sand is overlain by grey clay and silt accumulated largely in the oxbow lake on the floodplain.

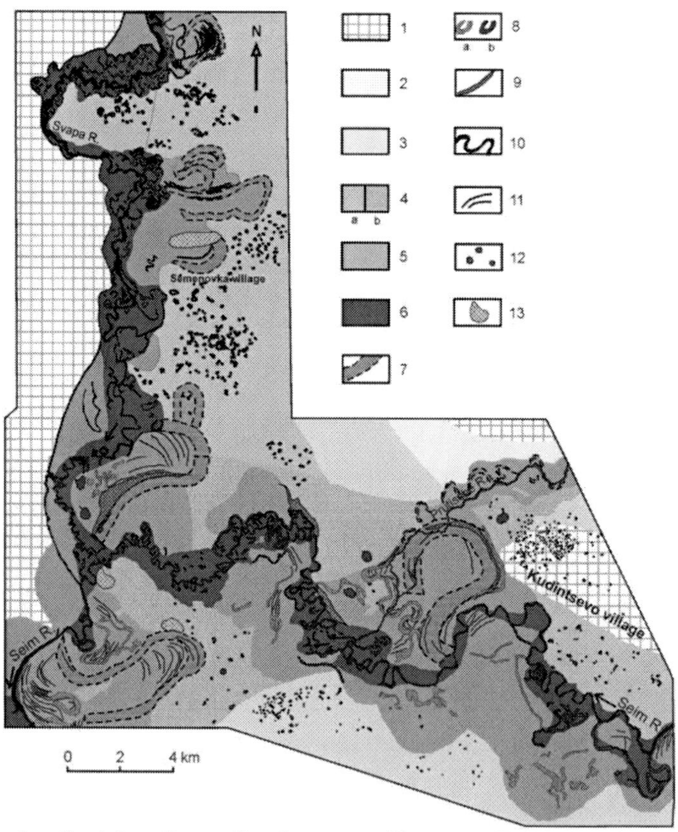

Keys: (1) watersheds; alluvial surfaces of various age: (2) terrace II (18–25 m), (3) terrace Ib (12–15 m), (4) terrace Ia (levels a and b), (5) lateglacial floodplain, (6) Holocene floodplain; fluvial forms: (7) large paleomeanders, (8) small oxbows (a – lateglacial, b –Holocene), (9) modern channel, (10) big alluvial levees, (11) floodplain levees; relic non-fluvial forms: (12) small thermokarst depressions; (13) aeolian dunes.

Figure 7. The geomorphological map of the Seim and Svapa valley bottoms.

Deposits at the base of this unit were radiocarbon-dated to 12630±70 and 13800±85 yr B.P., based on the bulk organic matter (samples Ki-6985 and Ki-6984). By the beginning of the Holocene the oxbow lake transformed into a fen, and mineral deposition was followed by peat accumulation. The thickness of the peat layer is up to 2 m.

In the Svapa River valley, the paleochannel formed steeply curved meanders with the mean half-wavelength of 1.4 km and the mean length along the channel of 3.8 km. The paleochannel fragment near Semenovka village is clearly expressed in the modern topography (Figure 9). Its surface lies only 4 m above the modern LWL, so that it is submerged during high floods. Coring reveals a box-shaped profile of the channel. The top of the lowers layer of fine and medium-grained channel alluvial sands lies 1.5-2.5 m below the modern LWL.

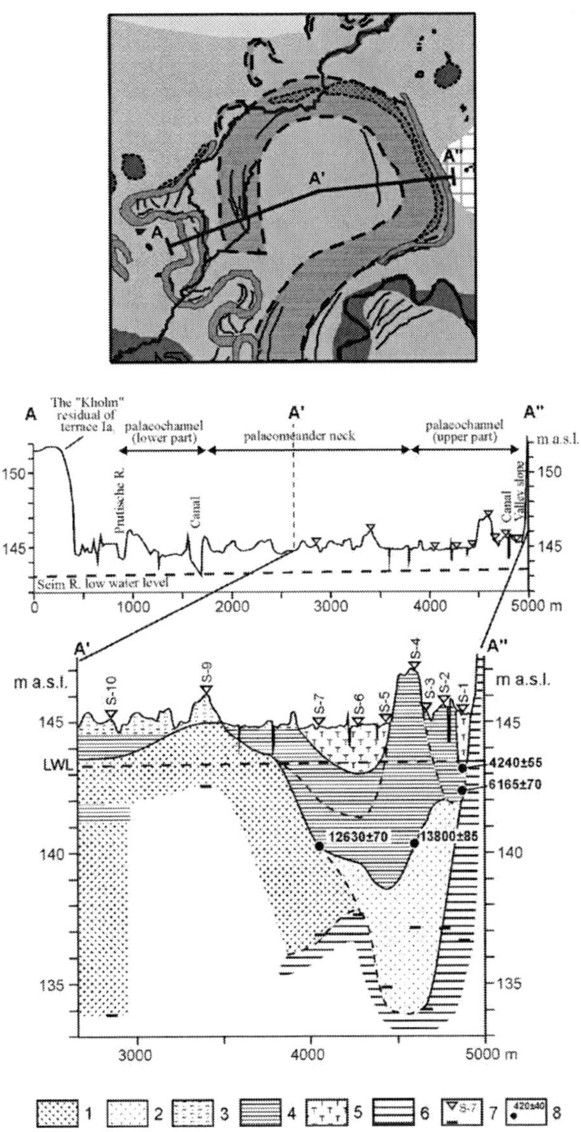

Figure 8. Topographic and geological profiles across the large meandering channel of Paleo-Seim River near Kudintsevo village. Keys: lithology: (1) medium sand, (2) fine sand, (3) silty sand and sandy silt, (4) loam and clay, (5) peat; (6) eluvium of parent rock; (7) cores, (8) radiocarbon dates.

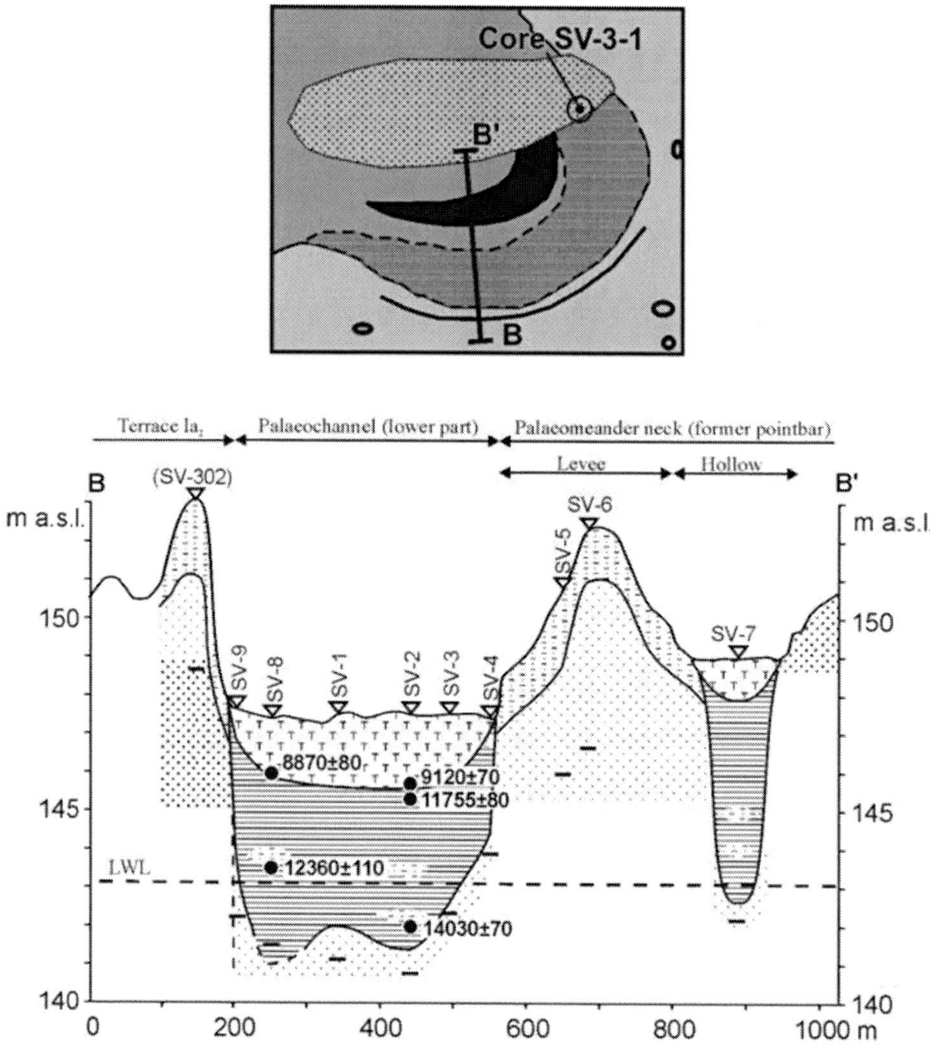

Figure 9. Geological section across the large meandering channel of the Paleo-Svapa River near Semenovka village. Keys as in Figure 8.

Silt and clay deposits with lenses of clayey sand fill in the trough. The accumulation of the fine-grained sediments has begun because of the abandonment of the paleochannel in the Oldest Dryas (14030±70 ^{14}C yr B.P., Ki-6997), and continued during the Bølling and the Allerød (12360±110, Ki-6999 and 11755±80, Ki-6996). In the end of the Late Glacial – beginning of the Holocene the paleochannel almost entirely dried up, and the rate of deposition became very low. Peat formation started in the late Preboreal (9120±70, Ki-6995; 9300 ± 120, GIN-11951).

D. Paleochannels of the Khoper River (the Don River basin)

The Khoper River basin area near Povorino is 19100 km^2. The mean annual discharge is 67.8 m^3 s^{-1} with mean flood maximum 991 m^3 s^{-1}. About 75-80% of the annual flow is drained during the flood period in April-May, with the typical levels 3-4 m above the low water mark and with the maximum about 6 m. The recent meandering river channel is 30-100 m wide with meander half-wavelength 100-500 m.

The river floodplain is 10-13 km wide here and its composition is heterogeneous (Sidorchuk, 1997; Panin et al., 2013). The youngest floodplain band is only 1.0 - 2.5 km wide and it follows the recent river channel. This floodplain was formed by the activity of the meandering channel of the recent size and morphological type. Its surface is covered by the numerous curved natural levees and ox-bow lakes of the same radii and width, as of the recent river channel. Mean floodplain height is 3-4 m above the low water mark with natural levee tops at 5-5.5 m.

Several radiocarbon dates shows that the youngest floodplain band was formed during the second half of the Holocene during the Subboreal – Subatlantic time (Figure 8). According to these dates the deposition rate is about 1.2-1.7 mm a^{-1} at the near-channel parts of floodplain and decreases to 0.3-0.32 mm a^{-1} at the central sections of the Holocene floodplain.

The main part of the Khoper River floodplain was formed by much larger meandering channels, than the recent one. Well preserved fragments of these paleochannels have different size and age at the different parts of the ancient floodplain. The paleochannel which follows the right bank of the valley is about 800 m wide with meander half-wavelength 2.5-3.5 km. This paleochannel has been rarely flooded after it was abandoned, so that its former bottom is locally exposed. The maximum thickness of the paleochannel filling deposits there does not exceed 1.5-3.0 m (Figure 10). The macromeander neck is about 6 m high (above the low-water mark) and is flooded only during the extreme floods. Radiocarbon and radio-thermo-luminescence dating of the bottom deposits (Figure 10) shows that the large paleochannel was formed during the Late Glacial and abandoned about 14 thousand radiocarbon years ago.

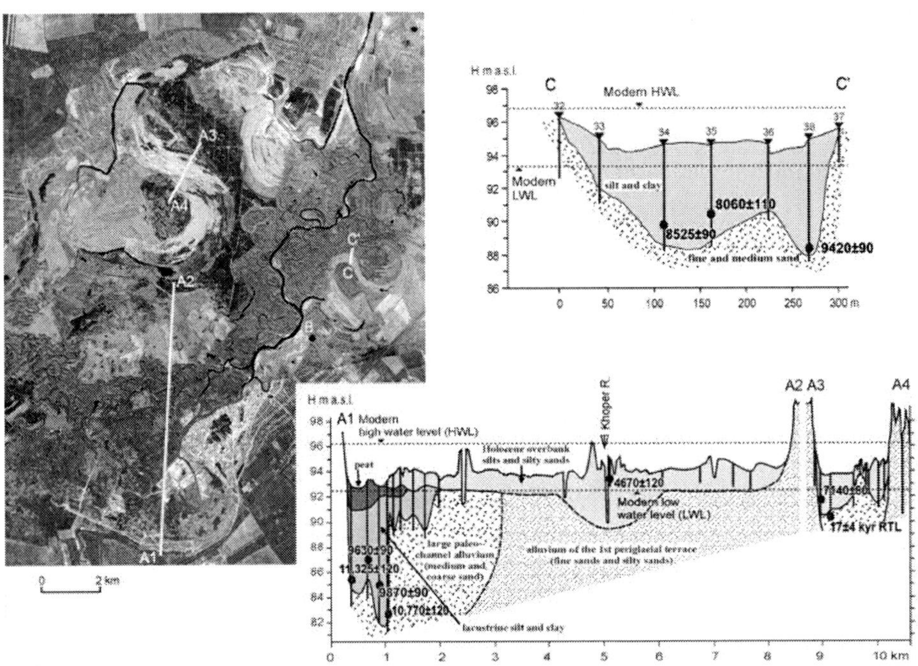

Figure 10. Space image and geological sections across the large meandering channels of the paleo–Khoper River near Povorino.

The meandering paleochannel along the left bank of the valley is of the same size but is completely filled in by fine-grained alluvium, beginning from 11.3-10.8 ^{14}C kyr B.P. The pools with the maximum depth of 9-11 m are now represented by shallow wide trough and are nearly invisible in floodplain relief, being discovered only by the coring.

A system of smaller paleochannels (although still larger, than the recent Khoper River channel) was preserved at the outspread of floodplain in the valley bend (see Figure 10, cross-section C-C'). The channel width was about 200 m, meander half-wavelength – about 1200 m. These channels were active before the Boreal time, presumably during the Younger Dryas.

The typical thickness of the floodplain alluvium of the first stage of floodplain formation is about 16 m, while the modern alluvium thickness is only 6 m. A greater alluvium depth corresponded to higher discharges of the rivers at the first stage, especially during the spring floods.

The Paleolandscapes during the First Stage of Floodplain Evolution

To reconstruct the main climatic indexes in the region with the large river remnants the method based on the composition of fossil floras derived from palynological data (method of "arealograms") was employed. Fossil floras used for paleoclimatic reconstructions were derived from six sites (Figure 11). At four of them, palynological studies were conducted on fluvial sediments filling in segments of large paleochannels found on the floodplains of the Moskva, Seim, Svapa, and Khoper rivers (Sidorchuk & Borisova, 2000; Borisova et al., 2006; Sidorchuk et al., 2009, 2011). Another of the localities is an Early Man site Yudinovo, where loamy sediments containing a Late Paleolithic cultural layer were subjected to detail pollen analyses (Borisova, Novenko, 1999). One more fossil flora includes both pollen and plant macrofossils identified in the so-called 'Usvyacha' deposits exposed in several outcrops in the Dvina River valley near the villages Sloboda and Drichaluki (Velichkevich, 1982). In the course of pollen analysis, an attempt was made to achieve the highest possible taxonomic resolution. Pollen identifications have been made to species or genus levels for arboreal plants and for certain groups of herbaceous plants. Where identifications were possible only to the family level, they were not included in the resulting lists of fossil floras. On the whole, these paleobotanical data proved to be sufficient to locate modern geographical analogues to 15 fossil floras and, therefore, to estimate climatic changes, which occurred within the East European plain since approximately 18 ^{14}C kyr B.P (Table 2).

The accuracy of the match of a modern geographical analogue to a fossil flora depends not only on the richness of the latter, but also on the knowledge of the modern geographical distribution of each plant genus or species. In some cases the area of a region-analogue remains rather large and includes different types of vegetation associations. Therefore in this study all such regions were located on the map (see Figure 11) and additionally checked afterwards against the types of plant communities, reconstructed from the pollen assemblage of a given sample. Sometimes such comparison enabled us to reduce the area of the region-analogue, but even then the resulting areas remained large enough to show a substantial variability of hydro-climatic characteristics within them. The reconstructed ranges of climatic parameters are shown in Figure 12 as shaded boxes, their vertical size corresponding to the time intervals characterized by each fossil flora.

Table 2. Main climatic indexes in the East European Plain in the Late Glacial and the Holocene

Radiocarbon ages of the fossil floras kyr B.P.	Indices and site names	Temp. Jan. °C	ΔTemp. Jan. °C	Temp. July °C	ΔTemp. July °C	Precipitation mm per year	Δ Precipitation mm per year	Runoff depth mm per year	Δ Runoff depth mm per year
17.5	Fl_sd1 Sloboda/ Drichaluki	-21 ÷ -22	-13÷ -14	8÷10	-7 ÷ -9	400÷600	-200÷0	350÷500	140÷290
(15.0)	Fl_mr1 Moskva	-18÷-20	-7÷-9	11÷13	-5÷-7	850÷950	350÷450	440÷500	210÷370
(14.2)	Fl_yu1 Yudinovo	-14 ÷ -18	-5.5÷-9.5	16÷17	-1.5÷-2.5	700÷800	125÷225	500÷550	350÷400
13.8	Fl_sm1 Seim	-22 ÷ -26	-14÷-18	15.5÷16.5	-3.5÷-2.5	425÷475	-125÷ -75	350÷400	225÷275
(13.5)	Fl_mr2 Moskva	-31÷-33	-20÷-22	17÷19	-1÷1	320÷480	-180÷-20	240÷360	10÷130
12.6	Fl_sm2 Seim	-23 ÷ -27	-15÷-19	17÷19	-2 ÷ 0	375÷425	-175÷-125	300÷350	175÷225
12.4	Fl_sv1 Svapa	-21 ÷ -25	-13÷-15	15÷19	-4 ÷ 0	425÷475	-125÷ -75	350÷400	240÷290
12.2	Fl_sm3 Seim	-15.5÷-16.5	-7.5÷-8.5	16÷18	-3 ÷ -1	600÷700	50÷150	150÷250	25÷125
(12.0)	Fl_mr3 Moskva	-15÷-17	-4÷-6	13÷15	-3÷-5	600÷1000	100-500	170÷290	-60÷60
11.9	Fl_kh1 Khoper	-16 ÷ -18	-6 ÷ -8	16÷17	-3 ÷ -4	500÷600	40÷140	150÷200	40÷ 90
(11.7)	Fl_mr4 Moskva	-27÷-29	-16÷-18	15÷17	-1÷-3	320÷480	-180÷-20	240÷360	10÷130
11.5	Fl_sm4 Seim	-15.5÷-16.5	-7.5÷-8.5	17.5÷18.5	-0.5÷-1.5	675÷725	125÷175	200÷250	75÷125
(11.3)	Fl_mr5 Moskva	-13÷-15	-2÷-4	12÷14	-4÷-6	650÷1100	150÷600	200÷360	-30÷130
(10.8)	Fl_mr6 Moskva	-29÷-31	-18÷-20	17÷19	-1÷1	320÷480	-180÷-20	240÷360	10÷130
(10.5)	Fl_sv2 Svapa	-17 ÷ -19	-9 ÷-11	14.5÷15.5	-3.5÷-4.5	400÷500	-150÷-50	150÷300	40÷190

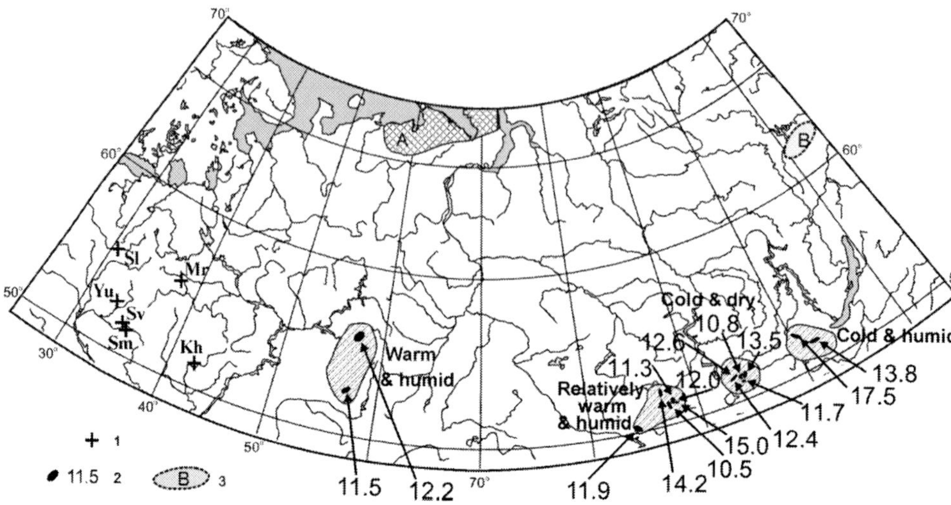

Figure 11. Location of the modern region analogues to the Last Glacial Maximum (LGM) and Late Glacial time (LGT) fossil floras. Keys: (1) key sites (Sm – Seim, Sv – Svapa, Kh – Khoper, Yu – Yudinovo, Sl – Sloboda/Drichaluki, Mr – Moskva River); (2) modern region analogues of the fossil floras and its ^{14}C age in kyr B.P.; (3) corresponding hydrological region analogues in the lowland areas (A – Bol'shezemel'skaya Tundra and Yamal Peninsula, B – Central Yakutian Plain).

1) The earliest fossil flora (Flora sd1 in Table 2) derived from the palynological and plant macrofossil studies of the Sloboda and Drichaluki sections belongs to the relatively cold stage of the Late Pleniglacial, dated by radiocarbon method to 17,460±210 yr B.P., Tln-309 (San'ko, 1987). The flora includes several of the typical Arctic and Arctic-mountain species, which at present grow in various European and West-Siberian tundra and forest-tundra communities, along with some xerophytes tolerant to low winter temperatures, such as *Ephedra distachya*. It also includes trees and shrubs, growing at present in the regions with cold and highly continental climate, mainly in Siberia (*Larix* sp., *Alnaster fruticosus*). Such a complexity of flora is highly typical of the glacial floras in northern Eurasia (Grichuk, 1969). The region currently inhabited by the species of fossil flora sd1 lies in the south of the East Sayan Mountains, in the upper part of the Oka River basin. The area has a cold climate with the mean January air temperature -21 ÷ -22°C and that of July 8÷10°C, which is characteristic for mountain tundra in this region. Because of permafrost, the runoff coefficient should be very high (more than 0.8). The mean annual precipitation is 400-600 mm, and the calculated runoff depth is 350÷500 mm (Table 2).

2) Pollen flora mr1 of the sediments, corresponding to the initial stage of filling in of the large paleochannel of the Moskva River dated by radiocarbon to app. 15.0 kyr B.P. characteristically combines cryophile (*Alnaster fruticosus*, *Betula nana*) and xerophile (*Kochia scoparia*, *Eurotia ceratoides*) plants, inhabitants of the boreal forest, steppe, meadow, and riverine communities. The closest modern floristic analogue for this assemblage can be found within the Altai Mountains south of the Teletskoye Lake, in the lower reaches of the Chulyshman River basin (Figure 11). Within this small area, sub-alpine and alpine meadows of grass, sedge and forbs exist next to mountain tundra and larch-Siberian pine open forest. The area is characterized by a cold and relatively humid climate. The mean January temperature is −19°C and that of July is about 12°C. The mean annual precipitation varies between 800 and 1000 mm. The mean annual surface runoff depth is about 450 mm. The

region is situated near the boundary of the permafrost zone, within the area with discontinuous spread of permafrost. It is characterized by high annual runoff coefficient close to 0.5.

3) An Early Man site Yudinovo is located on the first terrace of the Sudost' River, a tributary of the Desna River in the Dnepr basin. Both the geomorphological position of the site and a series of radiocarbon dates based on the materials from the cultural layer indicate that the site was inhabited approximately from 14.5 to 14 kyr B.P. (Svezhentsev, 1993). According to palynological data, fossil flora yu1 of this interval consists mainly of forest and meadow plants. It includes Boreal and Arcto-Boreal trees, forest club-mosses (*Lycopodium clavatum*, *L. selago* and other), *Pteridium aquilinum* and other mesophile plants. Such a composition of the flora suggests relatively humid conditions during the interval in question. A region-analogue for fossil flora yu1 lies at the Altai Mountains, in the middle reaches of the Biya River, on the east-faced slopes (Figure 11). In this region, pine and birch forests and spruce-fir-Siberian pine mountain taiga forests occur along with wet meadows. The area is characterized by milder and wetter climatic conditions compared to the region-analogue sd1: the mean January air temperature there is –16° and the mean July temperature is 16-17°C. The mean annual precipitation is 700-800 mm, and the runoff depth is 500-550 mm (see Table 2).

4) Sediments of the initial stage of filling in of the large paleochannel of the Seim River (core S-4 in Figure 8; 13,800±85 yr B.P., Ki-6984) generally correspond to the beginning of the Oldest Dryas. Fossil flora sm1 derived from these sediments combines cryophile (*Alnaster fruticosus*, *Selaginella selaginoides*) and xerophile (*Ephedra* sp.) plants, inhabitants of the boreal forest, steppe, meadow, and riverine communities. The closest modern floristic analogue for this assemblage can be found in the forest steppe in the middle reaches of the Irkut River basin, west of the Lake Baikal (see Figure 11). Within this small area, larch and pine forest grow next to south Siberian meadow steppes, with patches of spruce forest in the river valleys. The area is characterized by a cold semi-arid and extremely continental climate. The mean January temperature is -24°C and that of July is about 16°C. The mean annual precipitation varies between 425 and 475 mm. The region is situated near the boundary of the permafrost zone with high annual runoff coefficient. The mean annual surface runoff depth is 350-400 mm (see Table 2).

5) Fossil flora mr2 from the paleochannel of the Moskva River also corresponds to the Oldest Dryas stage and is dated ~13.5 yr. B.P. The flora contains a variety of xerophytes and xerohalophytes, which can tolerate cold winters. Such a composition suggests a rapid drying of the topsoil in the watershed areas during relatively warm summers. Similar conditions occur at the present time in the intermountain depressions in southern Siberia. A region-analogue for fossil flora mr2 lies at the headwaters of the Yenisei River, within a depression situated south from the confluence of the Biy-Khem and Ka-Khem rivers (Figure 11). The region is characterized by extremely cold winters with the mean January air temperature about −32°C, while the summer is warm, the mean July temperature being about 18°C. The annual magnitude of air temperature changes is app. 50°, which is at least 20° greater than at the study site at the present time. The mean annual temperature is close to −7°C, providing favorable conditions for the permafrost development. Therefore the runoff coefficient is estimated at 0.75. The mean annual precipitation is app. 400 mm, and the calculated annual runoff depth is about 300 mm.

6) Another sample of the fluvial deposits filling in the large paleochannel of the Seim River was obtained from the basal part of core S-7 taken in the same profile as core S-4

(Figure 8). According to the radiocarbon date (12,630±70 yr B.P., Ki-6985), it corresponds to the final part of the Oldest Dryas. The flora of this sample (Flora sm2 in Table 2) is distinctive for the diversity of xerophytes and xerohalophytes, which suggest a rapid drying of the topsoil in the watershed areas in relatively warm summers. Similar conditions occur at the present time in southern Siberia. A region-analogue for fossil flora sm2 lies at the headwaters of the Yenisey River, within an inter-mountain depression downstream from the confluence of the Biy-Khem and Ka-Khem rivers (Figure 11). In this area south Siberian grass dry steppes are found next to mountain forest of *Pinus sibirica* and *Larix sibirica*. The area is characterized by extremely cold winters with the mean January air temperature from -23° to -27°C, while the summer is warm, the mean July temperature being about 18°C. The annual magnitude of air temperature changes is about 43°, and that is app. 15° greater than at the study site at the present time. The mean annual precipitation is about 400 mm and calculated runoff depth is about 325 mm.

7) Fluvial deposits filling in the paleochannel of the Svapa River were cored near Semenovka village (see Figure 9). According to the ^{14}C date (12,360±110 yr B.P., Ki-6999), fossil flora sv1 derived from core SV-1-8 belongs to the beginning of the Bølling Interstadial. It includes species of dark coniferous taiga forest (*Picea abies*, *Abies sibirica*, *Pinus sibirica*), light coniferous and mixed boreal forest (*Pinus sylvestris*, *Pteridium aquilinum*, *Betula alba*), along with species of riverine shrub associations, meadows, psammophytes (e.g., *Spergula*), and xerophytes. A region-analogue for fossil flora sv1 lies at the headwaters of the Yenisey River. (see Figure 11). In this area south Siberian grass dry steppes are also the main vegetation type, though the role of dark coniferous mountain forest of *Picea*, *Pinus sibirica* and *Abies sibirica* there is slightly greater compared to region-analogue sm2. The area is characterized by milder climatic conditions with the mean January air temperature -23° and that of July about 17°C. The mean annual precipitation is app. 450 mm, and the calculated runoff depth is about 375 mm.

8) Palynological data for the fluvial deposits infilling the paleochannel of the second generation in the Seim River valley were obtained from borehole S-11. Flora sm3, derived from a sample dated to 12,250±70 yr B.P. (Ki-6987), includes broad-leaved temperate tree species (*Quercus*, *Ulmus*, and *Tilia cordata*), as well as *Alnus glutinosa* – tree alder, growing on the waterlogged ground. Its composition indicates that in the late Bølling the area was covered by a complex vegetation of the forest steppe type, with birch and pine copses with minor participation of the broad-leaved trees. The presence of the broad-leaved trees in the region implies a complete degradation of permafrost in the end of the Bølling. It is confirmed by generally mesophile character of the flora and by presence of pollen of relatively heat-demanding aquatic plants, such as *Nymphaea alba*. A diversity of the hygro- and hydrophytes and an absence of xerohalophytes in this flora suggest an increase in the rainfall. A region-analogue for fossil flora sm3 lies at the headwaters of the Ufa River at the Southern Ural Mountains (see Figure 11). In this region, meadow steppes come into close contact with southern Urals pine and birch forests and spruce-fir sub-taiga forests with minor presence of the broad-leaved trees. The area is characterized by significantly milder climatic conditions compared to the region-analogue of the early Bølling flora sv1: the mean January air temperature there is –16° and the mean July temperature is about 17°C. The mean annual precipitation is 650 mm, and the runoff depth is about 200 mm with 70% of the flow passing during the spring flood. Annual runoff coefficient is 0.31, and that for the flood period is about 0.6.

9) Fossil flora mr3 from the paleochannel of the Moskva River also corresponds to the Bølling interstadial (~12.0 yr B.P). It includes mesophile shrubs (*Salix* spp., *Alnus incana, Viburnum opulus, Ribes* sp.), inhabitants of various forest biotopes – dark coniferous taiga forest (*Picea abies, Pinus sibirica*), light coniferous and mixed boreal forest (*Pinus sylvestris, Larix, Betula alba*), along with a variety of meadow plants, helophytes and aquatics. The presence of xerophytes and plants growing on the eroded ground in this flora indicates that the periglacial steppe communities were preserved at that time in favorable conditions, although their landscape role probably decreased compared to the Oldest Dryas. A region-analogue for fossil flora mr3 is found at the headwaters of the Biya River, south of the Teletskoye Lake (see Figure 11). In this area open *Larix* forests with steppe species in the ground cover occur next to *Larix-Pinus sibirica* mountain forest and mixed woodlands of *Betula alba* and *Pinus sylvestris*. The region is more forested compared to region-analogue 1. It is characterized by relatively mild climatic conditions with the mean January air temperature −16° and that of July about 14°C. The mean annual temperature is close to −1°C, which is not favorable for the permafrost preservation. At present, the region-analogue is characterised by sporadic spread of permafrost. Probably, in the Moskva River basin permafrost degraded during the Bølling interstadial warming. The mean annual precipitation within the region-analogue is app. 800 mm, and the runoff depth is about 300 mm. The annual runoff coefficient is close to 0.3.

10) The alluvium deposits infilling the meandering Khoper paleochannel were studied near the town of Povorino (see Figure 10). At the base of core B, a radiocarbon date was obtained for the layer enriched with organic matter – 11,900±120 yr BP (Ki-5305). Flora kh1 of this horizon includes some arboreal species with broad ecological tolerances, such as Scots pine (*Pinus sylvestris*) and tree and shrub birch (*Betula alba, B. humilis*), but it also includes Siberian pine (*P. sibirica*), growing in the regions with continental climate, xerohalophytes and xerophytes (*Atriplex pedunculata, A. verrucifera* and other). A region-analogue of fossil flora kh1 is located in the north-eastern Kazakhstan, in the Bukhtarma River basin (Irtysh River basin) at the boundary between dry steppe and semidesert and close to the region where communities of shrub birch and open dark coniferous forest are spread on the western slopes of the Altai Mountains (see Figure 11). The mean January air temperature in this region is -17°C; mean July temperature is about 16.5°C. The mean annual precipitation varies between 500 and 600 mm. The annual runoff depth is 150-200 mm. The region-analogue is situated beyond the permafrost zone but close to its boundary.

11) The short Older Dryas cold stage (~11.7 yr B.P.) is represented by fossil flora mr4 from the paleochannel of the Moskva River. It belongs to the periglacial forest steppe type, combining typical cryophytes, xerophytes, and heliophytes along with arboreal species with broad ecological tolerances. A region-analogue for fossil flora mr4 is found within the depression south from the confluence of the Biy-Khem and Ka-Khem Rivers (see Figure 11). With respect to the region-analogue for flora mr2, it is shifted to the south, towards the northern piedmont of the Eastern Tannu-Ola ridge. In this region dry steppes occupy the depression floor, while the lower parts of the mountain slopes are covered by larch and birch open forests, and Pinus sibirica participate in the larch woodlands at higher altitudes. The mean January air temperature in the region-analogue 4 is app. −28°C, and that of July is 16°C. Therefore, the annual magnitude of air temperature changes reaches 44°. The mean annual temperature is about −6°C. As at the region-analogue mr2, the runoff coefficient is estimated

at 0.75. The mean annual precipitation is app. 400 mm, and the calculated runoff depth is 300 mm.

12) Fossil flora sm4 of the dated sample from section S-11 (see Figure 8, 11,450±60 yr B.P., Ki-6986) includes many helophytes – plants, growing on wet meadows, near the water margin, and in shallow water (*Sparganium* spp., *Menyanthes trifoliata*, *Sagittaria sagittifolia*, and others). Composition of this flora indicates that the Allerød warm interval was favourable for the expansion of dark coniferous trees (*Picea abies*, *Pinus sibirica*, and *Abies*) in the Seim River basin. Forest communities were dominated by birch and Scots pine, and larch participated in the pine forest at this period. Of the relatively thermophile trees and shrubs, flora of the Allerød included *Ulmus*, *Tilia*, *Corylus*, and *Viburnum*. A modern analogue for fossil flora sm4 is located at the headwaters of the Belaya River near the Southern Urals (see Figure 11). The region is currently occupied by forest steppe, where meadow steppes are associated with birch and aspen woods and grow next to larch-pine open forests with steppe elements in the herbaceous cover, and to the broad-leaved (oak-linden) forests of Southern Ural type. This area is characterized by the mildest climatic conditions for the entire Late Glacial, with the mean January air temperature -16° and that of July 18°C. The mean annual precipitation there reaches 700 mm, the runoff depth is about 225 mm, with 60% of the flow passing during the spring flood. Annual runoff coefficient is 0.32, and that for the flood period is about 0.56.

13) Flora mr5 from the Moskwa River paleochannel show that the Allerød interstadial was the most favourable time for the expansion of *Pinus sylvestris* in the forests for the entire Late Glacial. *Abies* participated in dark coniferous forests of *Picea abies* and *Pinus sibirica* in the end of the interstadial. Rare pollen grains of *Larix* indicate that larch also occurred in the forest at that time. Fossil flora mr5 features a great diversity of the meadow forbs and helophytes. Of the relatively thermophile species, pollen of *Viburnum opulus* is registered. A modern analogue for fossil flora mr5 is located south of the Teletskoye Lake, in the eastern part of the Chulyshman River basin in its lower reaches (see Figure 11). The region is currently occupied by fir and Siberian pine mountain taiga, which grow next to larch-pine south Siberian mountain forest. This area is characterized by relatively mild winter conditions with the mean January air temperature –14°C. The summer is cool, the mean July temperature being 13°C. The mean annual precipitation reaches 900 mm, the runoff depth is about 270 mm. Annual runoff coefficient is 0.3.

14) Flora mr6 is correlated to the beginning of the Younger Dryas cold stage. Changes in the pollen assemblages, concentrations, and the flora are indicative of both colder and more arid climatic conditions compared to the previous interval, with probable re-establishment of permafrost in the region. A region-analogue for fossil flora mr6 is similar to that for flora mr2, being slightly shifted to the west (see Figure 11). In the vegetation of this area patches of cold mat-grass steppes are spread next to open larch woodlands with steppe elements in the ground cover. The area is characterized by extremely continental and relatively dry climate with the mean annual precipitation of 400 mm and the runoff depth app. 300 mm. An annual runoff coefficient is estimated at 0.75. The mean January air temperature is –30°C, and that of July is 17°C. These values suggest that in the Younger Dryas the mean annual air temperature in the Moskva River basin was as low as –6°C, which would promote permafrost development.

15) Fossil flora sv2 derived from a loam and clay sediment layer in the core SV-1-8 is correlated to the Younger Dryas on the basis of pollen composition and radiocarbon dating.

This horizon is distinguished for a high content of non-arboreal pollen, as well as for a distinct maximum of Artemisia pollen and a smaller peak of Chenopodiaceae. Among trees, *Pinus sylvestris* and *Betula alba* were predominant. Of the dark coniferous trees, *Picea abies* and *Pinus sibirica* are registered in this flora. Of the cold-tolerant shrubs, *Betula humilis* and *Alnaster* are present. Relatively thermophile trees once again disappear from the local flora. Changes in the flora are indicative of both colder and more arid climatic conditions compared to the previous time interval, with probable re-establishment of permafrost in the region. A region-analogue for fossil flora sv2 is situated in the middle reaches of the Biya River in the Altai Mountains (see Figure 11). In the vegetation cover of this area, open larch woodlands with steppe elements in the ground cover and patches of steppes are spread next to fir and Siberian pine forests on the mountain slopes. Birch and Scots pine communities occur on sandy soil. The area is characterized by a cold, continental and relatively dry climate with the mean annual precipitation of 450 mm and the runoff depth 150-300 mm. The mean January air temperature is -18°C, and that of July 15°C.

On the whole, changes in geographical position and hydro-climatic characteristics of the paleofloristic region-analogues reflect a complexity of climate change during the Late Glacial against the background tendency towards warmer and wetter climate (Sidorchuk et al., 2011). Secondary oscillations can be seen on this general trend. Air temperature was the lowest in the LGM, in the late part of the Oldest Dryas and in the Younger Dryas. Relatively warm intervals precede the Oldest Dryas and correspond to the Bølling and the Allerød (Figure 12). Extremely low winter temperatures indicate existence of permafrost during the Late Pleniglacial, the Oldest Dryas – early Bølling and in the Younger Dryas.

Precipitation changes in LGM and LGT generally followed temperature changes, the cold stages being relatively dry, and the warm stages relatively humid, with the maximum precipitation achieved in the pre-Oldest Dryas warming (app. 14.5-14 ^{14}C kyr B.P.), and in the Bølling-Allerød Interstade (app. 12.5-11 ^{14}C kyr B.P.). The surface runoff is not determined entirely by annual precipitation, as the water losses depend on air temperature. Runoff coefficient is inversely correlated with the temperature changes showing a strong overall decrease during LGM-LGT. Generally following changes in the runoff coefficient, the surface runoff also decreased considerably during this period (Figure 12). Secondary oscillations of precipitation and water losses led to significant variations in the surface runoff. The greatest runoff was characteristic for the pre-Oldest Dryas warming, because of relatively high precipitation and preserved permafrost. Second runoff maximum was achieved at the beginning of the Bølling interval, when the temperatures still remain low while precipitation began to rise. During the cold and dry Younger Dryas an increase in the runoff was caused by low water losses to evaporation. During the warm and relatively humid Bølling and Allerød interstadials, higher losses to evaporation caused a relative decrease in surface runoff. Nevertheless, the entire LGM-LGT interval was characterized by high surface runoff compare to the present-day level, which was achieved only at the beginning of the Holocene (see Figure 12).

For relatively mild periods of the Lateglacial (pre-Oldest Dryas warming, Bølling and Allerød interstadials) the modern region-analogues are situated in the basins of the Biya River (the Altai Mountains) and of the Belaya River (southeastern Ural Mountains). For the cold periods (LGM, the Oldest, Older and Younger Dryas) they are located in the southeastern sub-Baikal region and at the headwaters of the Yenisei River (the Sayan Mountains). These territories can be used as hydrological analogues only partly (for example, for the estimation of runoff coefficient) because of the mountainous relief, high inclinations and small areas of the river basins.

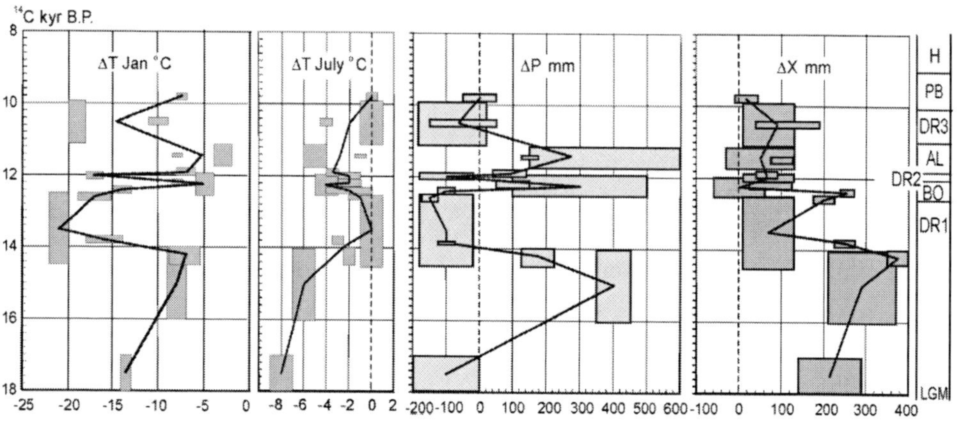

Figure 12. Estimations of the main climatic indexes (deviations from modern values) based on the composition of fossil floras.

For the estimation of such characteristics as the discharge variability, lowland territories with similar climatic indexes (mean temperature of January and July, annual precipitation) and landscapes, close to those of the LGT (sparse open vegetation, widespread permafrost), must be selected. The lowland with climatic characteristics similar to those at the Altai and Ural Mountains, is the Bol'shezemel'skaya Tundra in the north-east of the European Russia and Yamal Peninsula at the north of the Western Siberia. The lowland with climatic characteristics similar to the Yenisei River headwaters is located within central Yakutia (see Figure 11). Hydrological regimes of large rivers at these two territories are quite similar: spring flood is high and sharp, and the most of annual flow passes during the flood. Low water period during the summer, autumn and winter is long, but some years only 10-12% of annual runoff passes during this period. Coefficient a in formula (7), which depends on hydrological regime, is therefore very similar for these two territories: it is 2.25 for the Bol'shezemel'skaya Tundra and 2.23 for the central Yakutia.

Former annual discharges for the large rivers of the first stage of floodplains evolution (Table 1) were calculated with the formulae (5) - (7) (Sidorchuk et al., 2008a). At that stage, water runoff from the northern megaslope of the Russian Plain was about 380 km^3 a^{-1}, or 150% of the recent runoff from the same catchment area. In the Volga River basin annual runoff was about 500 km^3, despite the catchment area of the Upper Volga being significantly reduced then. The main part of this runoff was contributed by the Oka and Kama Rivers, as their discharges were 3 to 3.5 times greater than the modern ones. This high surface runoff in the Volga River basin caused an extreme Khvalyn transgression of the Caspian Sea up to the mark 50 m a.s.l. In the basins of the Dnieper and Don Rivers the discharges at that stage were also 3 to 4 times greater than the modern ones. In the Ob River basin the runoff was two times its modern value, in the Irtysh River basin – three times its modern value (Sidorchuk et al., 2008b).

The duration of the first stage of the floodplain formation and the age of the large paleochannels in Northern Eurasia were estimated using the radiocarbon and pollen analyses of the oxbow deposits. Maximum water flow and formation of the broad-floor river valleys filled in with alluvium date back to the period of app. 13-16 kyr B.P. At that time, in cold periglacial climate with low evaporation and virtual absence of groundwater supply, winter precipitation increased, which caused a sharp increase in the spring flood runoff. Further on both the average, and maximum discharges in the rivers gradually decreased, but remained greater than the modern ones until the Early Holocene. During the cold stages of the Oldest, Older and Younger Dryas, the runoff increased due to the reduction of losses through evaporation and infiltration (though the rainfall at these stages was reduced). In the Bølling and Allerød warm interstadials the runoff decreased despite an increase in precipitation due to an increase in the water losses. A significant temporal variability of the water flow brought about substantial changes in the dimensions of paleochannels in many catchments. For instance, on the floodplains of the Seim, Khoper, and Moksha rivers the paleochannels of the intermediate size and age can be traced. They are smaller than the largest paleochannels found in the same catchments, but larger than the modern channels of these rivers. With the reducing height of floods the highest parts of the floodplains formed during the first stage became no longer flooded in spring. As a result they transformed into fragments of low terraces, often located within the floodplains. Some researchers refer to such low terraces as "the first terrace above the floodplain", and other call them "intermediate" between the floodplain and the first terrace. These terraces, first described by Volkov (1963) in West Siberia, are entirely climatic by their origin, since their formation does not require rivers incision.

Thus, in the river valleys on the plains of Northern Eurasia at the first development stage the main morphometric characteristics of floodplains took shape. Together with the intermediate terraces, these floodplains formed the morphological and lithological basis of the modern river valley bottoms. Their relief and lithology were largely determined by the activity of the maximum-size meandering paleochannels 16-13 Kyr B.P. Large paleochannels of the intermediate size, which were formed 13-9 Kyr B.P., are less widespread. In the next stages of the floodplain formation, the large levees of the first stage in some places were destroyed by the stream shifting in the course of the erosion and accumulation, but the fragments of deeply incised large paleochannels filled in with the lateglacial and Holocene sediments are preserved both in relief and lithology and can be used as a source of information about this important stage in the development of the river valleys.

Over large areas (e.g., in the Upper Volga basin) the morphology of modern channels of small and medium-sized rivers is almost completely determined by the configuration of large paleochannels which belong to the first stage of floodplain formation. The channels of such modern rivers follow large relict bends with the dimensions which do not match their current discharges. At the end of the first stage, when the water flow reduced, these rivers did not rework their former floodplain, which partly became low terraces. Therefore, the fluvial topography of the first stage has been well preserved there.

The Recent Analogues of the Large Periglacial Rivers

The rivers of the vast territories in the Northern Eurasia covered with deep permafrost are still at the first stage of floodplain development. This stage channel processes are typical in the rivers of the Yamal Peninsula in Western Siberia (Sidorchuk & Matveev 1994).

There are 223 days a year with the temperature below freesing at the Yamal Peninsula, which leads to permafrost development. In the west-central part of the peninsula the mean annual air temperature is -8.3°C. The runoff for the snowmelt period is 220-250 mm.

Precipitation in the form of rain occurs generally in June-September, with an average of 470 rainy hours. The mean rainfall for this period is 140 mm. The area represents a typical tundra landscape with a complex of Quaternary marine terraces with heights between 20-45 m a.s.l. The mainly fine-grained marine deposits contain lenses and sheets of ice with thickness up to 50 m and spatial dimensions of individual ice bodies up to several kilometers. Lower alluvial terraces occupy smaller areas. The main rivers of the area – the Mordy-Yaha River and its right tributary Se-Yaha River – form sharply curved meanders on the broad floodplain. The floodplain is 5-6 km wide, while the river width is only 100-150 m.

According to interpretation of the aerial photographs and radiocarbon dating, several levels of the floodplain have been defined. The convex banks of meanders are outlined by the young grass-shrubs floodplain with the age up to 200 years. This level is characterized by primary fluvial relief with clearly identified natural levees. The next level is mature grass-moss-shrubs floodplain with the age 200-2300 years. Here the fluvial relief is slightly transformed by permafrost polygonal crusting and by thermo-karst. The highest floodplain level with the age more than 2300 years is completely reworked by thermo-karst processes and has relief and vegetation pattern that shows different stages of thermokarst lakes development.

The dendrochronological and radiocarbon dating show very high rates of the vertical growth of the floodplain. The surface of the youngest grass floodplain grows with the average rate up to 50 mm/year due to nearly annual flood sedimentation. The lowest parts of the grass-shrubs floodplain grow with the average rates 3-19 mm/year. The average rate of accumulation is about 0.8 mm/year at the mature floodplain. This is more than two-fold of above-mentioned rates of the Khoper River floodplain grows. The main cause of such floodplain grows rates is the high contribution of non-mineral additional accumulation of peat and ice, which is about the half of the floodplain deposits.

The rates of bank erosion are much less than 1 m/year for the majority of river reaches. The average rates of erosion of the floodplain banks for 200-year period are 0.5-0.7 m/year along the lower Se-Yakha River. The maximum observed rate of erosion of the high terrace bank is 0.9 m/year. The cut-offs occur at the meanders with the coefficient of sinuosity more than 5-6, that is, when the opposite sides of the bends almost touch each other. The main cause of this type of channel meander dynamics is completely frozen surface of the floodplain at the period of spring flood.

At the same time there is clear evidence of the relatively recent (200-250 years ago) long (several kilometers) blow-outs on the mature floodplain of the lower Se-Yakha River. The channels of these blow-outs follow along the numerous linear depressions, and old ox-bows, and connect depressions of thermo-karst lakes on the surface of the floodplain. These blow-outs can form during the catastrophic floods. Two-D hydraulic modeling shows (Sidorchuk, 1996) that flow velocities on the floodplain are large enough to erode its surface when it is flooded to the depth 1.5-2.0 m.

During the summer low-water period the rivers of the central-western Yamal Peninsula are feed with the waters from large lakes of the central Yamal. The rivers of the eastern Yamal Peninsula are not connected with these lakes. In the conditions of deep permafrost ground water supply is very low. The channels of these rivers are nearly empty and broad sandy point bars are exposed to wind action (Figure 13). Therefore large aeolian dunes often develop on the floodplains with typical water-full tundra landscape (Figure 14).

Figure 13. Nearly empty channel of the Nurmayakha River (the eastern Yamal Peninsula) during the summer.

Figure 14. Aeolian dunes formation on the floodplain of the Nurmayakha River.

The Second Stage: Floodplain Formation by the Rivers Smaller Than Modern Ones

Many oxbow systems identified on floodplains have smaller channel width and size of bends than present-day channels of the same rivers. It often happens due to the floodplain formation by braided channels, or by meandering stream branches which were smaller than the main channel and could have a wide range of sizes. In some cases such smaller oxbows can be formed by the tributaries meandering within the floodplains of the main rivers. However, in many cases small sizes of the oxbows and natural levees are difficult to explain by these causes, especially when these fragments are found at the valleys with relatively narrow floodplain, where the anastomosing channels are not typical. Sometimes remains of a single small meandering paleochannel can be traced on the floodplain segments on both sides of the modern channel, the latter having larger bends and occupying an axial part of the floodplain (Figure 15). In other cases, a single meandering paleochannel of small dimensions can be traced at one-sided floodplain and the channel is pressed to the opposite bank of the valley (Figure 16). At the sites where both large and small oxbows and/or levees occur, their relative position indicates a younger age of the floodplain with small oxbows, thus attributing them to the second stage of the floodplain formation.

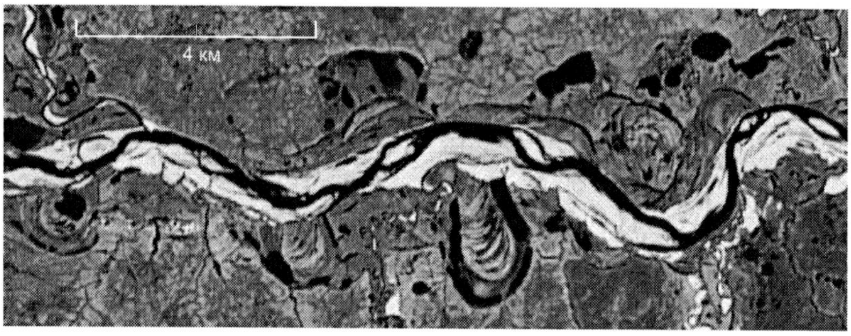

Figure 15. Channel transformation of the Khadutte River (lower Pur's tributary, No. 19 in Table 3) from steeply meandering with small meanders into gently meandering with large meanders and point bars. A segment with double-sided floodplain.

Figure 16. Channel transformation of the Upper Ob' River (No. 53 in Table 3) from steeply meandering with small meanders into gently meandering with large meanders and point bars. A segment with one-sided floodplain.

Currently 55 sections of river valleys with fragments of small paleochannels, which belong to the second stage of floodplain formation, are identified (Figure 17, Table 3). In the tundra no such sites were found, though in the forest tundra small paleochannels sometimes occur (as, for example, on the Khadutte River floodplain, see Figure 15). The greatest number of such sites is located in the forest zone, mainly in the coniferous forests of the north-eastern European Russia, and especially in the basins of the Vychegda and Vyatka rivers. Small paleochannels also occur in the zone of broad-leaved forests (*e.g.*, in the Desna River basin), in the forest-steppe (the Irtysh River basin near Omsk), and in the steppe (the Don and Seversky Donets basin). Most of the sites with fragments of small paleochannels on the floodplains are located on the East European Plain and the West Siberian Plain.

Table 3. The mean geometric characteristics of the river channels (meander half-wavelength $\overline{\lambda}$ and amplitude $\overline{A}$, channel width $\overline{W}$) at the sections with well-defined paleochannels of the second stage of evolution

Section index	The river	Lat. N	Long. E	Recent channel			Old channel			$\frac{\overline{W_o^*}}{\overline{W_r^*}}$	$\left(\frac{\overline{A}}{\overline{\lambda}}\right)_o \Big/ \left(\frac{\overline{A}}{\overline{\lambda}}\right)_r$	$\frac{\overline{Q_o}}{\overline{Q_r}}$
				$\overline{\lambda}$, m	$\overline{A}$, m	$\overline{W}$, m	$\overline{\lambda}$, m	$\overline{A}$, m	$\overline{W}$, m			
1	Padenga	61.85	42.50	433	167	41	123	122	30	0.44	0.39	0.36
2	Peza	65.72	45.65	2262	—	215	908	—	157	0.52	—	0.45
3	Uftyuga	61.63	46.62	767	301	165	468	479	105	0.62	0.38	0.59
4	Vychegda (mouth)	61.28	46.65	6000	2000	1100	3500	2500	600	0.55	0.33	0.49
5	Yug	60.33	47.08	2540	—	313	993	—	195	0.49	—	0.42
6	Mezen'	63.97	47.20	1517	370	286	1060	857	208	0.71	0.30	0.70
7	Yel'	63.97	47.20	1489	—	262	1071	—	208	0.76	—	0.76
8	Viled'	61.33	47.38	575	—	114	216	—	44	0.38	—	0.30
9	Luza	60.60	48.07	1280	388	206	733	673	155	0.66	0.33	0.63
10	Yarenga	62.20	48.97	1034	276	182	636	461	107	0.60	0.37	0.56
11	Sysola	63.52	50.70	1093	350	178	403	546	120	0.51	0.24	0.45
12	Pizhma	65.05	52.02	792	336	143	398	412	123	0.68	0.41	0.66
13	Vychegda (Lokchim)	61.87	52.05	2677	785	398	1692	1562	319	0.71	0.32	0.70
14	Vychegda (Nem)	61.58	54.63	1396	497	197	841	961	129	0.63	0.31	0.59
15	Kama	60.23	55.28	1863	284	321	1076	1020	196	0.59	0.16	0.55
16	Kolva	66.03	57.23	2434	293	327	1582	1470	246	0.69	0.13	0.68
17	Pechora	63.87	57.28	6662	—	603	2900	—	425	0.53	—	0.46
18	Irtysh (mouth)	61.03	69.03	4498	—	555	3810	—	—	0.85	—	—
19	Khadutte	67.45	76.33	2515	707	586	1451	1367	290	0.53	0.30	0.47
20	Ngarka-tabyakha	66.63	76.58	945	426	190	682	749	145	0.74	0.41	0.74
21	Taz	63.70	84.42	630	—	110	431	—	—	0.68	—	—
22	Tyung	63.92	121.52	1690	551	265	1114	1101	270	0.83	0.33	0.86
23	Vilyuy	63.77	121.58	5671	—	913	6742	—	706	0.99	—	1.10
24	Yana	70.00	135.53	5150	2055	765	2780	1677	523	0.60	0.66	0.56
25	Zapadniy Bug	52.68	21.87	1042	269	107	522	497	105	0.68	—	0.65

Table 3. (Continued)

Section index	The river	Lat. N	Long. E	Recent channel			Old channel			$\frac{\overline{W^*_o}}{\overline{W^*_r}}$	$\left(\frac{\overline{A}}{\overline{\lambda}}\right)_o / \left(\frac{\overline{A}}{\overline{\lambda}}\right)_r$	$\frac{\overline{Q_o}}{\overline{Q_r}}$
				$\overline{\lambda}$, m	$\overline{A}$, m	$\overline{W}$, m	$\overline{\lambda}$, m	$\overline{A}$, m	$\overline{W}$, m			
26	Dnestr	49.18	24.59	1135	305	125	577	547	105	0.63	0.28	0.60
27	Neman	53.52	25.35	819	216	122	444	403	78	0.59	0.29	0.54
28	Berezina	53.23	29.20	648	251	94	447	353	80	0.76	0.49	0.77
29	Dnepr	54.97	32.98	740	310	113	320	426	93	0.61	0.32	0.57
30	Desna	52.63	34.00	1060	212	129	423	384	75	0.47	0.22	0.40
31	Severskiy Donets	48.33	40.23	1434	418	173	694	589	114	0.55	0.34	0.50
32	Don (Noviy Liman)	49.93	40.85	1650	461	162	949	806	164	0.73	0.33	0.72
33	Don (Khoper)	49.58	42.63	1676	248	240	1443	941	189	0.83	0.23	0.86
34	Don (Kalach na Donu)	48.90	43.67	2793	640	405	1909	1285	358	0.77	0.34	0.78
35	Sura	55.90	46.10	1918	511	285	758	942	178	0.50	0.21	0.43
36	Vetluga	56.38	46.32	1635	—	242	892	—	225	0.72	—	0.71
37	Moloma	58.72	48.45	1159	324	157	352	493	94	0.43	0.20	0.35
38	Velikaya	58.88	49.03	454	113	54	260	223	53	0.74	—	0.73
39	Bystritsa	58.55	49.25	513	79	75	247	209	46	0.54	0.18	0.48
40	Belaya Kholunitsa	58.87	50.58	439	172	77	148	138	36	0.40	0.42	0.32
41	Kilmez'	56.97	51.10	1032	286	136	329	362	43	0.32	0.25	0.23
42	Vyatka	55.70	51.42	5175	—	494	2384	—	331	0.53	—	0.47
43	Buy	56.23	54.30	403	115	89	222	267	57	0.60	0.24	0.55
44	Belaya (Blagoveschensk)	55.25	55.58	3225	1164	414	1444	2017	399	0.66	0.26	0.64
45	Belaya (Aptrakovo)	52.88	56.20	842	239	165	482	580	102	0.60	0.24	0.55
46	Tobol (Yalutorovsk)	56.70	66.42	954	354	110	413	584	117	0.68	0.26	0.66
47	Tobol (Tavda)	57.77	67.30	1642	—	253	1145	—	—	0.7	—	—
48	Irtysh (Uvat)	59.58	69.32	4596	1628	538	1897	2499	491	0.61	0.27	0.57
49	Irtysh (Tevriz)	57.55	70.05	3207	1723	343	1397	1530	304	0.60	0.49	0.56
50	Irtysh (Omsk)	55.00	73.23	2937	—	361	1664	—	315	0.69	—	0.67
51	Vasugan	59.12	76.90	750	308	110	367	412	111	0.72	0.37	0.71

Section index	The river	Lat. N	Long. E	Recent channel			Old channel			$\frac{\overline{W}_o^*}{\overline{W}_r^*}$	$\left(\frac{\overline{A}}{\overline{\lambda}}\right)_o \Big/ \left(\frac{\overline{A}}{\overline{\lambda}}\right)_r$	$\frac{\overline{Q}_o}{\overline{Q}_r}$
				$\overline{\lambda}$, m	$\overline{A}$, m	$\overline{W}$, m	$\overline{\lambda}$, m	$\overline{A}$, m	$\overline{W}$, m			
52	Charysh	52.08	82.63	1093	468	196	665	878	166	0.73	0.32	0.72
53	Ob'	53.47	83.10	3951	772	674	3924	2576	648	0.98	0.30	1.08
54	Chumysh	53.80	83.67	1067	479	171	641	686	119	0.65	0.42	0.61
55	Peschanaya	52.33	84.98	560	263	122	363	373	82	0.66	0.46	0.64
56	Chulym	56.45	90.45	1128	—	200	644	—	183	0.74	—	0.74

They are less common in the plains of East Siberia (for example, some such fragments are found in the Tyung River valley) and in the Yana lowland. In the western part of the East European Plain small paleomeanders are located on the floodplains of the tributaries of the Upper Dnepr, on the Neman, the Western Bug and the Upper Dnester floodplains. In Western Europe, small paleochannels on floodplains are not distinguished. The ratio of the widths of the small and modern channels of different rivers varies from 0.32 to 0.99. Although the number of studied sites is not large, a tendency is rather obvious towards an increase in this ratio both to the north and to the south of latitude 57-58°N, as well as both to the east and to the west of longitude 50°E. The dimensions of the small paleochannels differ most significantly from those of the present-day channels in the catchment areas of the Vyatka River and the middle Irtysh River. The curvature of the paleomeanders of the second generation was throughout greater than the modern one.

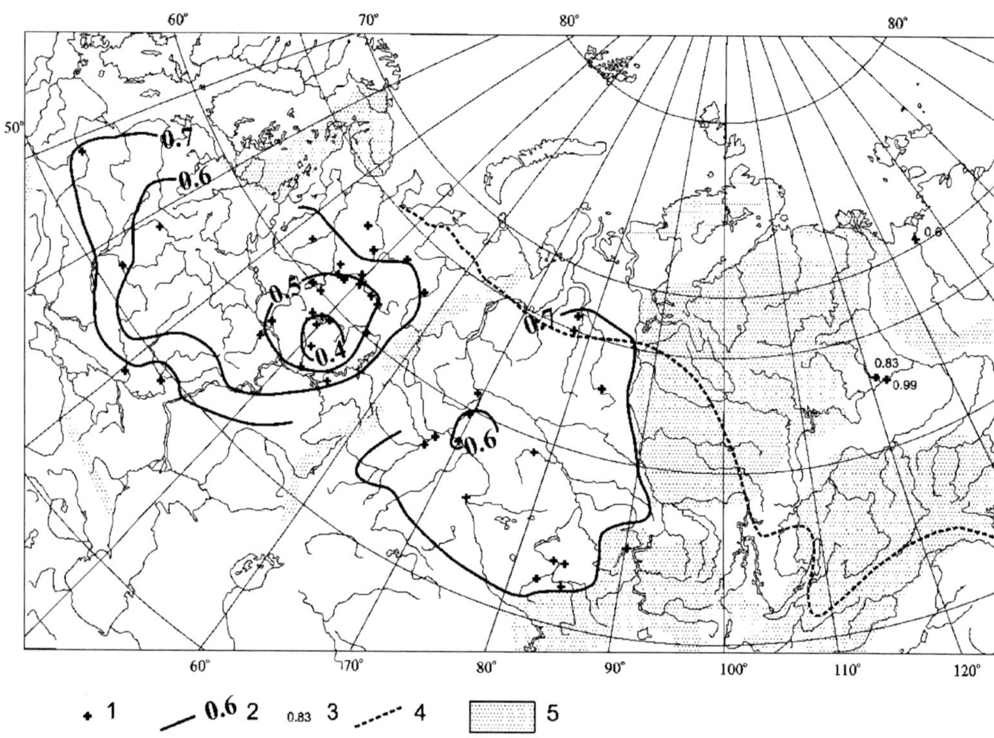

Figure 17. The distribution of the ratio between middle Holocene and modern mean channel widths in northern Eurasia. Keys: (1) positions of the sites with small oxbows, (2) contour lines of the ratios of ancient and modern channel widths, (3) individual values of the ratios of ancient and modern channel widths, (4) the southern boundary of the continuous permafrost, (5) areas of limited channel transformation.

The duration of the second stage of the floodplain formation and an age of the small paleochannels in Northern Eurasia were estimated by radiocarbon dating and pollen analysis of the oxbow deposits. The formation time of the meandering channel of the upper Ob River (see Figure 16) is estimated as 6.5-3 kyr B.P. Later the meandering channel there has been transformed into the anastomosing one (Chernov & Harrison, 1981). Radiocarbon dates obtained within the Lower Yana floodplain downstream from the Kular Ridge give a similar age estimate of the small meandering channel which existed there 5-7 kyr B.P. (Sidorchuk et al., 2003). The most complete chronology of the channel transformation in the middle and late Holocene is obtained for the Vychegda river catchment, both for its lower reaches (Sidorchuk et al., 1999, 2000) and for the upper and middle reaches (Zaretskaya et al., 2007, 2010).

A. The Lower Vychegda Key Site

The Vychegda River is the main right tributary of the Northern Dvina River. Its length is 1,130 km and the basin area is 121,000 km^2. The river flow begins on the Timan Ridge at the altitudes 200 - 300 m above sea level. The main part of the basin is a dissected plain with mean altitudes 140 - 160 m. The precipitation is 700 mm per annum, of this 350 mm falls during the winter-spring period. The mean annual air temperature is 1.2 °C. Mean annual water discharge at the mouth of the Vychegda is 1160 m^3 s^{-1} with the mean maximum value of 7500 m^3 s^{-1}. The runoff depth is 300 mm, including 200 mm for the winter - spring period (including the flood). On average about 56% of the river flow passes during the flood.

The morphology of the lower Vychegda valley was investigated within a 120 km reach near the confluence (Sidorchuk et al., 2001b). Valley bottom width increases here from 8 - 10 up to 35 - 50 km at which point valley of the Vychegda joins that of Northern Dvina. Several terraces and a floodplain with complicated morphology, and a system of paleochannels are distinguished in the area.

After recession of the Last Glacial ice sheet about 12,500 years ago and dam-lake blow-out the base level for Vychegda River lowered considerably. River incision had begun, and the sequence of erosion terraces was formed. There are several steps of different age on the surface of the terraces. The primary fluvial forms are quite clearly seen on the surface of these steps: there are paleochannels, mid and alternating bars, natural levees and chutes. The curved bluff of the third terrace forms the right margin of the oldest paleochannel A (Figure 18) on the highest step of the second terrace. The paleochannel is completely filled with very fine deposits and peat. The surface is now 6-7 m above the level of flooding. Its width is being about 1200 m (Table 4). A chain of aeolian dunes traces the left margin of the oldest paleochannel. This part of paleochannel A was not dated directly. One of the secondary channels on the lower step of the terrace cuts paleochannel A, which indicates that the latter is older. This secondary channel was active more than 9200 years ago, because peat and plant remnants in filling deposits have a radiocarbon age 9255±65 (Ki-6406) and 8950±50 (MGU-1454) years BP.

The age of peat filling a similar large paleochannel on the terrace near Lokchim River mouth and near the village of Gam (middle Vychegda valley) is 10500 - 10900 years BP (Potapenko, 1975, Arslanov et al., 1980).

A well preserved meandering paleochannel B with islands and alternating bars is found on the lowermost step of the second terrace near Sol'vychegodsk. Its mean width is 1300 m (1.5 km in broader sections). Meander half-wavelength is about 6-7 km. Elongated aeolian dunes 7-10 m high on the older step of the second terrace mark the right margin of this paleochannel. Coring at the riffles of the paleochannel shows that the bankfull depth was about 8 - 10 m. Riffles are composed of fine sand. Plant remnants in the top layers of these alluvial sands are related to the initial phase of paleochannel filling, and have an age of 8655±60 (Ki-6413) (Figure 18), 8630±60 (Ki-6405) and 8400±70 years BP (Ki-6407). Paleochannel B was filled with very fine silty sand, sandy loam and peat, but the deposition was not sufficient to cover completely the primary fluvial relief. This very fine material covers well sorted fine and medium-grained alluvial sands. It is 1.5 - 2.0 m thick on the ancient bars, and about 4.0 - 5.0 m thick in the channels at the riffle sites. The surface of the second terrace within the paleochannel is generally 3 – 4 m higher than the modern floodplain, but at the lowermost sections the river can still flood it.

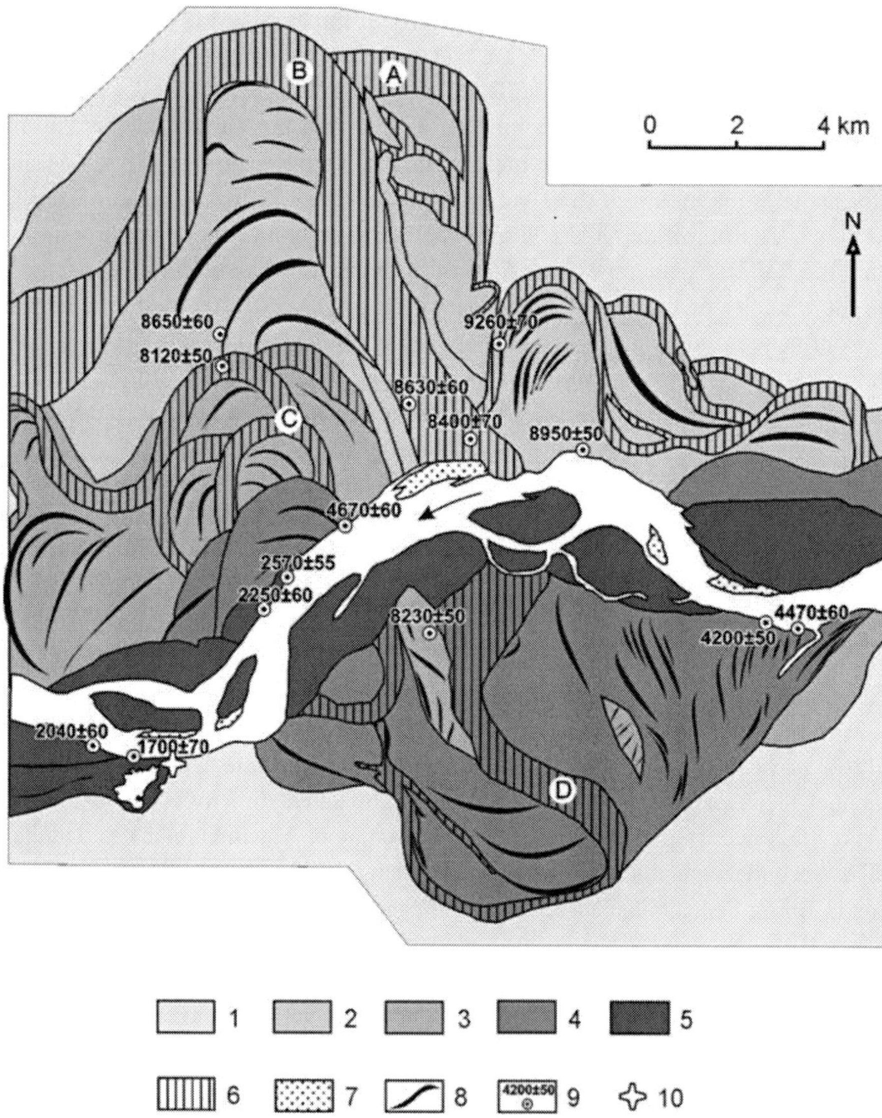

Figure 18. Fluvial relief of the lower Vychegda River. Keys: (1) the 3rd terrace, (2) the 2nd terrace, (3) the 1st terrace, (4) the higher level of floodplain, (5) the lower level of floodplain, (6) paleochannels of various age, (7) modern high sands, (8) natural levees of various age, (9) radiocarbon dates, (10) location of the section Baika-2.

Table 4. Morphology of the channels in the lower Vychegda valley

Surface	Relative altitude (m)	Channel index	Channel pattern	Bankfull width (m)	Bankfull depth (m) at the riffle/pool	Meander half-wavelength (km)
Second terrace high step	13 - 14	A	meandering with the braids	1200	?/~16	~6.5
Second terrace low step	10 - 11	B	meandering with the braids	1300	9/?	6.5
First terrace	9 - 10	C	meandering	600	6/?	3.5
Floodplain higher step	6 - 7	D	meandering	800	6/12	4.5
Floodplain lower step	5 - 6	modern	sinuous with the braids	1100	7/14	6

The first terrace is about 2-4 km wide and has a relative height above low flow level of 9-10 m. Several systems of meandering paleochannels (C in Figure 18) with natural levees and chutes are well preserved on its surface. Their mean widths are about 600 m; meander half-wavelengths are about 3.5 km. The longitudinal slope within these paleochannels is about 0.07 — 0.08 $^0/_{00}$. The paleochannels are filled with very fine silty sand, loam and peat. The filling layer covering the fine sand alluvium is about 3.5 m thick. Plant remnants in the top layers of alluvial fine sand (Figure 18) have a radiocarbon age 8120±50 years BP (Ki-6404).

The floodplain of the lower Vychegda River is 5-7 m high above the low water level. Its total width is about 8 km. The floodplain can be subdivided into two main steps of different age. Well preserved remnants of meandering paleochannels (D in Figure 18) with natural levees, chutes and oxbows exist on the higher step. The paleochannel is 800 m wide, meander half-wavelength is 4.5 km, and the bankfull depth on the riffles is 6-8 m. Water surface slope within the channel was 0.05 — 0.06 $^0/_{00}$. The paleochannel is filled by sandy loam 1.0-2.0 m thick at the point bars and by sandy loam and silty fine sand up to 9 m thick at the former pools. Some of the former pools remain in the modern relief as oxbow lakes up to 12 m deep. Peat in the depressions between adjacent natural levees on the floodplain has an age of 4200±50 (Ki-6401), 4470±60 (Ki-6402) and 4670±60 (Ki-6409). Plant remnants and charcoal found in the deposits filling oxbows are 3980±60 (Ki-6395) and 1900±50 (Ki-6390) years old.

The modern channel of the Vychegda River forms the lowermost step of the floodplain. The river has a sinuous pattern with numerous braids within the main channel. The total width of the river is about 1100 m at the crossings, meander half-wavelength is about 6 km and the water surface slope is 0.07 — 0.08 $^0/_{00}$. Bankfull depth is about 7 m at the riffles and up to 12-14 m at the pools. Channel alluvium is composed of fine and medium sand with the lenses of fine gravel. The floodplain is composed of fine and very fine sand with loam cover 0.5-1.0 m thick. Peat in the depressions between adjacent natural levees on the floodplain has an age 1700±70 (Ki-6391), 2040±60 (Ki-6392), 2250±60 (Ki-6393) and 2570±60 (Ki-6394).

The Paleolandscapes at the Lower Vychegda Region during the Holocene

A series of 15 samples of fluvial deposits was subjected to detail palynological study. An attempt was made to achieve the highest possible taxonomic resolution: pollen identifications have been made to species or genus levels for arboreal plants and for certain groups of herbaceous plants. Where identifications were possible only to the family level, they were not included into the resulting lists of fossil floras. For three pairs of successive samples the floristic lists were combined to obtain more complete fossil flora. It was possible in the case of close resemblance between pollen spectra of successive samples. Landscape and climatic reconstructions using paleofloristic method were accomplished for 12 resulting fossil floras with the radiocarbon ages ranging from 52,000 to 1400 yr BP (Sidorchuk et al., 2001b).

Fossil floras from 6 to 12 characterize consequent layers of the section Baika-2 (Figure 19). Alluvial deposits there are represented by loam and sand with two layers of peat in the upper part of the sequence. The deposits accumulated during the Holocene. Pollen assemblages were dominated by spruce, pine and birch pollen throughout the whole sediment sequence, but nevertheless substantial changes took place both in the composition of flora and vegetation. These changes are reflected by "migration" of modern floristic analogue regions over the East European Plain beginning from the Middle and Southern Urals, as far west as the Sukhona River basin, and back to the site under study at the downstream of the Vychegda River (Figure 20).

The modern floristic analogue of flora 6 (app. 8800 yr B.P.) is located at the headwaters of the Kolva River, at the boundary between dark coniferous mountain taiga forests (*Picea abies*, *Pinus sibirica*) and middle- and southern taiga pine forests (*Pinus silvestris*). Present-day climate of this region is much more continental and severe than that at the lower Vychegda. Mean January temperature was 3.5°C lower and mean July temperature 0.5°C lower than at present (Table 5, Figure 21). The mean annual temperature was slightly below 0°C. Annual precipitation exceeded the modern value by more than 150 mm.

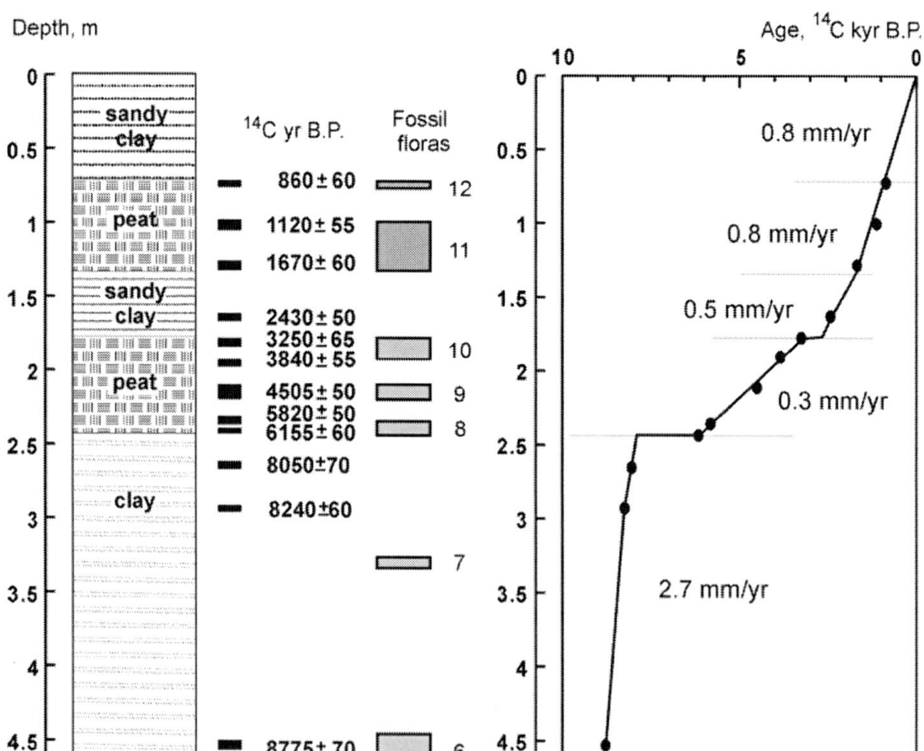

Figure 19. Lythology, age-depth curve, and estimated sedimentation rates at Baika-2, the floodplain section 14 km from the Vychegda River mouth.

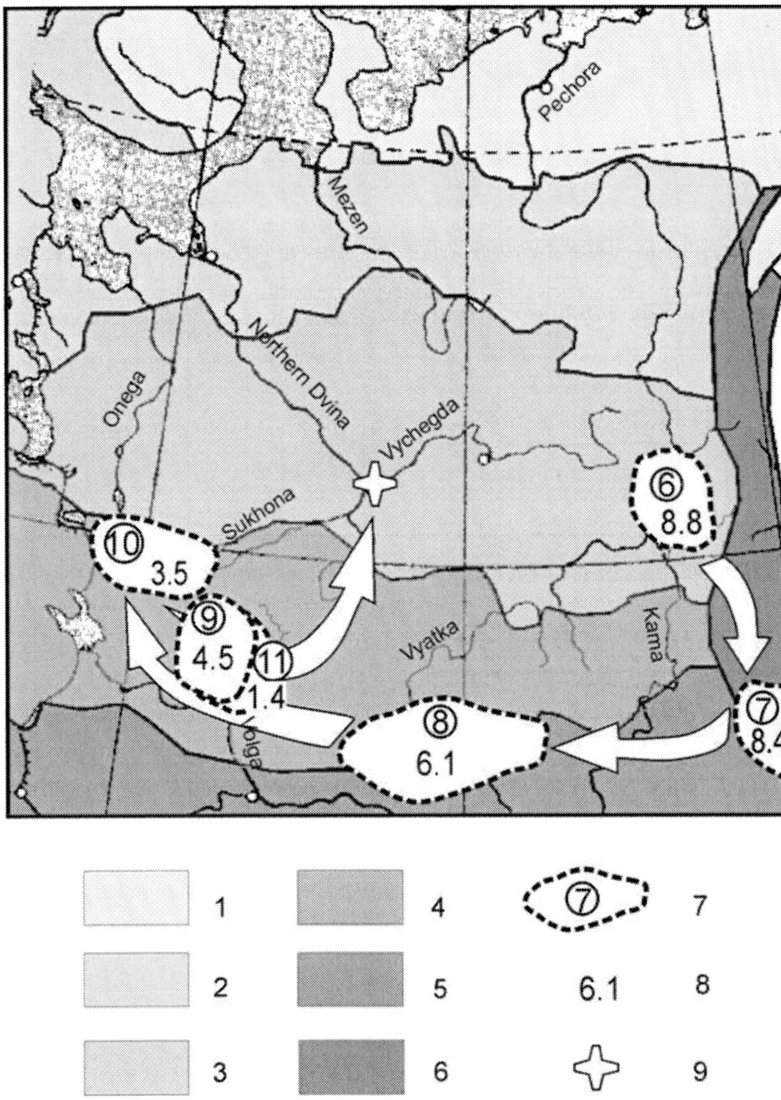

Figure 20. Locations of present-day region-analogues for the fossil floras from Baika-2 section. Keys: (1) tundra and forest tundra, (2) northern taiga, (3) middle taiga, (4) southern taiga, (5) broad-leaved/coniferous forest, (6) mountain forests, (7) the number of fossil flora and its region-analogue, (8) radiocarbon age of the fossil flora, kyr B.P., (9) Baika-2 site.

The region-analogue for flora 7 with the age estimated by interpolation between the closest ^{14}C dates as 7500 yr B.P. lies near the headwaters of the Chusovaya River. In this area broad-leaved-dark coniferous mountain forests of the Ural type are spread in the vicinity of dark coniferous mountain taiga forests (*Picea abies* with *Pinus sibirica*). The continentality of climate was then very high, as the temperature of the coldest month was still 2.5°C lower than at present, while that of the warmest month was almost 1°C higher. Mean annual temperature was already above 0°C. The duration of the frostless period slightly exceeded the present-day one. Annual precipitation was slightly less than the present-day one (630 mm).

Table 5. Positions of the paleofloristic analogues and contemporary hydro-climatic characteristics of these regions

№ of fossil flora	Region	Region-analogue coordinates		fossil flora age, ^{14}C yr BP	Recent air temperature		Recent flow depth, mm		Recent precipitation, mm		Recent evatranspiration, mm	
		Northern latitude	Eastern longitude		January	July	annual	Winter-spring	annual	Winter-spring	annual	Winter-spring
6	Basin of Kolva River	61	57	8800	-17.5	16.5	380	260	850	410	470	150
7	Basin of Chusovaya River	57.5	59	8400	-17	18	190	140	630	310	440	170
8	Middle Vyatka River	57.5	49	6000	-14	19	200	160	640	360	440	200
9	Basin of Unzha River	58.5	44	4500	-13	18	250	180	720	360	470	180
10	Headwaters of Sukhona River	59.5	39.5	3500	-13	18	330	220	750	400	420	180
11	Basin of Unzha River	58.5	44	1400	-13	18	250	180	720	360	470	180
12	Basin of Vychegda River	61.3	47	860 - recent	-14	17	300	200	700	350	400	150

Table 6. Water flow and precipitation at the lower Vychegda River basin at the Late Glacial and Holocene, reconstructed from paleochannel (and recent channel) morphology and contemporary hydro-climatic characteristics of regions – analogues

region-analog	age, years BP	Paleo channel index	Paleo channel width, m	Coefficient *a* in formula (7)	Discharge m^3/s		Flow depth, mm		Precipitation, mm	
					Mean	Mean maximum	annual	winter-spring	annual	winter-spring
input					reconstruction					
Basin of Kolva River	8800	B	1300	3.48	1490	9910	390	265	860	415
Basin of Chusovaya River	8400	C	600*	2.80	440	3670	115	90	555	260
Middle Vyatka River	6000	C	700*	1.80	400	5110	105	85	545	285
Basin of Unzha River	4500	D	800	3.99	840	4970	220	150	690	330
Headwaters of Sukhona River	3500	D	900*	2.32	670	6710	175	115	595	295
Basin of Unzha River	1400	recent	1100*	3.99	1270	7720	330	225	800	405
Basin of Vychegda River	recent	recent	1100	3.66	1170	7950	300	205	700	355

* - interpolated values.

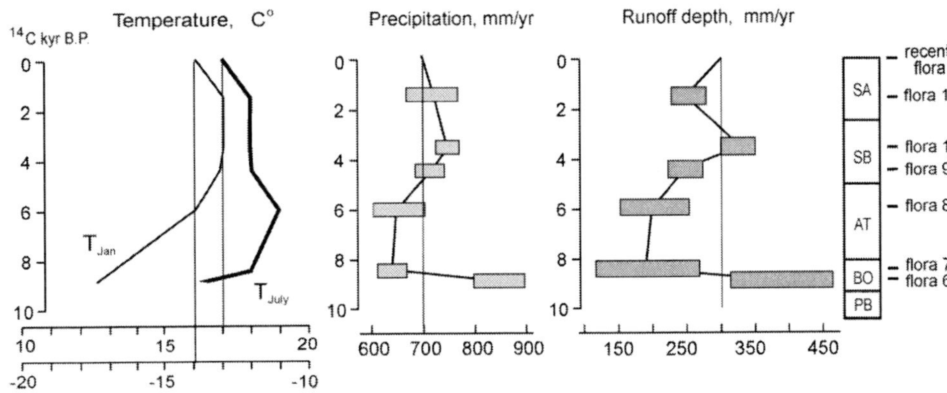

Figure 21. Climatic and hydrological characteristics of the Holocene at the lower Vychegda based on the analyses of the floodplain deposits near the Baika village. The vertical lines correspond to the modern values. Rectangles show the ranges of the reconstructed values and the appropriate radiocarbon ages.

A considerable shift of the modern analogue for flora 8 (about 6000 yr B.P.) to the west corresponds to the rise of the summer temperature, which exceeded the present-day one by 2°C, while the winter temperature reached the present level. This region in the middle part of the Vyatka River basin is covered at present by broad-leaved-dark coniferous sub-Volga forests (*Picea abies* with *Quercus robur* and *Tilia cordata*). The frostless period at that stage was almost one month longer than at present, though the annual amplitude of air temperature was still larger than now. Precipitation was about 640 mm. The floodplain sedimentation rate was extremely low at this period (see Figure 19), showing very rare flooding of this surface.

The modern analoque for the fossil flora 9 (4500 yr. B.P.) occupies the northern part of the southern taiga dark coniferous forest in the Unzha River basin. The boundaries of the present-day ranges of such species as *Quercus robur* and *Thalictrum angustifolium* form its eastern boundary. Present-day climatic conditions of the region show that the mean temperature of January for the first time exceeded the present one by about 1°C, while that of July decreased by 1°C compare to the previous stage. A substantial increase of precipitation (by app. 80 mm per year) was also reconctructed. The continentality of climate reached its minimum value for the entire Holocene. Therefore, the time interval characterized by fossil floras 8 and 9 can be regarded as the climatic optimum of the Holocene.

A region-analogue to flora 10 (about 3500 yr B.P.) is found at the Sukhona River basin, near the northern limit of the southern taiga dark coniferous forest, where large peat bogs are widespread. At that stage the mean January temperature was still about –14°C (1°C higher than the present-day one), that of July was about 18°C. Annual precipitation reached 750 mm, that is, it was 50 mm greater than at present.

The modern analogue for the fossil flora 11 is shifted again towards the east, into the Unzha River basin, where the southern taiga dark coniferous forests are spread along with pine middle- and southern taiga forests of the North European type. As it is evidenced by the climatic conditions of the region, the climate at the lower Vychegda basin about 1400 yr B.P. was still warmer and less continental than at present. The precipitation was about 720 mm per year.

Fossil flora 12 of the upper peat layer in the Baika-2 section dated by ^{14}C as 860±60 yr B.P. (KI-7026) does not include any taxa which are absent from the modern flora of the region, where the studied section is situated. It suggests that the climatic conditions at that time were essentially similar to the present-day ones.

Lower Vychegda River Channel Evolution During The Holocene

The paleochannel B on the lower step of the second terrace was abandoned about 8400-8600 years ago. It was still active during early Boreal time 8500-9000 years ago. According to paleofloristic reconstruction, the region-analogue for that time is situated at the south - eastern part of the middle taiga. The territory was already free of permafrost, and the depth of seasonal freezing of the ground was close to the contemporary one. The annual distribution of the water flow during the early Boreal was also close to the present-day one. Large and wide paleochannel in the conditions of low variability of the river flow corresponds to relatively high mean annual discharge 1500 $m^3\ s^{-1}$. Mean maximum discharge was about 9900 $m^3\ s^{-1}$. The depth of mean annual runoff was 390 mm, the depth of spring runoff was 265 mm. In conditions of seasonal ground freezing the runoff coefficients were close to recent ones: they were about 0.45 for the whole year and up to 0.63 for a flood period. The air temperature and evatranspiration at the lower Vychegda valley were close to those in the region-analogue. That gives a rather high estimate for the mean annual precipitation: 860 mm. Precipitation in the winter-spring period was 415 mm, and that of summer - autumn period was up to 445 mm (Table 6).

Paleochannel C on the first terrace of the lower Vychegda was abandoned about 8200 years ago. It was active during late Boreal time 8200-8500 years ago. According to the paleobotanical reconstruction the region - analogue is situated at the eastern part of the southern taiga. The depth of seasonal freezing of the ground was close to the contemporary one. Annual distribution of the water flow during the late Boreal was more variable than at present, and humidity at that time was significantly lower. The paleochannel on the first terrace is smaller than all other paleochannels. Its width and curvature corresponds to mean annual discharge about 440 $m^3\ s^{-1}$, the mean maximum discharge was about 3700 $m^3\ s^{-1}$. The depth of mean annual runoff was 115 mm, the depth of spring runoff was 90 mm. This corresponds to the minimum value of the winter-spring precipitation depth (260 mm) for the whole period studied. Mean annual precipitation was about 555 mm.

On the Viled' River (a tributary of the Vychegda River) an average width and the meander wavelength of the paleochannel were 2.6 times less than those of the present-day channel. The radiocarbon age of this paleochannel is about 7700 yr B.P. The minimum water flow occurred at the lower Vychegda at the late Atlantic, about 6000 years ago.

The latest paleochannel D at the lower Vychegda valley is well preserved on the oldest steps of the floodplain. This channel evolved during the Subboreal period and was abandoned at the beginning of the Subatlantic, about 2500 years ago. The geometry of natural levees and chutes on the floodplain shows the main paths of the meandering channel migration. The rate of channel migration down the valley was estimated as 1.6 m per annum near Durnitsino village for the period 4200-4500 years ago, according to radiocarbon dating of a sequence of natural levees and depressions between them. Migrating channel reworked deposits of the first terrace. Several remnants of this surface were preserved within the floodplain. Development of curved omega-shaped meanders shows low erosivity of the flow on the floodplain. It probably indicates a relatively low annual variability of the water discharge.

Subboreal paleochannels were formed during a time of significant decrease in climate continentality. The last paleochannel morphology corresponds to these climatic conditions. Its width and curvature were formed by a mean annual discharge about 840 - 670 $m^3 s^{-1}$, and the mean maximum discharge was about 5000 – 6700 $m^3 s^{-1}$. Annual distribution of the water flow was characterized by low variability at the beginning of the period. It increased at the end of the paleochannel formation. The depth of the mean annual runoff was 175-220 mm, the depth of winter – spring runoff was 115-150 mm. Together with the evatranspiration value for the region–analogue this gives a mean annual precipitation depth equal to 600-690 mm. Precipitation of the winter-spring period was about 290-330 mm.

At the beginning of the Subatlantic period the morphological type of the river channel was altered again: it became braided-meandering. The stream then abandoned the omega-shaped meanders of paleochannel D, which are preserved now as oxbows. This was caused by an increase of the maximum discharge at the end of Subboreal and by the general humidity of the climate in Subatlantic. Calculations show that the water flow and precipitation reached their maximum about 1500 years ago, at the period of lowest variability of the flow during the year (see Table 6). By recent times both the calculated mean annual discharge and precipitation depth decreased to 1170 $m^3 s^{-1}$ and 700 mm correspondingly, and the mean maximum discharge increased to 7950 $m^3 s^{-1}$due to a greater discharge variability within a year.

B. Upper and Middle Vychegda Key Sites

In the Upper Vychegda valley downstream from the Nem River mouth the floodplain is 2.0-3.5 km wide. At this site three age generations of the floodplain are identified (Zaretskaya et al., 2007): (1) a flat floodplain with poorly preserved troughs-and-islands primary relief, (2) floodplain with troughs and levees formed by meandering channel migration with much smaller and steeper bends compare both to the older and modern channels, and (3) floodplain formed during the development of the modern channel bends. Modern meander bends are larger than those of the previous stage, and have much less curvature. The radiocarbon dating suggest that the formation of small channels of the second stage of floodplain formation at the Upper Vychegda has begun before 3600 and ended after 2800 ^{14}C yr B.P.

At the stretch of the Vychegda River near the Lokchim River mouth (see Figure 1) the main part of the floodplain was formed by the meandering channel with the bends substantially smaller and steeper than those of the modern channel. One of the oldest floodplain fragments with small paleochannels, on which the Neolithic site Pezmog-4 is situated, was formed 6700-6800 ^{14}C yr B.P. (Zaretskaya et al., 2010).

Thus, throughout the Vychegda River valley the second stage of the floodplain formation can be clearly distinguished. The floodplain of the second generation was formed in the process of erosion and accumulation by a meandering channel with smaller dimensions than both the older and modern channels. In the lower reaches of the Vychegda River the width of this channel was 600 m (1.8 times less than today) early in the second stage and 800 m (1.4 times less than today) at the end of the stage. This stage characterized by a substantial reduction of the channel size and an increase in the curvature of meanders included the entire Atlantic period and an early part of the Subboreal period of the Holocene, with a minimum size of the river channel during the Holocene Optimum.

The Paleohydrology during the Second Stage of Floodplain Formation

According to the sizes of paleochannels, the floodplain of the second stage was formed by a fairly low water flow (Table 3). On the northern megaslope of the Russian Plain the runoff at the time was about 180 $km^3 a^{-1}$, which is 30% less than the modern runoff from the same catchment area (Sidorchuk et al., 2012). In the Volga River basin an annual runoff was about 134 km^3, which is almost half of today flow. The runoff from the Vyatka River catchment area was even lower – app. 60% less than the present-day one. Such a dramatic reduction of the runoff from the Volga River drainage basin fully explains a deep Mangyshlak regression of the Caspian Sea which followed the Khvalyn transgression. At the maximum of this regression the Caspian Sea level fell to the mark about minus 50 m. In the basins of the Don and the Dnieper rivers the runoff in the Holocene optimum was 40% less than at present. In the basin of the Ob' and Irtysh rivers the runoff was 30% less than at present (Sidorchuk, 2010).

The Third Stage: Floodplain Formation by the Modern Rivers

About 4000 years ago the general cooling of the climate and some increase in precipitation commenced. These processes developed during the Subboreal and Subatlantic periods of the Holocene. A significant (1.5-2 times) increase in runoff resulted from the climatic changes has led to a change in the river channels size over large areas. In the upper reaches of the Ob' River a single channel with sharp bends transformed into an anastomosing channel with several very gently meandering main branches. At the Vychegda River basin where a sufficient number of radiocarbon dates makes it possible to distinguish all the stages of the channel evolution, such transformation has begun about 3600 yr B.P. At that time small and steep meanders were cut off from the main channel and turned into small oxbows. At the lower reaches this process ended about 2500-2700 yr B.P., when the channel acquired its modern branched-sinuous pattern. Over time, the channel width and the meander wavelength increase, while the width of the meandering belt decreases. More often, as on the rivers Khadutte, Belaya, middle Vychegda, upper Dnestr, Vilyui, and Tyung, channels with steep small meanders became converted into wider gently meandering channels with point and braid bars. The amplitude of meandering was reduced, and the meander wavelength increases (as at the Dnestr, Desna, Don, Seversky Donets, Irtysh, Tobol, and other rivers).

The history of floodplain evolution and the geometry of the paleochannels are often crucial in the modern river dynamics. The floodplain width was mostly formed by the large rivers of the first stages of floodplain evolution, and the modern channel often didn't rework this floodplain, being too narrow. This mismatch between the floodplain and the channel width of the rivers was referred to by Dokuchaev (1878). Davis (1896) suggested the term "misfit rivers" for similar rivers in Western Europe. The wide point bars of former large meandering paleochannels are still flooded by the high waters of smaller recent rivers. Thus, the great variety of the river valleys keep a clear morphological “memory” of past events, and this “memory” significantly influences the modern dynamics of erosion/sedimentation system. Two key-study examples illustrate this thesis (Sidorchuk, 2003).

Water Flow and Sediment Fluxes in the Moksha River Basin

The Moksha River (a tributary of the Oka) is 656 km long. Together with its main tributaries, Tsna (451 km, 21,500 km^2) and Vad (222 km, 6,500 km^2), it has a catchment area of 51,000 km^2. The river headwaters are situated at the Privolzhskaya Upland, and the southeastern part of the basin has altitudes 200–350 m above sea level. The northwestern part of the basin is mainly swampy and forested lowland, crossed by the elongated Oka-Tsna ridge, where the lower Tsna and Moksha rivers form narrow canyon-like valleys. Snowmelt waters mainly feed the river, and the flood season occurs in April/May. The annual water discharge near the river mouth is about 180 $m^3 s^{-1}$, and sediment discharge at the Shevelovskiy Maidan station is 11 kg s^{-1}.

Fragments of the lateglacial paleochannels of the Moksha and Tsna Rivers are well preserved on the river floodplains and the lower terrace. In the upper and middle sections of the valley the paleochannel has a meandering pattern with the channel width, meander wavelength and amplitude gradually increasing down the river. The main geometrical characteristics of the floodplain in the Moksha basin are determined by the morphology of the paleochannels. The width of the floodplain is more or less equal to the total width of the paleochannels, including some point bars and braid bars. The floodplain margins follow the shape of the paleochannel. Former paleochannel pools, partly filled with sediments (now swamps and ox-bow lakes), and not reworked by subsequent fluvial activity, occupy a significant part of the floodplain. During the Holocene, a meandering channel was active within a relatively narrow belt, often in the middle part of the wide inherited floodplain. The lowland swampy conditions of the recent meander belt development caused multiple blowouts of the natural levees, formation of numerous ox-bows, chutes and distributaries, often of anastomosed pattern.

There are 68 hydrological stations at the Moksha River basin with the different periods of observation, 11 of them measured the suspended load during a 10–50 year period. Paired correlation coefficients of the annual water discharges, and of the mean annual suspended sediment concentrations, were obtained for those stations, each pair of stations being situated on the same river or on the main river and its tributary. When related to the distance between the stations, these correlations show rather high level of the water regimen similarity in the Moksha River basin (Figure 22). Correlation coefficients for sequences of annual water discharges, significant at a 0.005 level, are equal to 0.82–0.96 for the distances of 150-400 km, and decrease to 0.68 for distances over 500 km.

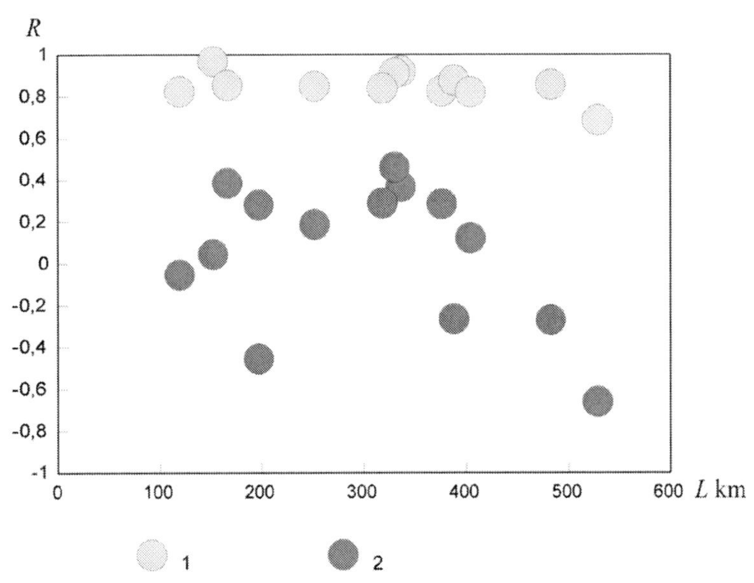

Figure 22. Paired correlation coefficients R for the rows of the mean annual discharge Q (1) and sediment concentration Q_s/Q (2) measured at different hydrological stations in the Moksha River basin. L is the distance between the hydrological stations.

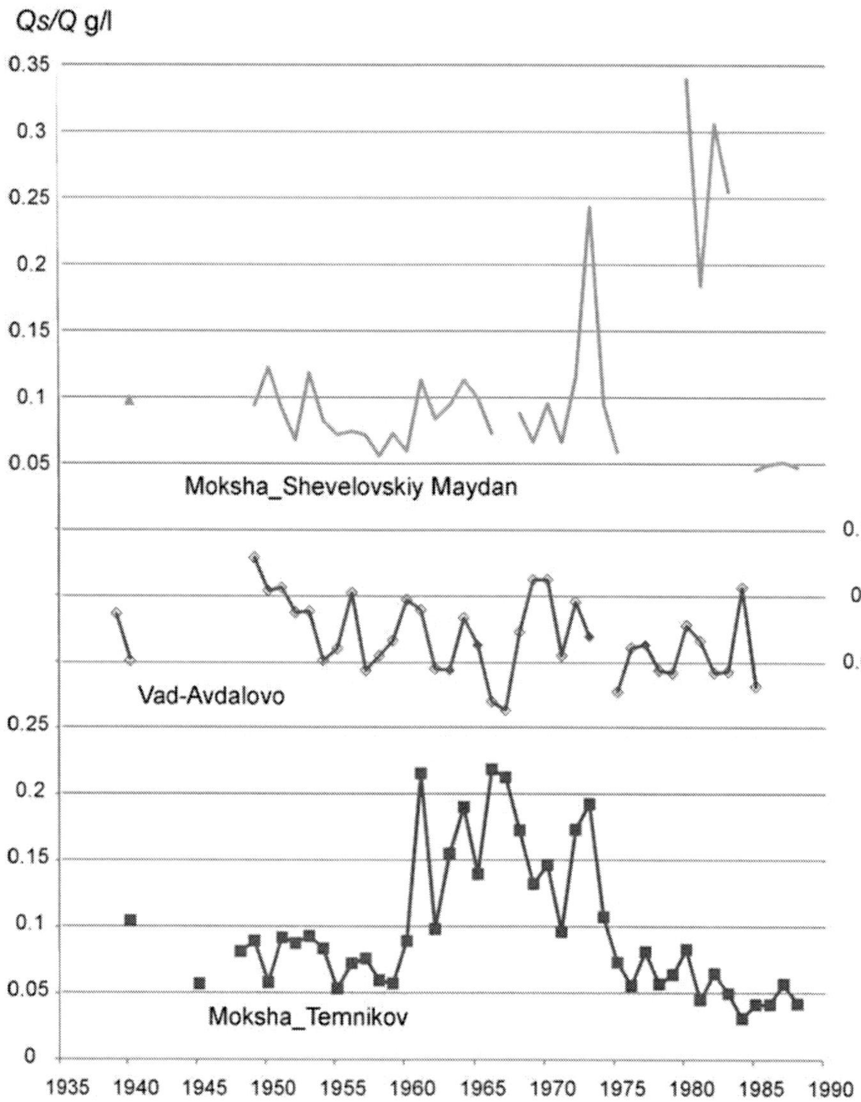

Figure 23. Changes of the mean annual sediment concentration Q_s/Q through time at the main hydrological stations in the Moksha River basin.

The correlation coefficients for the suspended sediments show significant disconnectivity in the sediment regimen in the Moksha River basin (see Figure 22): they are as low as 0.05–0.46 for the distances of 150–350 km, and negative for distances over 400 km. The very low connectivity of the suspended sediment regimen in the Moksha River valley is a characteristic feature of valleys of this type, where ancient alluvial forms bring about an unconformity between the floodplain and the modern channel.

The longest records of suspended sediment are available for three stations: Temnikov (T) on the Moksha River, 224 km from the river mouth, contributing basin area 15,800 km^2; Avdalovo (A) on the Vad River, 105+86 km from the Moksha River mouth, 1930 km^2; and Shevelovskiy Maidan (SM) on the Moksha River, 72 km, 28,600 km^2 (Figure 23). At site T, a period of relatively low suspended sediment load in the ‘40s and ‘50s was followed by a period of relatively high suspended sediment load in ‘60s and early ‘70s. In 1975 another period of relatively low suspended sediment load began. At site A, a general tendency towards decreasing sediment load persisted during entire measurement period. At site SM, on the contrary, the annual sediment load increased until the beginning of the '80s, when sediment concentration decreased sharply. Stations T and A collect data from 56% of the basin up to SM. Total annual water discharge at the upper stations accounts for 96.8% of the annual discharge variability at the lower station. The annual sediment load at the upper stations accounts for only 2% of the annual sediment load variability at the lower station.

The main reason for sediment load discontinuity is the absence of a link between the slopes in the basin and the river channel (Sidorchuk, 2003). The floodplain is much wider than the river channel, and is not completely submerged during an average flood. Most of the tributaries of the Moksha River are not connected with the main river channel, their f;ow being dispersed on the floodplain and forming sediment fans there, and only the largest tributaries with basin areas over 1300 km^2 reach the main channel. The Moksha River erodes high slopes along only 5% of its valley length. In the Tsna River valley such sections amount to approximately 11% of the total valley length. The proportion of such eroded slopes varies significantly along the valleys, being 25–30% at the confined valley sections, and nearly 0 at the valley sections with the wide floodplain. Such river valley morphology, resulting from evolution under the Late Glacial/ the Holocene environmental change, leads to the deposition of the bulk of the sediments eroded from the slopes and gullies onto the floodplains. This sediment from the slopes and gullies is only delivered to the main river channel within confined valley sections. For example, the SM station is situated within such a section, where the floodplain is only 14 times wider than the river channel (2800 m and 200 m, respectively). Within this 20-km section of the valley, the river follows a high bluff dissected by gullies. The sediment load, measured at SM, is formed mainly by local erosion within this confined valley section, being non-correlated with the sediment regime in other parts of the river basin.

Water Flow and Sediment Fluxes Connectivity in the Khoper River Basin

The Khoper River (one of the largest tributaries of the Don) is 1008 km long. With its main tributaries, the Vorona, Savala and Buzuluk, it drains a basin of 61,100 km^2. The Khoper headwaters are situated at the Privolzhskaya (Kerensk-Chembarsk) Upland. The major part of the basin has altitudes of 200–350 m above sea level. The river feeds mainly from the snowmelt, and the main flood period occurs in the April-May. The annual water discharge at Dundukovskiy station is 152 $m^3 s^{-1}$, the sediment discharge is 13.4 kg s^{-1}.

Fragments of the lateglacial paleochannels of the Khoper and Vorona Rivers are well preserved on the river floodplains and on the lower terrace in the upper and middle parts of the basin (see Figure 10). At the lower section of the main valley (downstream from the Savala River mouth) the paleochannel was mainly reworked by the subsequent fluvial activity of the river.

The morphology of the floodplains in the upper and middle sections of the Khoper River basin is controlled by the remnants of paleochannels. The width of the floodplain is approximately equal to that of the paleochannels together with their point bars and the paleochannel pools, partly filled with sediments (present-day swamps and lakes), which occupy a significant part of the modern floodplain. During the Holocene, the meandering channel was active within a relatively narrow belt, often situated in the middle of the wide inherited floodplain, with numerous natural levees, oxbows, chutes and distributaries (see Figure 10).

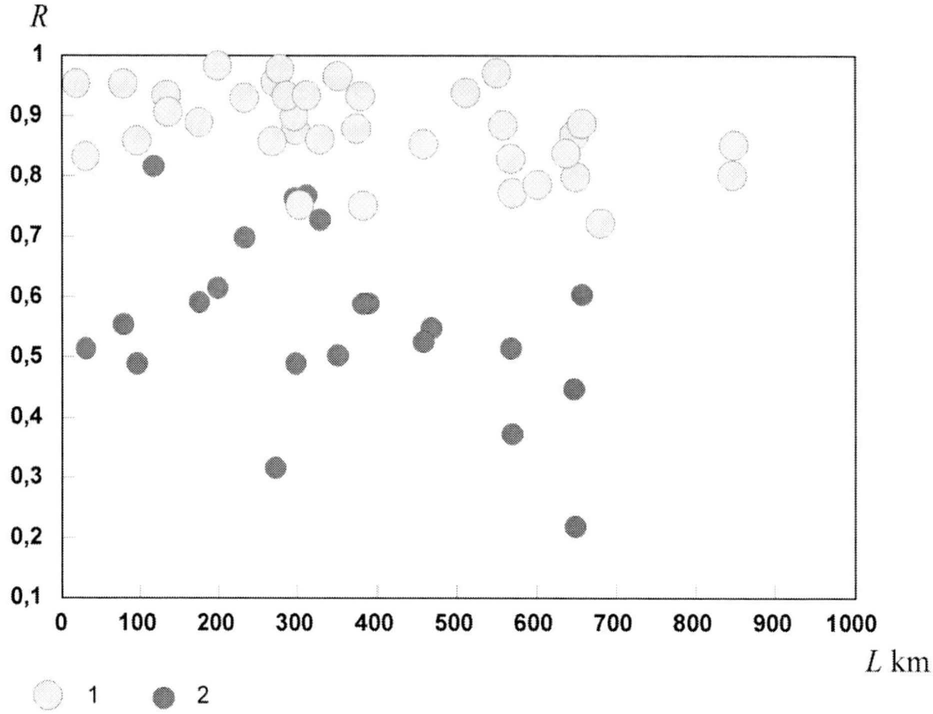

Figure 24. Paired correlation coefficients R for the rows of the mean annual discharge Q (1) and sediment concentration Q_s/Q (2) measured at different hydrological stations in the at the Khoper River basin.

In the Khoper River basin there are 78 hydrological stations with the different length of the records, and 19 stations measured the suspended load for a period of 10–50 years. Comparisons of the measurements at these stations show similar trends. A period of relatively high sediment load in the '30s and early '40s was followed by a period of relatively low sediment load at the end of the '40s. At the beginning of the '50s the sediment load increased and was stable in the '60s (with some trend to decrease at several stations). Between 1973 and 1975 a new period of relatively low suspended sediment load started. The above-mentioned trends can be explained by historical changes in the land use in the Khoper River basin. A rapid increase of the area of ploughed land took place in the '20s and '30s, followed by a dramatic decrease of agricultural activity during World War II. The pre-war level of agriculture was restored in the '50s and '60s. A slight decrease of the arable land area and an intensification of soil conservation work in the '70s and '80s brought a decrease in suspended sediment load. The general correlation between the sediment load of the Khoper River and the land use in its basin indicates a rather high degree of connectivity of sediment fluxes in this fluvial system.

The paired correlation for sequences of annual water discharges and the mean suspended sediment concentration also indicate high flow connectivity in the Khoper River basin (Figure 24). The correlation coefficients for water discharges are to 0.83–0.98 for the distances less than 100 km, and gradually decrease to 0.72–0.85 for distances over 600 km. The correlation coefficients for sediment load vary from 0.48 to 0.81 for distances less than 150 km, and decrease to 0.22–0.51 for the distances over 600 km. These correlations of the suspended sediment flow are much higher in the Khoper River basin than in the Moksha River basin.

There is a rather strong link between the basin slopes and the river channel in the Khoper River basin: about 19% of the high slopes are eroded by the rivers. This proportion increases from 10% at the upper sections to 28% at the lower ones. In general, it also increases (with some scatter) with the channel/floodplain width ratio. The floodplain at the Vorona River valley is wider compared with the channel width, and the main channel erodes only about 12% of high banks. The ratio is distributed along the river more uniformly than in the Khoper River valley.

Most of the tributaries of the Khoper River are connected with the main river channel. Only small tributaries with basin areas under 800 km^2 do not reach the main channel, and their flow are dispersed on the floodplain. Slope and gully sediment is delivered to the main river channel in the confined valley sections, but is deposited on the wide sections of the floodplain with remnants of lateglacial relief, e.g., deposition prevails in the lower section of the Vorona River, where the channel represents a chain of lakes and swamps. The sediment load measured at the hydrological stations on the Khoper River system, reflects both the variability of local hydraulic and erosion conditions and the sediment regime of the slopes in the river basin as a whole.

The relatively high connectivity of suspended sediment fluxes in the Khoper River valley, where the remnants of the ancient alluvial forms on the floodplain were partly reworked by channel activity in the Holocene, can be explained mainly by closer slope – channel links.

Sedimentation on the Inherited Floodplains

The main role of swampy inherited floodplains is as a reservoir for deposition of sediments delivered from the river basin, thus forming a barrier between the erosion section of the fluvial system (rills and gullies on the slopes) and its transportation section (permanent river channels). The great width of a floodplain of this type is out of proportion to the modern channel size. As a result, the probability that meandering and anastomosed river channels would erode the valley slopes and link directly with the erosion features on the slopes is quite low and correlation of sediment regimes at different, often neighboring, segments of the river channel is very poor.

The inherited floodplains at the lowland Moksha and Khoper River valleys were formed in the Late Glacial/Holocene transition, and their subsequent evolution took more than 10,000 years. In the lower Khoper River basin (downstream from the Savala River mouth), the continuous channel is already formed, as the floodplain has been reworked by the channel during the Holocene. In the upper part of the basin, the main channel and its tributaries still often have the chain-of-lakes pattern. The sediment yield at different stations in the Khoper River basin varies from 64 to 6 t km^{-2}, generally decreasing with the contributing area. The calculated rates of annual erosion of arable land within the basin vary from 100 to 300 t km^{-2}. Both the slope erosion and the sediment yield in the channels increased dramatically as a result of human activity during the last 150-200 years, as 95% of the basin is ploughed. In natural conditions forest-steppe vegetation covered the upper part of the Khoper River basin. The estimated natural sediment yield in the rivers of the region was about 1.2 t km^{-2} (Dedkov & Mozzherin, 1984, Table 2.11), and an annual deposition rate on the inherited floodplain could be ~10-20 t km^{-2}. The lower part of the basin is within the steppe zone, and the natural sediment yield there could be about 4.4 t km^{-2} (Dedkov & Mozzherin, 1984, p. 45). Annual deposition on the floodplains within the zone could reach 150 t km^{-2}. In their main length the channels in the Moksha River basin contain chain-of-lakes and anastomosed patterns. The modern sediment yield at the different stations varies from 58 to 8 t km^{-2}, and the calculated rate of accelerated erosion from arable land (75% of the basin) varies from 100 to 2000 t km^{-2}. The Moksha River basin is mainly situated in the broad-leaved forest zone, where the sediment yield under natural vegetation (Dedkov & Mozzherin, 1984, Table 2.10) could be about 1.8 t km^{-2}. An annual deposition on the floodplain could therefore be about 30 t km^{-2}. These estimates show that for a medium-size river the transition from an inherited floodplain to a typical one with a continuous channel occurs after the deposition of about 1000-1500 10^3 t of sediments per 1 km^2 of the floodplain area. When the total deposition amounts to 200-740 10^3 t km^{-2} of the floodplain area, the river channel has a transitional chain-of-lakes and anastomosed pattern, and the floodplain still represents a significant barrier between the erosion and transport sections of the fluvial system.

4. Spatial Morphometric Patterns in the Floodplain Relief

The spatial distribution of the widths ratios between the channels of the first and third stages (Figure 25), as well as those of the second and third stages of the floodplain formation (Figure 17), shows three well-defined areas where the size of the bends of river channels underwent maximum changes during the Late Glacial – Holocene – up to 20 times. These areas are located in the south-west of the East European Plain (in the eastern part of the Dnepr River basin), in the north-east of the Plain (in the Upper Vychegda and Vyatka river basins), and in the south-west of the West Siberian lowland (in the Tobol and middle Irtysh basins).

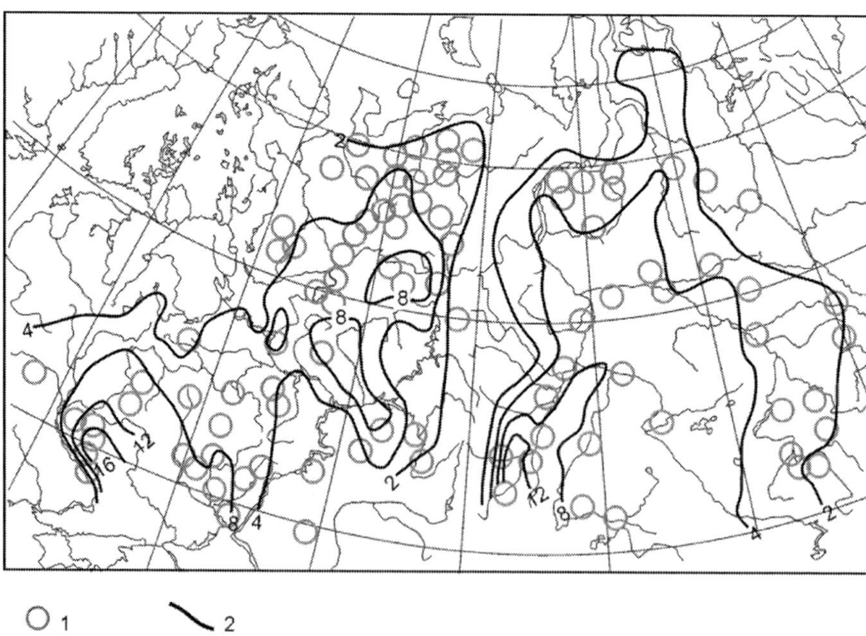

Figure 25. The distribution of the ratio between lateglacial and modern mean channel widths in the northern Eurasia. Keys: (1) the sites with large paleochannels, (2) contour lines of the ratios of the ancient and modern channel widths.

In the Late Glacial, on the first stage of the floodplain formation, the left (eastern) tributaries of the Dnepr River were characterized by high discharges, wide channels and meanders with large wavelengths and amplitudes. Accordingly, in the valleys with relatively small catchment areas broad floodplains were formed. At the beginning of the Holocene, the conditions of the surface runoff formation underwent major changes in the process of permafrost degradation and development of the modern-type landscapes. Accordingly, the dimensions of the river channels on the second stage of floodplains formation decreased 15-20 times as compared to the first stage. At the third stage of floodplains formation (in the Subboreal time of the Holocene) channel sizes have increased ones again, although by no more than 30%. At that stage many rivers formed floodplains with anastomosing patterns, so that the differences in the floodplain morphology between the second and third stages are hard to detect.

As a result, the region has the largest widths ratio between the floodplains formed on the first stage by large lateglacial channels and the modern floodplains of the same rivers. In the south-west of Western Siberia similar morphology is characteristic for floodplains in the Tobol River basin, where all the valley bottoms are occupied by broad floodplains with large natural levees of the first (lateglacial) stage of formation, while modern branched or meandering channels transform only narrow band of floodplains forming small natural levees.

In the north-east of the East European Plain another area of maximum transformation of river channels at different stages of the floodplain is located. Here, in the basins of the Vyatka and Upper Vychegda, large lateglacial rivers worked out wide valley floors on the first stage of the floodplain formation. By the Holocene climatic optimum, on the second stage of the floodplain formation, the channel sizes in this region decreased by 15-20 times, that is about the same ratio as in the southern areas of the maximum transformation of rivers. But in contrast to the southern regions, in the north-east of the East European Plain the increase in the meander wavelength and water discharges was significant between the second and third stages of the floodplain formation.

At the same time at every next stage the width of the meandering belt and, accordingly, the width of the floodplain was smaller than at the previous stage. As a result, in this region all three generations of floodplains can be discerned within the long stretches of river valleys.

In the tundra changes in the morphology of river channels are not observed and floodplains are relatively homogenous.

A similar situation exists for the Eastern Siberia where the changes in the river channel morphology are minimal. This is primarily due to the high stability of incised river channels. However, the sizes of rivers with broad floodplains (for example, in the Vilyui and Lower Yana basins) also experienced little changes. During the period of the main changes (the middle Holocene) channel transformation here was limited due to small variation of discharges. Homogeneous floodplains are also typical for the areas of limited channel transformation on the northwestern East European Plain and the Urals.

5. Applied Aspects of the Study of the Main Stages in the Floodplains Evolution

Applied aspects of the floodplains evolution studies are numerous. The need to distinguish the evolutionary stages of floodplains for the analysis of riparian forest successions was mentioned above (Braslavskaya, 2004). Evolutionary factors play an important role in the soil formation on the floodplains. Further only the water flow regime changes under the conditions of the possible anthropogenic global warming are discussed.

A considerable part of territory of the Northern Eurasia is occupied by continuous or discontinuous permafrost. The present-day hydrologic regime of rivers in the permafrost areas is similar to the hydrologic regime of the Late-Glacial large rivers in the East European Plain and Western Siberia. Under the conditions of anthropogenic global warming, the landscapes of these areas will experience very fast and radical changes. It is quite probable that the changes in the hydrologic regime of rivers and the morphology of river channels within the permafrost territories will follow the scenario of transformation of the Late-Glacial large rivers. The annual runoff volume will decrease due to greater losses through evaporation; the runoff will become more uniform due to a significant decrease in runoff during spring flood and increase in water discharges during the low-water period. The river channel sizes in central Siberia, Yakutia, and northeastern Russia will considerably decrease; large braided rivers may transform into a single channeled meandering rivers.

The same type of scenario arises in the analysis of the second stage of floodplain formation in the Atlantic period of the Holocene. According to paleoclimatologists (Borzenkova, 2002), the Holocene optimum can be used as an analogue of the anthropogenic climate warming, with an increase in average global air temperature by approximately 1°C. If we accept this hypothesis, the results of the morphological studies of small paleochannels of the Middle Holocene can be used to develop scenarios of water availability and river regime in the middle part of the 21st century. The results of the study of the evolution of river floodplains of Northern Eurasia shows that the Holocene Optimum rivers have been much smaller and the water flow was more uniform. If this scenario can be used for the future, change in the channel size, width and position of the meandering belt will affect all sectors of the ecology and economy associated with the rivers. For example, this scenario proposes a considerable decrease of the Volga River discharge, which will cause the Caspian Sea level drop up to the complete draining of the modern northern shelf of the Caspian Sea.

References

Arslanov, Kh.A., Lavrov, A.S., Lyadov, V.V., Nikiforova, L.D., Potapenko, L.M., & Tertychnaya, T.V. (1980). Radiocarbon geochronology and paleogeography of Middle Valdai interval and last glacial sheet at the north – east of the Russian Plain // *Geochronologiya chetvertichnogo perioda.* Moskva: Nauka. P. 68-81 (in Russian).

Baryshnikov, N.B. (1984). Morphology, hydrology and floodplain dynamics. Leningrad: *Gidrometeoizdat*, 280 p. (in Russian)

Borisova, O.K., & Novenko, E.Yu. (1999). Environmental conditions at the Late Paleolithic site Yudinovo (based on palynological data) // Bolikhovskaya, N.S., & Rovnina, L.V.

(eds). *Topical Problems in Palynology at the Beginning of the Third Millennium.* Moscow: IGIRGI Press. P. 54–60 (in Russian).

Borisova, O.K., Sidorchuk, A.Yu., & Panin, A.V. (2006). Palaeohydrology of the Seim River basin, Mid-Russian Upland, based on palaeochannel morphology and palynological data // *Catena*. Vol. 66. P. 53-78.

Borzenkova, I.I. (2002). Empirical Paleoclimatology: The State of the Problem and Research Methods // *Climate Change and Its Consequences*. St. Petersburg: Nauka. P. 75-92 (in Russian)

Braslavskaya T.Yu. (2004) Vegetation cover of the floodplains in the forest zone // *East European forests*. Vol. 2. Moscow: Nauka. P. 384-426 (in Russian)

Chalov, R.S., & Chernov, A.V. (1985). Geomorphic classification of floodplains of lowland rivers // *Geomorfologiya*. No. 3. P. 3-11 (in Russian)

Chalov, R.S., Zawadski, A.S., & Panin, A.V. (2004). *River meanders*. Moscow: Moscow State University Press, 371 p. (in Russian)

Chernov, A.V. (1983). *Geomorphology of the floodplains of lowland rivers*. Moscow: Moscow State University Press, 198 p. (in Russian)

Chernov, A.V. (2006). About typification and classification of floodplains and floodplain processes // *Floodplain and Floodplain Processes*. St. Petersburg: Izd. RSU. P. 12-30 (in Russian).

Chernov, A.V. (2009). *Geography and geo-ecological state of the channels and floodplains of northern Eurasia*. Moscow: Korona, 682 p. (in Russian)

Chernov, A.V., & Harrison, L.M. (1981). Paleogeographic analysis of channel deformation of the wide-floodplain rivers in the Holocene (an example of the Upper and Middle Ob') // *Bull. MOIP. Geol. Section*. Vol. 5. No. 4. P. 97-108 (in Russian)

Davis, W.M. (1896). The Seine, the Meuse, and the Moselle // *National Geographic Magazine*. Vol. VII. P. 189-202, 228-238.

Dedkov, A.P., & Mozzherin, V.I. (1984). *Erosion and sediment yield on the Globe*. Kazan': Kazan' Univ. Pres. 264 p. (in Russian).

Dokuchaev, V.V. (1878). *The Ways of the River Valleys Formation at the European Russia*. St. Petersburg: Dermakov Publ. 123 p. (in Russian).

Dury, G.H. (1954). Contribution to a general theory of meandering valleys // *Am. J. Sci. Vol.* 252. No. 4. P. 193-224.

Dury, G.H. (1964). Principles of underfit streams // *U.S. Geological Survey Professional Paper* 452-A. 67 p.

Dury, G.H. (1965). Theoretical implications of underfit streams // *U.S. Geological Survey Professional Paper* 452-C. 43 p.

Evstigneev, V.M. (1990). *River Flow and Hydrological Prognosis*. Moscow, Moscow University Press. 304 p. (in Russian).

Grichuk, V.P. (1969). Glacial floras and their classification // Gerasimov, I.P. (ed). *The Last Ice Sheet in the Northwestern European Part of the USSR*. Moscow: Nauka. P. 57–70 (in Russian).

Hintikka, V. (1963). Über das Groβklima einiger Pflanzenareale in zwei Klimakoordinatensystemen dargestellt // *Annales Botanici Societatis Zoologicae Botanicae Fennicae 'Vanamo'*. Vol. 34. P. 1-64.

Iversen, J. (1944). *Viscum, Hedera, Ilex* as climate indicators. A contribution to the study of the post-glacial temperature climate // *Geologiske Föreningen Stockholm Förhandlingar.* Bd. 66. P. 463-483.

Kondratyev, N.E., Popov, I.V., & Snischenko, B.F. (1982). *Fundamentals of Hydro-Morphological Theory of Channel Process.* Leningrad: Gidrometeoizdat. 272 p. (in Russian).

Leliavsky, S. (1955). *An Introduction to Fluvial Hydraulics.* London: Constable. 257 p.

Lewin, J. (1978). Floodplain geomorphology // *Progr. Phys. Geogr.* Vol. 2. No. 3. P. 408-437.

Makkaveev, N.I. (1969). *Experimental Geomorphology.* Moscow: Moscow State University Press. 179 p.

Makkaveev, N.I., & Chalov, R.S. (1970). Some features of the valley bottoms of large rivers, associated with periodic changes in the flow // *Voprosy Geografii.* Vol. 79. P. 156–167 (in Russian).

Nanson, G.C., & Croke, J.C. (1992). A genetic classification of floodplains // Floodplain Evolution (Eds Brackenbridge, G.R. & Hagedorn, J.). *Geomorphology.* Vol .4. No. 6. P. 460-486.

Page, K., Nanson, G., & Price, D. (1996). Chronology of Murrumbidgee River palaeochannels on the Riverine Plain, Southeastern Australia // *Journal of Quaternary Science.* Vol. 11. Issue 4. P. 311-326.

Panin, A.V., & Sidorchuk, A.Yu. (2006). Macromeanders: the problem of origin and interpretation // *Vestnik MGU. Series 5. Geografiya.* No. 6. P. 14-22 (in Russian).

Panin, A.V., Sidorchuk, A.Yu., & Chernov, A.V. (1992). Macromeanders of the rivers in the European Russia and problems of the paleohydrological reconstructions // *Water Resources,* No. 4, p. 93-97.

Panin, A.V., Sidorchuk, A. Yu., & Chernov, A.V. (1999). Historical background to floodplain morphology: examples from the East European Plain // *Floodplains: Interdisciplinary Approaches.* London, Geological Society Special Publications. Vol. 163. P. 217-229.

Panin, A.V., Sidorchuk, A.Yu., & Chernov, A.V. (2011). Main stages of floodplain formation in North Eurasian Rivers // *Geomorfologiya,* No. 3. P. 20–31 (in Russian).

Panin, A.V., Sidorchuk, A.Yu., & Vlasov, M.V. (2013). Large Late Valdai discharge in the Don River basin // *Izvestiya Akad. Nauk. Ser. Geogr.* No. 1. P. 118-129 (in Russian).

Popov, I.V. (1969). Deformation of river channels and engineering. Leningrad: *Gidrometeoizdat*. 363 p. (in Russian).

Potapenko, L.M. (1975). *Quaternary deposits and valley evolution of the lower Vychegda River.* Abstr. PhD thesis. Moscow University. 25 p. (in Russian)

Rotnicki, K. (1983). Modelling past discharges of meandering rivers // Gregory, K. (Ed.) *Background to Palaeohydrology.* Wiley. P. 321–354.

Rotnicki, K. (1991). Retrodiction of palaeodischarges of meandering and sinuous rivers and its palaeoclimatic implications // Starkel, L., Gregory, K.J., & Thornes, J.B. (Eds.) *Temperate Palaeohydrology.* Wiley. P. 431–470.

San'ko, A.F. (1987). *The Neopleistocene of North-East Byelorussia and Neighboring Districts of the RSFSR.* Minsk: Nauka i Tekhnika. 178 p. (in Russian).

Schantser, E.V. (1951). Alluvium of the plain rivers in the temperate zone and its implications for understanding the structure and formation of alluvial suites // *Proc. Geol. Institute of the USSR Acad. of Sci. Series geol.* No. 135. 275 p. (in Russian).

Sidorchuk, A.Yu. (1996). Channel processes on the Yamal Peninsula (Western Siberia) during the last 2300 years // *The Late Glacial Palaeohydrology and Modelling of Environmental Change*. Toledo, Spain. P. 23.

Sidorchuk, A.Yu. (1997). Palaeohydrology in the Southern Russian Plain: methods and results // VI Congresso da Associacao Brasilera de Estudos do Quaternario e Reuniao sobre o Quaternario da America do Sul. Resumos Expandidos. Curitiba, Parana, Brasil. P. 479-482.

Sidorchuk, A.Yu. (2003). Floodplain sedimentation: Inherited memories // *Global and Planetary Change*. Vol. 39. No. 1-2. P. 13-29.

Sidorchuk, A.Yu. (2010). The Middle Holocene stage of the low water flow of the rivers and its expression in the morphology of river meanders // *Ancient and Modern Valleys and Rivers: The History of Formation, Erosion and Channel Processes*. Volgograd: VGPU Publ. P. 106-117 (in Russian)

Sidorchuk A.Yu., & Borisova O.K. (2000). Method of paleogeographical analogues in paleohydrological reconstructions // *Quaternary International*. Vol. 72. No. 1. P. 95-106.

Sidorchuk A.Yu., Borisova, O.K., Kovalyukh, N.N., Panin, A.V., & Chernov, A.V. (1999). Paleohydrology of the Lower Vychegda in the Late Glacial and the Holocene // *Vestnik MGU. Series 5. Geografiya*. No. 5. P. 35-42 (in Russian).

Sidorchuk, A., Borisova, O., & Panin, A. (2001a). Fluvial response to the Late Valdai/Holocene environmental change on the East European Plain // *Journal of Global and Planetary Changes*. Vol. 28. P. 303-318.

Sidorchuk, A., Borisova, O., Panin, A. Elias S.A., & Syvitski J.P. (2000). The channel morphology and river flow in the northern Russian Plain in the Late Glacial and Holocene // *International Journal of Earth Science*. Vol. 89. P. 541-549.

Sidorchuk, A., & Matveev, B. (1994). Channel processes and erosion rates in the rivers of the Yamal Peninsula in western Siberia // *Variability in Stream Erosion and Sediment Transport.* IAHS Publ. No. 224. P. 197-202.

Sidorchuk, A., Panin, A., & Borisova, O. (2003). The Late Glacial and the Holocene palaeohydrology of the Northern Eurasia // Gregory, K.J., & Benito, G. (eds.). *Palaeohydrology: Understanding Global Change*. Chichester: Wiley. P. 61-76.

Sidorchuk, A.Yu., Panin, A.V., & Borisova, O.K. (2008a). Climate-induced changes in surface runoff on the north-Eurasian plains during the Late Glacial and Holocene // *Water Resources,* Vol. 35, No. 4, pp. 386-396.

Sidorchuk, A.Yu., Panin, A.V., & Borisova, O.K. (2008b). Late Glacial paleochannels of rivers in Western Siberia // *Izvestiya Akad. Nauk. Ser. Geogr*. No. 2. P. 67–75 (in Russian).

Sidorchuk, A.Yu., Panin, A.V., & Borisova, O.K. (2009). Morphology of river channels and surface runoff in the Volga River basin (East European Plain) during the Late Glacial period // *Geomorphology*. Vol. 113. P. 137–157.

Sidorchuk, A.Yu., Panin, A.V., & Borisova, O.K. (2011). Surface runoff to the Black Sea from the East European Plain during Last Glaciation Maximum–Late Glacial time // Buynevich, I., Yanko-Hombach, V., Gilbert, A.S., & Martin, R.E. (eds). Geology and Geoarchaeology of the Black Sea Region: Beyond the Flood Hypothesis. *The Geological Society of America Special Paper* 473. P. 1-25.

Sidorchuk, A.Yu., Panin, A.V., & Borisova, O.K. (2012). River runoff decrease in north Eurasian plains during the Holocene optimum // *Water Resources*. Vol. 39. No. 1. P. 69-81.

Sidorchuk, A., Panin, A., Borisova, O., & Kovalyukh, N. (2001b). Lateglacial and Holocene palaeohydrology of the lower Vychegda River, western Russia // Maddy, D., Macklin, M.G., & Woodward, J.C. (eds). *River Basin Sediment Systems: Archives of Environmental Change*. A.A. Balkema Publ. P. 265-295.

Svezhentsev, Y.S. (1993). Radiocarbon chronology for the Upper Paleolithic sites on the East European Plain // Soffer, O., & Praslov, N.D. (eds). *From Kostenki to Clovis*. New York: Plenum Press. P. 23–30.

Szafer, W. (1946–1947). Flora pliocenska z Krocienka nad Dunajcem (The Pliocene flora of Kroscienko on Dunajec River) // *Rozprawy Wydzialu matemat.-przyrodn*: Polska Akad. Umiejet. Krakow. 72 B. 1. 162 p. 2. 213 p. (in Polish).

Velichkevich, F.Yu. (1982). Pleistocene Floras of the Glaciated Districts of the East European Plain. Minsk: *Nauka i Tekhnika*. 239 p. (in Russian).

Volkov, I.A. (1963). Remnants of powerful flow in the valleys of the southern West Siberia // *Doklady AN SSSR*. Vol. 151. No. 3. P. 648-651 (in Russian).

Yelenevsky, R.A. (1936). *Problems of Floodplain Investigations and Reclamation*. Moscow: VASKHNIL Publ. 207 p. (in Russian).

Zagwijn, W.H. (1996). An analysis of Eemian climate in western and Central Europe // *Quaternary Science Reviews*. Vol. 15. P. 451-469.

Zaretskaya, N.E., Chernov, A.V., Karmanov, V.N., Panin, A.V., Buravskaya, M.N., & Volokitin, A.V. (2010). History and development of the Middle Vychegda valley during the Late Glacial and the Holocene // Geomorphological Processes and Their Applied Aspects. *Proceedings of the VI Schukin Workshop*. Moscow: MSU Geogr. Faculty. P. 421-424 (in Russian).

Zaretskaya, N.E., Panin, A.V., Sidorchuk, A.Yu., Buravskaya, M.N., & Chernov, A.V. (2007). Dynamics of the Upper Vychegda in the second half of the Holocene (based on the analytical and geomorphological data) // *Proceedings of the V All-Russian Workshop on the Quaternary Research*. Moscow, November 7-9, 2007. P. 130-133 (in Russian).

In: Floodplains
Editor: Enner Herenio Alcântara
ISBN: 978-1-62808-998-1

Chapter 5

THE IMPORTANCE OF SPATIAL AND LOCAL ENVIRONMENTAL FACTORS TO STRUCTURING PHYTOPLANKTON COMMUNITY IN THE FLOODPLAIN LAKES OF CUIABÁ RIVER (NORTHERN PANTANAL, BRAZIL)

Simoni Maria Loverde-Oliveira**, Simone Jaqueline Cardoso and Ibraim Fantin-Cruz
Universidade Federal de Mato Grosso, Departamento de Ciências Biológicas, Mato Grosso, Brasil

ABSTRACT

Ecological studies have shown that ecosystem properties depend greatly on biodiversity and how organisms are distributed over the landscape. Understanding how spatial and local environment factors interact structuring the diversity and distribution of organisms is very important, because it can subsidize ecological theories and concepts and also give bases to resource management and biodiversity conservation plans. This study aimed to evaluate the effects of spatial and local environmental factors structuring the phytoplankton abundance and species diversity in nine lakes along the Cuiaba River, Northern Pantanal, Mato Grosso, in different periods of flood pulse (high waters and low waters). Monthly samplings were made from December 2008 to December 2009. Analyses of variance partition were based on the abundance and the occurrence (presence and absence) of phytoplankton species and evaluated by partial redundancy (pRDA) analysis. The abundance and occurrence of phytoplankton species were explained both by spatial and local environmental factors during the different periods of the flood pulse. However, local environmental factors showed greater importance in relation to the spatial ones, except during the low waters, when the spatial factors explanation was slightly higher. Our results reveal that spatial and local environmental factors are partially

* Corresponding author: Universidade Federal de Mato Grosso, Departamento de Ciências Biológicas. Rodovia Rondonópolis-Guiratinga, km 06 (MT-270), Sagrada Família, Rondonópolis, MT, Brasil. CEP 78735-910. E-mail: loverde@terra.com.br.

responsible for phytoplankton abundance and occurrence in the floodplain systems studied.

1. INTRODUCTION

The relationship between species diversity and distribution and their driving factors in aquatic and terrestrial ecosystems has been a topic under intensive study (Magurran and McGill, 2011). The most frequent hypotheses tested are related to drivers and stressors of species diversity in different time and spatial scales, such as habitat size, heterogeneity and stability, resources availability, presence of grazers, competition and dispersal ability (Stendera et al., 2012).

In general, most of the studies have provided many insights on the species requirements and biodiversity patterns, but a combined comprehension of mechanisms are still lacking (Konar et al., 2013). The knowledge about how spatial and local environmental factors structure species diversity and distribution is very important, because it can subsidize ecological theories and concepts (Ward and Tockner, 2001), and also give bases to resource management and biodiversity conservation plans (Hooper et al., 2005).

In aquatic ecosystems the microorganisms and specially the phytoplankton has been used as model in many studies (Reynolds, 2006). The phytoplankton community can be driven by a set of multi-scale spatial-temporal environmental processes such as: the availability of resources and resources utilization (light and nutrients), algal morphological traits, resistance to grazers and competition, changes on climatic conditions (e.g. temperature), habitat size and isolation, and other mechanisms such as land use and losses of habitat (Sommer, 1993; Heino and Soininen, 2006). In spite of the increasing number of studies related to microorganism diversity patterns, there is still a huge lack of knowledge in tropical floodplain areas.

The Pantanal is among the most endangered ecosystems in the world. Currently it is been subject to many anthropogenic impacts, both in land and water (Gopal and Junk, 2000). Moreover, studies have point out numerous problems in Pantanal, such as loss of biodiversity, increase in the sedimentation processes and alteration in the flood pulse (Da Silva and Girard, 2004; Junk and Wantzen, 2004). Understanding the driving factors of species diversity and distribution in Pantanal will help the creation of better strategies of use and management of the local natural resources and the creation of conservation plans.

The objective of his study was to evaluate the effects of spatial and local environmental factors structuring the phytoplankton abundance and species diversity in nine floodplain lakes along the Cuiabá River in different periods of flood pulse (high water and low water).

2. STUDY AREA

The Pantanal is a large intracontinental depression (16 - 20° S and 56 - 58° W) drained by the Paraguay River basin. The annual precipitation alternating from dry to rainy seasons have resulted in formation of a heterogeneous wetland. The Pantanal is composed of different subsystems, however only 20-30% of the plain is occupied by water or permanently flooded. The remainder of its area belongs to aquatic terrestrial transition zone (ATTZ), which extends continuously between terrestrial and aquatic habitats (Junk et al., 2011). The region has many oxbow lakes characterized by fluvial hydrodynamics and subject to fluctuations in water level. The lakes are often shallow (< 3.0 m), with different dynamics (lentic vs. lotic), connectivity and degree of interaction with the main river (Cardoso et al., 2012) and the vegetation in the in marginal areas, which is mainly composed of flooded savanna (Cerrado vegetation).

The systems analyzed in this study are nine oxbow lakes and the Cuiabá River. They are located at a national natural reserve (RPPN SESC Pantanal) in the northern Pantanal, along the Cuiabá River, in the state of Mato Grosso (Figure 1 and Table 1). The regional climate is Aw (Tropical Savanna, Köppen classification), with a rainy season during the summer and a dry season during the winter. The historical data of total annual precipitation range from 800 to 1600 mm, and the annual mean temperature is around 28 °C (PCBAP, 1997).

3. Field Sampling and Analytical Methods

Samplings were made monthly in nine oxbow lakes and in the Cuiabá River from December 2008 to December 2009 (Figure 1 and Table 1). Samples for phytoplankton and nutrient analysis were collected with a van Dorn bottle at the subsurface (~ 0.30 m). Phytoplankton was fixed with Lugol's solution. The fractions for total nitrogen and phosphorus were fixed in the field with 0.5 mL absolute sulfuric acid. Water temperature was measured in situ using a multiparameter probe (YSI 6920A), pH using a Micronal portable meter (model B474), and water transparency using a Secchi disk (SD).

The total phosphorus concentration (TP) was analyzed by the spectrophotometric method (Golterman et al., 1978) and the total Kjedahl nitrogen concentrations (TN) by titration method (Mackereth et al., 1978) and total suspended solids (TSS) according to APHA (1998). Phytoplankton abundance was estimated using an Olympus CK40 inverted microscope, by the settling technique (Utermöhl, 1958). The units (cells, colonies, and filaments) were enumerated to at least 100 specimens of the most abundant species (Lund et al., 1958).

Table 1. Locations of the systems sampled

Sampling site	System	Coordinates
1	Assombrada	16°42'55'' S, 56°29'55'' W
2	Ilha dos Turistas	16°37'25'' S, 56°27'03'' W
3	Biguá	16°31'07'' S, 56°23'25'' W
4	Luzardo	16°37'06'' S, 56°27'18'' W
5	Santa Rosa	16°41'55''S, 56°28'35'' W
6	Cobras	16°43'59'' S, 56°30'09'' W
7	Moquém	16°34'06'' S, 56°24'01'' W
8	Trilha do Tatu	16°31'14'' S, 56°23'12'' W
9	Antônio Alves	16°31'17'' S, 56°23'39'' W
10	Rio Cuiabá	16°37'05'' S, 56°27'04'' W

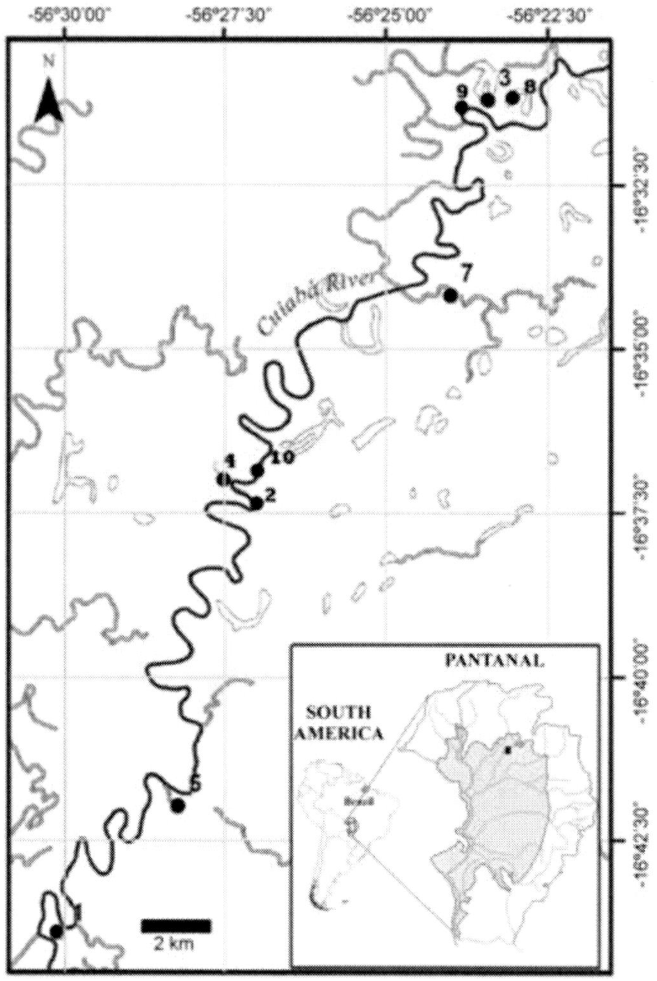

Figure modified from Cardoso et al. (2012).

Figure 1. Map of the study sites along the Cuiabá River: 1 – Assombrada, 2 – Ilha dos Turistas, 3 – Biguá, 4 – Luzardo, 5 – Santa Rosa, 6 – Cobras, 7 – Moquém; 8 - Trilha do Tatu, 9 - Antônio Alves and 10 - Cuiabá River.

3.1. Data Analysis

The euphotic depths (zeu) of the systems were estimated as three times the SD extinction depth (Cole, 1994). The phytoplankton was analyzed at the species level and also classified as major groups: Cyanobacteria, Cryptophyceae, Dinophyceae, Chrysophyceae, Bacillariophyceae, Euglenophyceae, Chlorophyceae, and Zygnematophyceae, according to Van den Hoeck et al. (1993), Round et al. (1990) and Komárek & Anagnostidis (1999, 2005).

Phytoplankton species diversity was calculated for each station as the total number of species found per sample. In order to compare species richness (species/sample) between different sampling periods, sample based rarefaction curves were computed (S_{MaoTau}). We calculated 400 times repeated re-samplings using EstimateS (Colwell et al., 2004; Gotelli and Colwell, 2001).

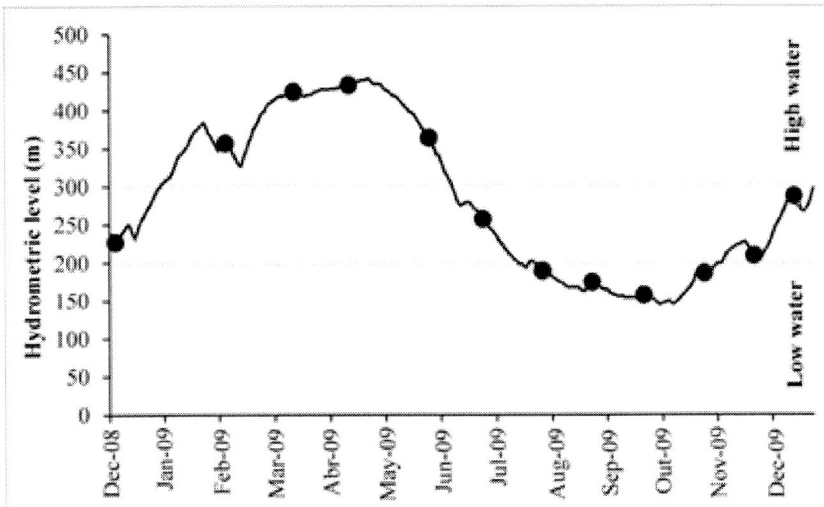

Figure 2. Changes in the water level of the Cuiabá River from December 2008 to December 2009. The circles indicate sampling dates and dotted lines the boundaries of the water level periods.

To distinguish between the different flood pulse periods, of high waters (HW) and low waters (LW), we used data of the hydrometric level of Cuiabá River (Figure 2). The threshold for each period was ≥ 3 m to HW, and ≤ 2 to LW (Figure 2). This criterion was adopted based on eco-hydrological studies developed in the region by Girard et al. (2010) and Fantin-Cruz et al. (2011). To test the relative influence of environmental and spatial factors on the phytoplankton species abundance and occurrence, we applied Partial Redundancies Analysis (pRDA) (Borcard et al., 1992; Legendre & Legendre, 1998). The outcomes of the RDA consists of four components: [E] the variation attributed solely to environmental variables; [S] the variation attributed solely to spatial variables; [SE] the combined variation explained by environmental and spatial variables (spatially structured environmental variation) and [U] the residual variance, portion not explained by any of the environmental and spatial factors (Borcard et al., 1992). The species occurrence data were standardized using the Hellinger function to preserve the Euclidean distance between sample units in an *n*-dimensional space (Legendre and Gallagher, 2001, Peres-Neto et al., 2006). The variations explained by each component were based on the adjusted fractions. The adjusted fractions are good indicators because they take into account the total number of predictors and the sample size. The significance of each fraction was tested using permutation tests with 999 randomizations. In this analysis, only the components E and S are testable (Peres-Neto et al. 2006). The analyses were performed using the VEGAN package and functions Varpart of the R Software (version 2.11.1; R Development Core Team, 2011).

4. Local Environmental Factors and Phytoplankton Diversity

The local environmental characteristics of the systems (except for water temperature and pH) varied widely among the systems, and between the different periods of the flood pulse (high and the low waters; Table 2). The lakes were characterized by similar mean values of pH (7.10, CV > 8%), water temperature 30 °C (CV = 11%) and phytoplankton richness (15 - 16 taxa / sample CV > 31%). During the high waters we found the greater depths (3.40 m, CV = 58%), water transparency (0.90 m, CV = 37%), and depths of euphotic zone (2.40 m, CV = 49%). During the low waters we found greater electrical conductivity 62 μScm^{-1} (CV = 16%), suspended solids (14 mgL^{-1}, CV > 100%), total phosphorus (0.06 mgL^{-1}, CV > 100%) and total nitrogen (0.28 mgL^{-1}, CV = 65%).

Table 2. Means and coefficients of variation (CV) of the main local environmental factors, phytoplankton abundance and phytoplankton species richness, sampled during the low and the high water periods in the systems studied

	High waters		Low waters	
	Mean	CV (%)	Mean	CV (%)
Water temperature	29.00	11	30.00	11
zmax (m)	3.40	58	1.90	81
Water transparency (m)	0.92	37	0.60	41
zeu (m)	2.40	49	1.40	61
Electrical condutivity (μScm^{-1})	58.70	21	62.00	16
pH	7.10	8	7.10	09
Suspended solids (mgL^{-1})	11.40	127	14.00	123
TN (mgL^{-1})	0.13	115	0.28	65
TP (mgL^{-1})	0.05	46	0.06	136
Phytoplankton abundance ($ind.mL^{-1}$)	2,159.00	278	2,448.00	259
Phytoplankton species richness	15.00	31	16.00	25

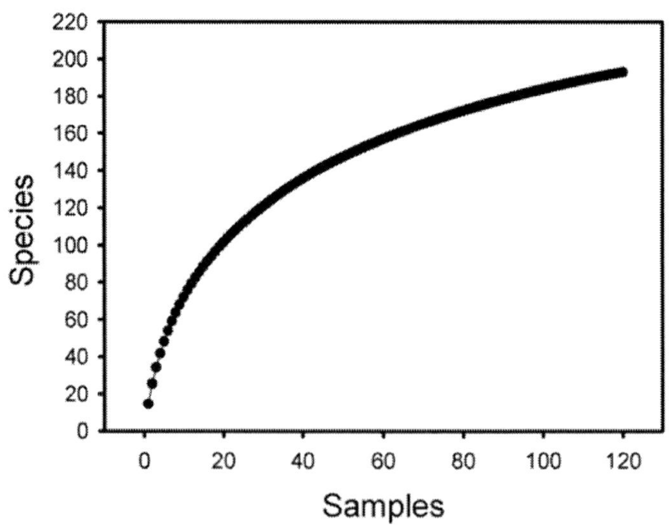

Figure 3. Sample-based rarefaction curve (S_{MaoTau}) of all phytoplankton species found in all systems studied from December 2008 to December 2009.

Large variation was found in the phytoplankton abundances among the systems (CV > 100%) and between the periods (Table 2). The most representative groups of phytoplankton abundances during the high and the low water periods, respectively, were Cyanobacteria (mean = 1,830.00 - 2,142.00 ind.mL^{-1}), Cryptophyceae (mean = 147.00 - 112.00 ind.mL^{-1}) and Chlorophyceae (mean = 117.00 - 121.00 ind.mL^{-1}). A total of 195 phytoplankton species were found considering all the systems and periods. The species rarefaction curve (Figure 3) indicated that the number of samples (n = 120) was able to capture most of phytoplankton species of the systems. Also, the number of species found per sample (15 - 16 spp.; Table 2) was similar to the average values reported by Cardoso et al. (2012) for the same region (14 ± 7 spp. per sample).

5. Contributions of Spatial and Local Environmental Factors to Phytoplankton Community

The phytoplankton abundance and occurrence of species found during the different flood pulse periods were explained both by environmental and spatial variables (Figure 4). The only exception occurred for the species occurrence data, where the fraction explained solely by the space was not significant (Figure 4a). It was also observed the environmental component showed greater importance in relation to the spatial component in structuring the phytoplankton community.

However, when comparing the contribution of the components between the different flood pulse periods (high and low waters), the spatial component explanation was a slightly higher during the low waters, which was not expected. We expected that during the low waters the isolation of the lakes in relation to the Cuiabá River would cause heterogeneity of habitats, consequently leading to differences in the local environmental characteristics of the systems, which in turn would promote differences in the spatial distribution of phytoplankton communities. Studies in North Pantanal are still few (Loverde-Oliveira et al., 2011; Loverde-Oliveira et al., 2012; Cardoso et al. 2012), and most of them also have shown that the isolation or connection of the lakes have not been sufficient to promote changes in phytoplankton community.

Further, despite of statistical effects, the proportions accounted by each component alone (environmental or spatial) or shared (when environmental and spatial components overlap) were low, with large portion of phytoplankton variation unexplained. Similar results of low explanation of phytoplankton variation were also found for lakes in Araguaia (Nabout et al., 2009) and Amazon (Lopes et al., 2011) river floodplains. We hypothesize that the low explanation of the environmental and spatial components may be related to the small scale of the study (regional). Also, it is been shown that in floodplain systems, the environmental gradients (limnological characteristics of the systems) may be less evident (Thomaz et al., 2007).

Moreover, limitations related to the phytoplankton dispersion may also have influenced the local structure of the assemblies (Tondato et al., 2013), together with other important variables for not accounted in this study, such as the magnitude and duration of hydrologic connectivity (Girard et al., 2010), trophic interactions (Loverde-Oliveira et al., 2009; Silveira et al., 2010) and temporal dynamics of organisms (Fantin-Cruz et al., 2010).

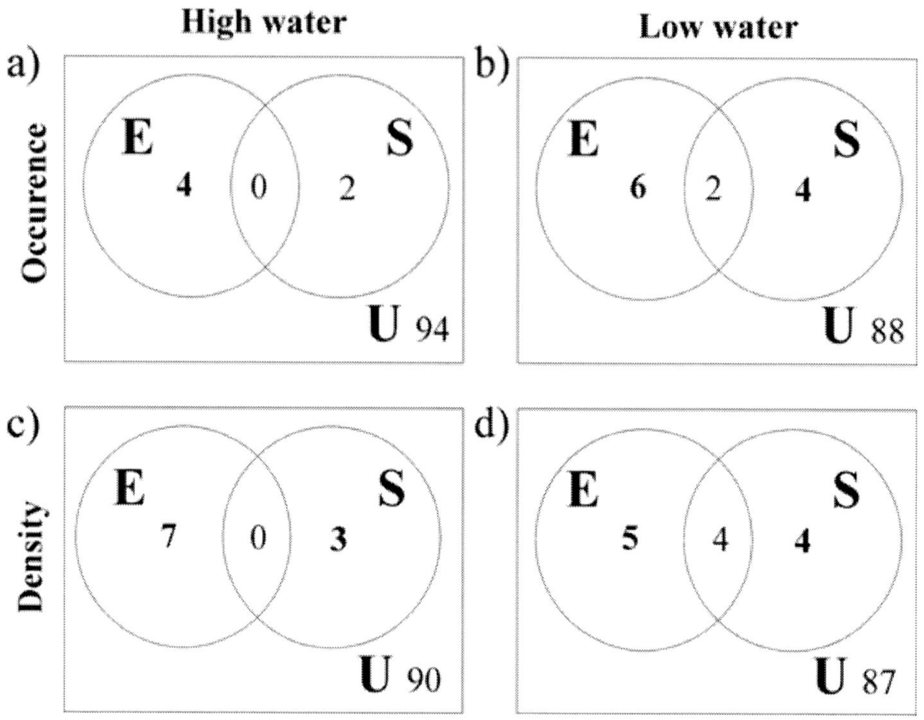

Figure 4. Relative contribution (%) of environmental (E), spatial (S) and shared components to the variability in the occurrence and density of phytoplankton species. U= component not explained. The zeros indicate values lower than 0.5%. The values in bold indicate significant values ($p < 0.05$).

Conclusion

The abundance and occurrence of phytoplankton species were explained both by spatial and local environmental factors during the different periods of the flood pulse. However, the local environmental factors showed greater importance in relation to the spatial ones in structuring the community along the different flood pulse periods.

Acknowledgments

We thank PELD SITE 12 for fieldwork support. This study was sponsored by Conselho Nacional de Desenvolvimento Científico e Tecnológico (CNPq), Brazil (fellowship to SJC).

References

APHA. (1985). *Standard methods for the examination of 604 water and wastewater.* APHA-AWWA-WPCF. 16th 605 edn. Washington.

Borcard D., Legendre P. & Drapeau P. (1992). Partialling out the spatial component of ecological variation. *Ecology* 73:1045-1055.

Cardoso, S.J., Roland F., Loverde-Oliveira, S.M. & Huszar, V.L.M. (2012). Phytoplankton abundance, biomass and diversity within and between Pantanal wetland habitats. *Limnologica* 42:235–241.

Cole, G.A. (1994). *Textbook of Limnology*, vol. 1. Waveland Press Inc., Illinois.

Colwell, R.K., Mao, C.X. & Chang, J. (2004). Interpolating, extrapolating, and comparing incidence-based species accumulation curves. *Ecology* 85:2717–2727.

Da Silva, C. J. & Girard, P. (2004). New chalenges in the managem of the Brazilian Pantanal and catchment area. Wetlands (Wilmington, N.C.). *Holanda,* v. 12, p. 553-561.

Fantin-Cruz , I., Pedrollo, O., Castro, N.M.R., Girard, P., Zeilhofer, P. & Hamilton, S.K. (2011). Historical reconstruction of floodplain inundation in the Pantanal (Brazil) using neural networks. *Journal of Hydrology* 399: 376–384.

Fantin-Cruz, I., Pedrollo, O., Bonecker, C.C., Motta-Marques, D., Loverde-Oliveira, S.M. (2010). Zooplankton Density Prediction in a Flood Lake (Pantanal – Brazil) Using Artificial Neural Networks. *International Review of Hydrobiology* 95: 330–342.

Girard, P., Fantin-Cruz, I., Loverde-Oliveira, S.M. & Hamilton, S.K. (2010). Small-scale spatial variation of inundation dynamics in a floodplain of the Pantanal (Brazil). *Hydrobiologia* 638: 223-233.

Golterman, H.L., Clymo, R.S. & Ohnstad, M.A.M. (1978). *Methods for Physical and Chemical Analysis of Freshwaters,* vol. 8. Blackwell Scientific Publications, Oxford.

GOPAL, B. & JUNK, W. J. (2000). Biodiversity in wetlands: an introduction. In: Gopal, B.; Junk, W.J.; Davis, J.A. (Org.). *Biodiversity in wetlands: assessment, function and conservation.* 1ed.Leiden: Backhuys Publishers, v. 1, p. 1-10.

Gotelli, N. & Colwell, R.K. (2001). Quantifying biodiversity: procedures and pitfalls in the measurement and comparison of species richness. *Ecol. Lett.* 4, 379–391.

Heino, J. & Soininen, J. (2006). Regional occupancy in unicellular eukaryotes: a reflection of niche breadth, habitat availability or size-related dispersal capacity? *Freshwater Biology* 51, 672–685.

Hooper, D. U., Chapin, F. S., Ewel, J.J., Hector, A., Inchausti, P., LavoreL,S., Lawton, J. H., Lodge, D.M., Loreau, M., Naeem, S., Schmid, B, Seta¨ la¨, H., Symstad, A. J., Vandermeer, J. & Wardle. D. A. (2005). Effects of biodiversity on ecosystem functioning: a consensus of current knowledge. *Ecological Monographs* 75 (1): 3–35.

Junk, W. J. & Wantzen, K. M. (2004). The Flood Pulse Concept: New Aspects, Approaches, and Applications - an Update. *Proceedings of the Second International Symposium on the Management of Large Rivers for Fisheries, Food and Agriculture Organizat* 2: 65-77.

Junk, W. J., Nunes da Cunha, C., Da Silva, C.J, & Wantzen, K.M. (2011). The Pantanal: a large South American wetland and its position im limnology. Pp. 2344. *In*: W. Junk; C.J. Da Silva, C.N. Cunha. K. Wantzen. (orgs.). *The Pantanal Ecology, biodiversity and sustainable management of a large neotropical seasonal wetland.* Pensoft Publishers, Moscou. 870p.

Komárek, J. & Anagnostidis, K. (1999). Cyanoprokaryota I. Teil Chroococcales. Pp. 1-548. In: H. Ettl, G. Gärtner, H. Heynig & D. Mollenhauer (eds.), *Süsswasserflora von Mitteleuropa,* 19(1). Gustav Fischer, Stutgart, BW. 548p.

Komárek, J. & Anagnostidis, K. (2005). Cyanoprocaryota. 2. Oscillatoriales. Pp. 1-759. In: B. Büdel; L. Krienitz L.; G. Gärtner & M. Schagerl. *Subwasserflora von Mittleuropa* 19 (2). Gustav Fisher, Stutgart, BW. 759p.

Konar, M., Jason Todd, M., Rachata Muneepeerakul, A.R. & Rodriguez-Iturbe, I. (2013). Hydrology as a driver of biodiversity: Controls on carrying capacity, niche formation, and dispersal. *Advances in Water Resources* 51, 317–325.

Legendre, P. & Gallagher, E. (2001). Ecologically meaningful transformations for ordination of species data. - *Oecologia* 129:271-280.

Legendre, P. & Legendre, L. (1998). *Numerical ecology*. Second English edition. Elsevier, Amsterdam, The Netherlands.

Lopes, P.M., Caliman, A., Carneiro L.S., Bini L.M., Esteves, F.A., Farjalla, V. & Bozelli, R.L. (2011). Concordance among assemblages of upland Amazonian lakes and the structuring role of spatial and environmental factors. *Ecological Indicators*, 11 (5): 1171-1176.

Loverde-Oliveira, S.M., Adler, M. & Silva, V.P. (2011). Phytoplankton, periphyton and metaphyton of the Pantanal floodplains: species composition and richness, density, biomass and primary production. Pp. 235-256. *In*: W. Junk; C.J. Da Silva, C.N. Cunha. K. Wantzen. (orgs.). *The Pantanal Ecology, biodiversity and sustainable management of a large neotropical seasonal wetland.* Pensoft Publishers, Moscou. 870p.

Loverde-Oliveira, S.M., Huszar, V.L.M., Mazzeo, N. & Scheffer, M. (2009). Hydrology-driven regime shifts in a shallow tropical lake. *Ecosystems* 12: 807-819.

Loverde-Oliveira, S.M., Pietro-Souza, W., Cardoso, S. J., Fantin-Cruz, I. & Mateus, L.M. (2012). Fatores Associados À Distribuição Espacial do Fitoplâncton em Lagos de Inundação (Pantanal Norte, Brasil). *Oecologia Australis* 16(4): 770-781.

Lund, J.W.H., Kipling, C. & Cren, E.D. (1958). The inverted microscope method of estimating algal number and statistical basis of estimating by counting. *Hydrobiologia* 11, 143–170.

Mackereth, F. J. H., Heron, J. & Talling, J. F. (1978). *Water analysis: some revised methods for limnologists.*

Magurran, A. E. & McGill, B. (2011). Biological Diversity. *Frontiers in measurement and assessment.* Oxford. 345pp.

Nabout, J.C., Siqueira, T., Bini, L.M. & Nogueira, I. S. (2009). No evidence for environmental and spatial processes in structuring phytoplankton communities. *Acta Oecol.* 35: 720-726.

PCBAP (1997). Diagnóstico dos meios físico e biótico. In: Projeto pantanal, programa nacional do meio ambiente, vol. II (1). *Plano de conservação da bacia do alto Paraguai*, Brasília.

Peres-Neto, P.R., Legendre, P., Dray, S. & Borcard, D. (2006). Variation partitioning of species data matrices: Estimation and comparison of fractions. Ecology 87: 2614–2625.

Reynolds, C. S. (2006). *The Ecology of Phytoplankton* 1. New York, Cambridge University Press.

Round F.E., Crawford R.M. & Mann, D.G. (1990). *The diatoms. Biology and morphology of genera.* Cambridge University Press, Cambridge, UK. 747p.

Silveira, R. M., Paiva, L. L.A.R. & Camargo, J.C. (2010). Top-down control in a tropical shallow lake of Northern Pantanal, Brazil. *Acta Limnologica Brasiliensia* 22(4), 455-465.

Sommer, U. (1993). Phytoplankton competition in Plubsee: a field test of the resource-ratio hypothesis. *Limnology and Oceanography* 38:838–845.

Stendera, S., Adrian, R., Bonada, N., Canedo-Arguelles, M., Hugueny, B., Januschke, F. & Pletterbauer, D. H. (2012). Drivers and stressors of freshwater biodiversity patterns across different ecosystems and scales: a review. *Hydrobiologia* 696:1–28.

Thomaz, S.M., Bini, L.M. & Bozelli, R.L. (2007). Floods increase similarity among aquatic habitats in river-floodplain systems. *Hydrobiologia* 579: 1-13.

Tondato, K.K., Fantin-Cruz, I., Pedrollo, O.C. & Suarez, Y.R. (2013). Spatial distribution of fish assemblages along environmental gradients in the temporary ponds of Northern Pantanal, Brazil. *Journal of Limnology* 72:95-102.

Utermöhl, H. (1958). Zur Vervollkommnung der quantitativen Phytoplankton Methodik. *Mitt. Int. Ver. Limnol.* 9: 1–38.

Van Den Hoeck, C.; Mann, D.G. & Jahns, H.M. (1993). *Algae: an introduction to Phycology.* Cambridge University Press, Cambridge, UK. 627p.

Ward, J. V. & Tockner, K. (2001). Biodiversity: towards a unifying theme for river ecology. *Freshwater Biology* 46: 807-819.

In: Floodplains
Editor: Enner Herenio Alcântara

ISBN: 978-1-62808-998-1

Chapter 6

SURVEY-BASED INFLUENCE OF STORAGE AND RELEASE OF WATER FROM FLOODPLAIN LAKES ON THE MAIN RIVER IN THE SPARSE-DATA AREA OF THE LOWER AMUR, RUSSIA

Vladimir V. Shamov*

Pacific Geographical Institute, Far-Eastern Branch, Russian Academy of Sciences
Laboratory of Hydrology & Climatology

ABSTRACT

The results of several field hydrometrical surveys of the Amur floodplain lakes are provided and discussed. Interrelationship between water level in the Amur and riparian lakes evidently controls the intensity and direction of water exchange within the river-lake hydraulic systems. It is found that at the time of a high flood decline the floodplain reaches including the lakes can ensure a total about 15% of the current discharge of the Amur flow nearby respective reaches. In periods of low and middle flood decline the lake outflow into the river is estimated to be low as it does not usually exceed 9-10% of the river discharge at the neighboring measuring section. When high river water level is sustained, the backwater (water lock) decreases sufficiently the outflow from lakes filled up by the river flood water – to 5-6% of the river discharge at the neighboring measuring section, and obviously lower. Under the conditions of the lowest water level in the Amur river as well as of the river backwater along the flood wave passing, discharges increase only slightly as within the limits of water flow rate measurement accuracy. For sure, at the time of a flood rise the Amur water inflows to the floodplain lakes through all the lacustrine channels that provide water exchange between the river and lakes.

* Corresponding author: Pacific Geographical Institute, Far-Eastern Branch, Russian Academy of Sciences; Laboratory of Hydrology & Climatology; Radio Street, 7 – 690041 Vladivostok, Russia Tel: +7 423 2312 857, Fax: +7 423 2312 159, E-mail: vlshamov@yandex.ru.

1. Introduction

The Russian part of the Amur R. basin contains 55,000 of lakes (Domanitskii et al., 1971), more than 19,000 of which are located within the Lower Amur area (Surface…, 1970). Most of these (99.5%) are small lakes with area of up to 1 km. Lakes with the largest area and volume are located in the Lower Amur valley (from Khabarovsk to the Amur Liman). Figure 1 depicts a map-scheme of location of the biggest lakes within the Lower Amur region, including the Lake Chlya (1), the Lake Dzhevdukha (2), the Lake Kadi (3), the Lake Kizi (4), the Lake Khavanda (5), the Lake Khummi (6), the Lake Padali (7), the Lake Kaltekheven, Bol'shaya and Malaya Sharga (8), the Lake Gassi (9) and the Lake Petropavlovskoye (10).

Water level behavior in the lakes is closely related with the Amur water regime. Their depths vary from 1 to 8 m at the mean depth of 2–3 m in summer. In winter, the lakes commonly freeze down to the bed everywhere, except for the deepest places and hollows.

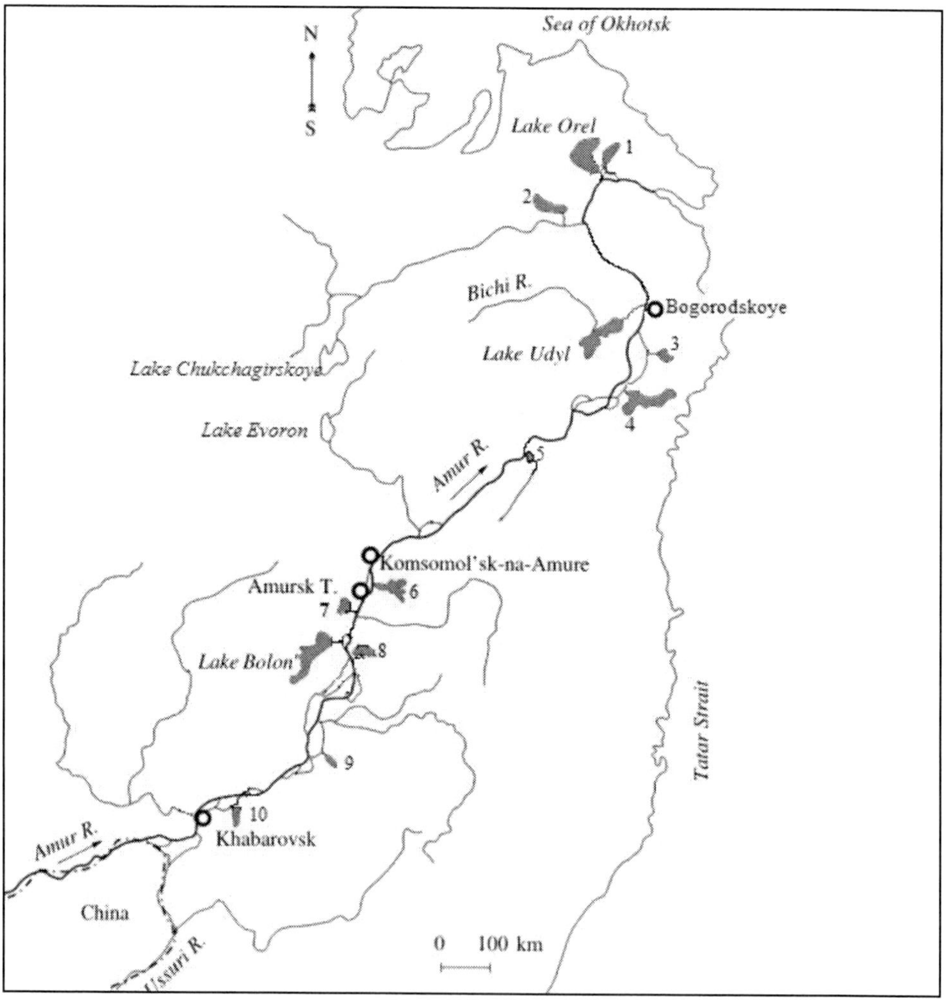

Figure 1. The biggest lakes in the Lower Amur valley.

The largest lakes of the Lower Amur were described by numerous researchers, who identified the lakes due to ponding (Bilibin, 1959), dam lakes (Mordovin & Sokhina, 1968), near-floodplain lakes (Glavatskii, 1961), etc. Most large lakes, which have different size and shape, are associated with the mouths of Amur tributaries in the periphery of its floodplain or are adjacent directly to the Amur channel or/and its distributaries.

The main cause of the development of lakes in the Lower Amur valley is the long-term sustained accumulation and, as a consequence, the increasing damming from the side of the Amur and the inundation of the mouths of tributaries and the structural elements of intermountain depressions adjacent to the Amur floodplain (Makhinov, 1992; Ufimtsev & Ivanov, 1984). The specificity of such water bodies, which are systematically inundated by the Amur water, determines two main types of their hydrological regime. During Amur low-flow periods when water levels do not exceed mean annual minimal values (normal annual low summer water level), the lakes are fed only from internal catchment areas, which are drained by rivers emptying into the lakes. In such cases, the lakes are flowing water bodies. In other seasons, Amur water enters these lakes through arms, i.e., vast external catchment areas contribute to lake nourishment, though these areas are difficult to reliably identify. In this case, the lakes become closed, i.e., local acceptors of the substances entering them.

Within the framework of research into the regular patterns of floodplain–Lower Amur channel interaction, the water exchange between the Amur valley tracts including the areas near the mouths of its tributaries is the subject for special study (Shamov, 2011). Under the conditions of directional valley accumulation, they have developed a great many lakes whose morphology is associated with geological structures occurring in their beds (Ufimtsev & Ivanov, 1984; Makhinov, 1992). The hydrological and sedimentation regimes of these lakes are largely determined by the regime of the Amur, while the regimes of the water bodies near the floodplain on the Amur-Amgun' lowland are also governed by the dynamics of the Amgun', a large tributary of the Amur, which is attributed to a tight hydraulic connection of these water bodies with the aforementioned rivers (Avaryaskin, 1975; Levshina et al., 2007; Shamov, 2003; Shamov et al., 1999). The objective of this paper is to gain a more penetrating insight into the regularities of water exchange by invoking evidence from field expeditions.

2. Objects and Methods

Lacustrine areas of the Amur valley are used in reference to the tracts within which there is a constantly or intermittently operating hydraulic connection of the channel of the main river with water bodies near the floodplain via lacustrine distributaries. At periods of the moderate and low water-level on the Amur, they are drainage water bodies and receive their water alimentation solely through their internal catchment which is a combination of rivers emptying into the lakes. In this case, the vast majority of shallow lakes disappear by turning into near-mouth tracts of lacustrine tributaries which flow along thalwegs of dried lake depressions being overgrown with luxuriant herbaceous vegetation and osier. Large lakes near the floodplain, however, become shallow, their water surface area is decreasing, and extensive tracts of their bottoms uncover themselves, which also rapidly become overgrown with shoal vegetation.

A rise in the water level of the main river is responsible for the backwater of the Amur tributaries in the tracts near the mouth. At the time of a rising level in the Amur, the lakes near the floodplain become closed and are rather rapidly impounded in their upper part with the water from the tributaries draining the internal lacustrine basins; in the lower part, they receive the water from the Amur which is usually more turbid, characteristically brownish-yellowish in color, which makes it possible to clearly delineate the boundary of the water masses of a different genesis.

Because of the regularly forming backwater under the conditions of a low local base-level of erosion, some places develop stable cascades of lakes near the Amur floodplain, the largest of which are: Darga – Toz, Kukharskoe – Petropavlovskoe, Al'bite – Gumen – Kiltasin – Bolon', Dzhalunskoe – Bol. Sharga – Baikal – Kaltakheven, Pereboevka Bay – Bol. Kizi, Mal. Kizi – Bol. Kizi, and Mal. Kadi – Kadi. At periods of the Amur's low water abundance, these cascades virtually disappear by turning into mouth tracts of its tributaries flowing along bottom furrows within the bare lake depressions.

High floods (with discharges more than 15 and 20 thou m^3/s at the beginning and end of the section between the city of Khabarovsk and the town of Bogorodskoe, respectively) are accompanied by a partial or complete impounding of the lakes near the floodplain with the Amur's water which "forces" the waters of the lacustrine tributaries to discharge into their mouth sections. The large water bodies near the floodplain: Bolon', Udyl' and Dudinskoe, which are connected with the Amur via one distributary, are characterized by a more complex mechanism of interaction with the main river. Thus, while Lake Bolon' and Lake Dudinskoe are considerably nearer to the Amur channel and more frequently experience the invasion of its waters and their subsequent discharge, Lake Udyl' is connected with the Amur by the Ukhta distributary about 36 km long, and by a relatively narrow massif near the floodplain which hamper the hydraulic connection of the lake with the main river. The water from the Amur enters Udyl' more rarely, during very high floods (usually, with peak flow rates in excess of 20 thou m^3/s) at the town of Bogorodskoe. Thus the volume of the Amur's discharge, regulated by the lakes near the floodplain, depends heavily on the water abundance of a flood as well as on its phase, and to determine the form of this dependence implies a separate problem that calls for systematic (fixed-site or routine) investigations.

The water regime of the lakes near the floodplain along the Amur valley is not monitored on a systematic basis. Hydrometric expedition-based summertime surveys in some measuring sections of the main river and in the lacustrine distributaries that were carried out by the Institute of Water and Ecological Problems, Far-Eastern Branch, Russian Academy of Sciences, in 1997 and during 2004–2007 furnish a means of making tentative estimates of the water exchange characteristics for typical lacustrine areas of the Amur valley in different phases of its regime. To accomplish this, a comparison was made of the lateral instantaneous afflux of the lacustrine area with the flow rate of the main river in the nearest (to the lake) outlet section – the local contribution from the lake discharge. It is particularly remarkable that the recorded discharge from Lake Udyl' was due solely to the inflow from the internal catchment; therefore, we omitted it from our estimations of the Amur discharge regulated by lake systems.

3. Results and Discussion

In August 1997, the investigations into the water regime of the Amur were made during the rise phase of a flood of a not high water abundance (a normal flood characterized by a maximum level of 366 cm as measured at the gauging station in the city of Khabarovsk, with the mean, season-highest, level of 391 cm). The flood wave was decreasing in height as it moved from Khabarovsk-city (966 km from the mouth) to the settlement of Takhta (75 km from the mouth) from 310 to 107 km, respectively. In this case, the measured water flow rates in the Amur amounted to 12–17 thou m^3/s, increasing from Khabarovsk to the measuring section at the town of Bogorodskoe (238 km from the Amur mouth). Under these circumstances, there emerged a backwater that blocked the discharge from the lakes near the floodplain into the Amur, while the floodplain and the lakes were impounded in minor amounts.

During the flood decline at the beginning of October 1997, a hydrometric survey was made of the outlet distributaries of Kaltakheven, Bolon' and Udyl'. In this case, the Amur discharge differed only slightly from its discharge in August, while the lakes that had been impounded during the flood discharged the surplus water into the gradually lowering river channel (see the table).

A more detailed hydrometric survey of the inlet and outlet lacustrine distributaries was made in June – July 2004 during the decline of the snowmelt-rainfall flood on the Amur (with the maximum level on June 6 amounting to 433 cm at the gauging station in the city of Khabarovsk). At that time, the near-floodplain water bodies of the Sredne-Amurskaya and Udyl'-Kizinskaya lowlands that were impounded in May – early June were discharging significant volumes of water into the Amur. The processed results from the surveys are provided in the table. To be precise, it should be remarked that the time of measurements the Lake Petropavlovskoe was, in effect, backwatered by the flood flow on the Amur, which was responsible for the exceptionally low value of flow rate in the outlet lacustrine Malyshevskaya distributary.

The velocity of the research vessel was somewhat higher, on the average, than the velocity of the flood wave, and in the second ten-day period of July 2004 the expedition reached the Amur-Amgun' lowland where the flood was only in its incipient stage of decline. Within the Orel'-Chlya water node (Fig. 2), at the time of hydrometric operations the situation was as follows. The flood waters of the Amur had impounded the Lake Chlya through a system of distributaries and started to be discharged into the Pal'vinskaya distributary, while the neighboring the Lake Orel' was backwatered by the outlet Pal'vinskaya distributary. The discharge from the Lake Chlya via the Chlinskaya and Chepchugan distributaries totaled 182 m^3/s on July 16–17, while at that time the Lake Orel' received about 80 m^3/s from the Chlinskaya distributary through the discharge of the Pal'vinskaya distributary.

Since both the internal catchment of the lake near the floodplain that is clearly identifiable on maps and the external catchment that can only be taken into account in the form of the inflow from the main river, are functioning in the flood conditions, it is only possible to use conditional specific characteristics of the discharge. By comparing the values of the conditional layer of discharge from the lakes, determined as the ratio of the measured discharge to the lake's internal catchment area with due regard for its water surface area, we can conclude that the water bodies with small catchment areas occurring at the floodplain expansions of the Amur (Lakes Kaltakheven and Sharginskoe, Dzhalunskoe and Dudinskoe) as well as in the intermontane depressions (Lakes Chlya, Kizi and Kadi) are distinguished by high values of specific stream discharge. In our view, this merely points to the dominant role of the Amur waters that had entered the lakes under investigation, in the formation of the discharge from them during the flood decline on the main river.

The conditions under which the hydrometric survey of the lacustrine areas were made in 2006 were characterized by a moderately low water-level on the Amur. The level of water as measured by the gauging station at the city of Khabarovsk in mid-June was fluctuating around a mark of 100 cm. At the survey period in the lacustrine distributaries the water reserves of the lacustrine areas along the Amur valley were, in effect, exhausted and started to be replenished in some places at the expense of the increasing discharge of the main river (see table). On the whole, the contribution from the Amur discharge regulated by the lakes is very small, or about 4%, which is as a rule within the discharge measurement error.

Floods of a moderate height on the main river, similar to the summer-autumn 2007 flood, are responsible for the supply of significant volumes of water via the lacustrine distributaries to the water bodies near the floodplain (except for Lake Udyl'). At the time of a high water level on the Amur, however, there is a strong backwater in the lacustrine distributaries. In these circumstances, the water reserves, accumulated in the lakes near the floodplain, are discharged slowly and make generally a low contribution to the measured/ instantaneous discharge of the Amur, or about 9–10% with due regard for the water bodies of the Amur-Amgun' lowland as shown in Table. Some of the small water bodies which are hydraulically connected with the Amur via one distributary (Lake Khavanda) are, in effect, closed ones because of the low runoff from their internal basins; a number of lacustrine areas of the Amur floodplain are characterized actually by a zero balance (Lake Kadi).

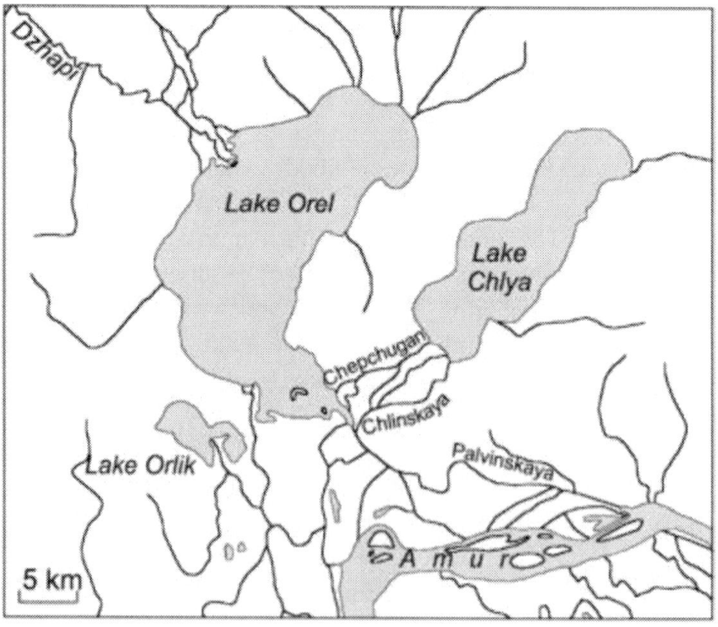

Figure 2. Schematic map of the Orel'-Chlya water node (Amur-Amgun' lowland).

Table. Water exchange characteristics in lacustrine areas along the Lower Amur

Lake	Area of lake, km²	Internal catchment area of lake, km²	Water flow rate in inlet distributaries, m³/s	Water flow rate in outlet distributaries, m³/s	Discharge to Amur river, m³/s	Conditional layer of flow, mm/day	Water flow rate in the Amur in the nearest measuring section, m³/s	Ratio of volume of discharge from lake to flow rate in the Amur, %
October 1997, end of decline in high rainfall flood on Amur								
Kaltekheven, Bol'shaya and Malaya Sharga	8.6	1290*	(8)	87.7	79.7	5.3	14200	0.56
Bolon'	338	12570	–	263	263	1.8	14460	1.82
Udyl'	330	11480	–	255	255	1.9	12700	2.00
July 2004, decline in high snowmelt-rainfall flood on the Amur								
Petropavlovskoe	55.4	3600	(0.3)	1.41	1.2	0.03	15100	0.01
Kaltekheven, Bol'shaya and Malaya Sharga	8.6	1290*	(15)	178	163	10.9	14200	1.14
Bolon'	338	12570	–	887	887	5.9	15090	5.88
Khavanda	3.7	1200	–	146	146	10.5	18000	0.81
Kizi	281	4890	159	980	821	13.7	18820	4.36
Kadi	67	905	16.2	–26.8	–43.0	–4.1	9690	–0.44
Irkutskoe	13.9	827	0	–21.0	–21.0	–2.2	9670	–0.22
Dudinskoe	39.4	267	–	–0.95	–0.95	–0.3	9670	–0.01
Udyl'	330	11480	–	165	165	1.2	11900	1.39
Orel'	314	4990	0	43.2	43.2	0.7	12800	0.34
Chlya	140	531	47.8	0	–47.8	–7.8	12800	–0.37
Total discharge from the lakes (except for Lake Udyl' and Lake Bolon')								3.30
Total runoff to the lakes								–1.06
Budget of lacustrine runoff								2.24
September 2007, decline in moderate flood on the Amur								
Petropavlovskoe	55.4	3600	(20)	105	80	1.9	14810	0.58
Kaltekheven, Bol'shaya and Malaya Sharga	8.6	1290*	(15)	112	107	7.2	12850	0.87
Bolon'	338	12570	–	406	406	2.8	13250	3.06
Khavanda	3.7	1200	–	9.5	9.5	0.7	15930	0.06
Kizi	281	4890	63	408	345	6.1	17280	2.00
Kadi	67	905	116	102	–14	–1.3	17270	–0.08
Irkutskoe	13.9	827	(1)	17.4	16.4	1.7	18310	0.09
Udyl'	330	11480	–	310	310	2.3	18530	1.66
Total discharge from the lakes (except for Lake Udyl')								6.5

Note. The dash in the table cell means the absence of pronounced inlet distributaries; the values in brackets were estimated with an error of 15–20%.

* We used the internal catchment as a combined set of catchments of Lake Bol. Sharga and Lake Mal. Sharga, Lake Dzhalunskoe, and Al'dabira Creek.

** Data were reconstructed from correlation with the values of layer of flow from counterpart lakes (see the text).

Some results of hydrometric surveys make it possible to roughly assess the correlations of the values of the measured/instantaneous flow rate from the lacustrine areas of the Amur valley. The estimated values of total measured/instantaneous layer of flow from lacustrine areas of the Lower Amur valley in different phases of the main river regime were shown in Figure 3. There one can see cases of the lakes: Petropavlovskoe (1), Kaltakheven, Bol'shaya and Malaya Sharga (2), Bolon' (3), Padali (4), Khavanda (5), Kizi (6), Kadi (7), Dudinskoe (8), Irkutskoe (9), Udyl' (10), Khilka (11), Orel' (12), Chlya (13). Negative values of discharge correspond to the amount of afflux.

There is some correlation of flows from Lake Bolon' and Lake Udyl' which are connected with the main river by one distributary. Figure 4 demonstrates such correlation in measured / instantaneous values of the flow layer (1) and a fitted curve (2).

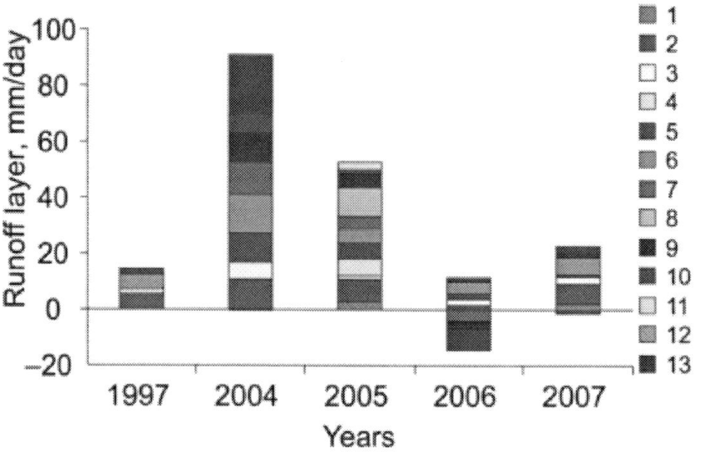

Figure 3. Estimated values of total measured/instantaneous layer of flow from lacustrine areas of the Lower Amur valley in different phases of the main river regime.

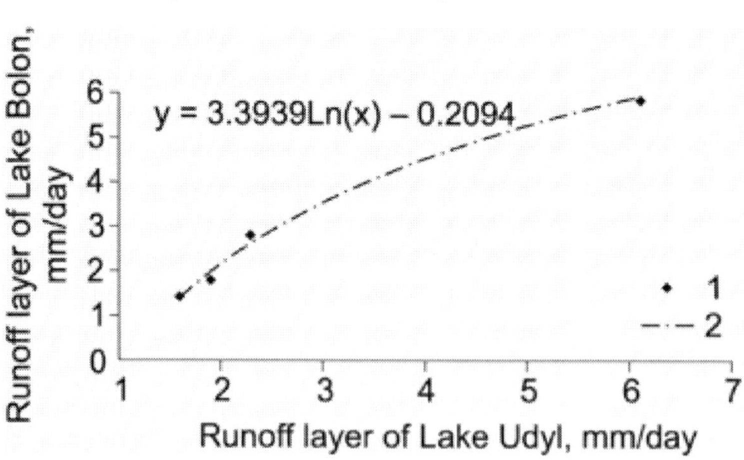

Figure 4. Estimated correlation of measured/instantaneous values of the flow rate from Lake Bolon' and Lake Udyl'.

Based on the connection between the measured discharges in the lacustrine distributaries during different phases of the regime, the current water balance of some tracts can be inferred through calculations by approximating regression equations. In view of the fact that the data obtained for the regime of the lacustrine areas of the Amur valley are highly limited and do not embrace the entire diversity of hydrological conditions, it only remains for us to suggest the presence of a complex connection between the measured/ instantaneous volume of discharge regulated by these areas and the current discharge (of the measured/instantaneous discharge) of the main river. Furthermore, the total contribution from the regulated discharge for the 1997 flood decline conditions is taken to be identical for the 2004 survey conditions, based on the equality of the partial contributions as calculated for Lakes Bolon', Kizi, Kaltakheven and Sharginskie from the data from the two surveys.

The plot in Figure 5 shows total proportion of the lake-regulated discharge of the Amur measured nearby Bogorodskoe-town, versus its current water flow rate. The Figure 5 also shows types of approximating functions, actual values for the impounding phase of lakes (1) and for the backwater phase of lakes due to the Amur flood (2) as well as fitted curves for the impounding phase of lakes (3) and for the backwater phase of lakes on the side of the Amur (4). In course of the floods on the Amur is accompanied, the lakes near the floodplain and of the mouth sections of the Amur tributaries are locked by river flow, ensuring, with a relatively rapid passage of the flood wave, a high contribution from the total instantaneous discharge regulated by the lacustrine areas (the increasing branch of the plot in Figure 5). At periods of a high discharge of the Amur there occurs a "locking" effect where the discharge on the lacustrine distributaries is difficult because of the backwater caused by the main river (the descending branch of the same plot).

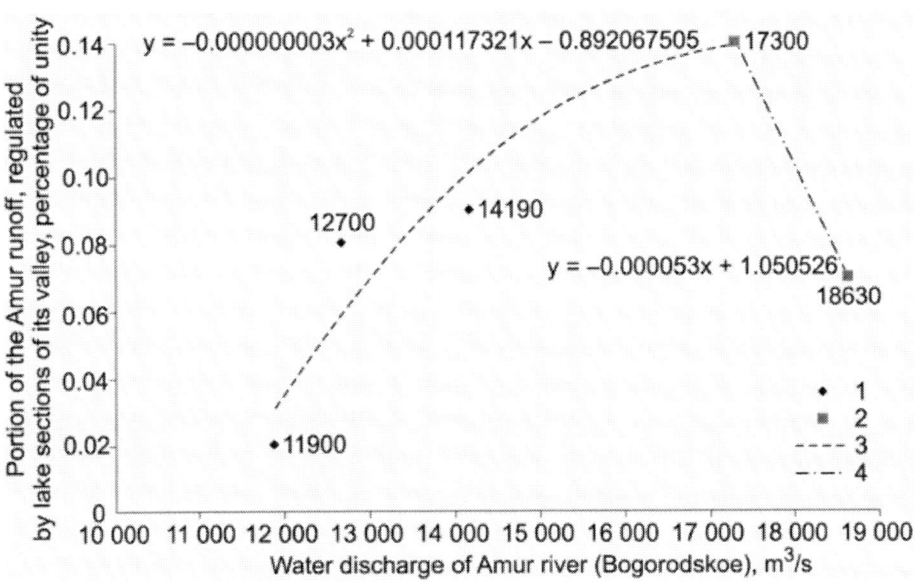

Figure 5. Total proportion of the regulated (by near-floodplain lakes) measured/instantaneous discharge of the Amur at Bogorodskoe-town versus its current water flow rate.

To estimate the volume of the Amur discharge, regulated by the lakes, for a flood or annual cycle calls for longer-lasting, systematic investigations, both in different tracts of the valley and in different cycle phases, having regard to the preceding humidification of the Amur floodplain. In a general form, the value of the volume of the discharge, regulated by the lakes near the floodplain and by the floodplain lakes can be expressed in terms of an integral:

$$W(t) = \int_T f(V, H)dt$$

where T is the length of the computational cycle (the period of time), $f(V, H) = dW$ is the instantaneous volume of the regulated discharge (water flow rate) as a nonlinear function of instantaneous inflow volume, V(t) (or the Amur water flow rate, Q(t)), and of the preceding humidification — the initial water capacity of the computational area, H(t) (or the lake's volume $W_{lake}(t)$).

Thus the problem of estimating the volume of the portion of the Amur discharge, regulated by the lakes, reduces to determining the function $f(V, H)$. In doing this, a promising approach would imply using the computational model for the interaction between the channel flow and the floodplain flow, based on the Saint-Venant equations where the channel and floodplain parts are originally separated from each other, and the floodplain has the role of an accumulating reservoir with a horizontal level of the free surface (Voevodin et al., 2008). In the absence of an inflow to the lakes from the inlet distributaries, in the general case we have $W(t) = W \cdot T = Q_{bas} \cdot T - Q_{lake} \cdot T$, where $Q_{bas} \cdot T$ is the volume of the inflow to the lake depression from the internal lacustrine basin for a computational period of time T, and $Q_{lake} \cdot T$ is the volume of discharge from the lake via the outlet distributaries for the same period. In this case, $W \cdot T$ stands for the total lateral afflux of the Amur from the lacustrine areas along its valley for a period T. When the flood cycle and the annual cycle are used in calculations at the initial and final times of the computational interval, to a first approximation it can be assumed that $Ho \approx H \cdot T$, where H_o is the water reserve in the lacustrine area at the initial time of the computational time interval, and $H \cdot T$ is the water reserve of the lacustrine area at the end of this period.

Conclusion

Occasional hydrometric surveys of the lacustrine branches along the valley of the Lower Amur under different water abundance conditions permit an assessment of the instantaneous water budget of the lacustrine areas thereby making it possible to assess the instantaneous discharge volume regulated by the lacustrine areas of the valley. The relationship of the water levels in the Amur and of the lakes adjacent to its floodplain determines the rate and directedness of a current water exchange within the lacustrine areas of the valley of the main river.

The areas, impounded at the time of high flood on the Lower Amur, which include the lakes near the floodplain, can ensure in the process of discharge of water reserves accumulated during floods, a total of more than 15% of the instantaneous discharge of the main river in the adjoining measuring sections. This increment in the discharge volume, regulated by the lacustrine areas, amounts to a significant value comparable with the value of total lateral afflux of the Amur in the long-term average context along a section nearly 1000 kilometers long from the mouth of the Ussuri to the mouth of the Amur. Under the high water abundance of the Amur, its backwater decreases substantially the amount of instantaneous discharge from the water bodies impounded by flood waters by as much as 5–6% or, perhaps, to a greater extent. Of course, the phase of water level rise on the river involves a further impounding of the lacustrine areas of the floodplain where the outflow distributaries operate as inflow distributaries.

During a decline in low and moderate floods, the total runoff in the lacustrine areas is estimated as low, due to the low water reserves, and appears to make up not more than 9–10% of the discharge of the Amur during the phase of flood decline.

In conditions of the steady-state summertime lowest-water level when the alimentation of the water bodies near the floodplain on account of their internal basin water reserves, the contribution of the lacustrine areas along the Amur valley to the regulation of the discharge of the main river decreases to zero. Furthermore, the total instantaneous discharge from the lakes will most likely make up only a few percent of the instantaneous discharge of the Amur, or within the limit of measurement accuracy of water flow rate.

REFERENCES

Avaryaskin, L. P. (1975). The mouths of the Lower Amur tributaries. *Questions of the Geography of the Far East.* Russia: Khabarovsk Book Publishers, issue 16. [in Russian].

Bilibin, Yu. A. (1959). Nekotorye geomorfologicheskie nablyudeniya v predelakh Dal'nego Vostoka (Some Geomorphological Observations in the Far East). *Russia: Akad. Nauk SSSR Publ.* Vol. *2*. [in Russian].

Domanitskii, Dubrovina, & Isaeva. (1971). Reki i ozera Sovetskogo soyuza (spravochnye dannye) (Rivers and Lakes of the Soviet Union (Reference Data)), Russia: Gidrometeoizdat. [in Russian].

Glavatskii, S. N. (1961). Groups and Types of Lakes in the Lower Amur Region, in Geologiya, geomorfologiya, poleznye iskopaemye Priamur'ya *(Geology, Geomorphology, and Useful Minerals of the Amur Region).* Vol. *1*. Iss. 72. Russia: Khabarovsk Book Publishers. [in Russian].

Levshina, Shamov, & Kim. (2007). *Organic matter in the water of lakes near the Lower Amur floodplain.* Water Resources.

Makhinov, A. N. (1992). *Near-mouth lakes of the Amur River tributaries.* Izvestiya RGO (Bulletin of Russ. Geograph. Society). [in Russian].

Mordovin, & Sokhina. (1968). Some features of the formation of lakes in the southeastern part of Udyl–Kizinskaya lowland. Geomorfologicheskie, landshaftnye i biogeokhimicheskie issledovaniya v Priamur'e (Geomorphological, Landscape, and Biogeochemical Studies in the Amur Region). Russia: Nauka. [in Russian].

Shamov, V. V. (2003). Landscape-hydrological typing of plain lakes in the Lower Amur region. *Geography and Natural Resources* (Geografiya i Prirodnye Resursy). [in Russian].

Shamov, V. V. (2011). Water exchange in lacustrine areas along the valley of the Lower Amur (as deduced from data of expedition-based investigations). *Geography and Natural Resources*.

Shamov, Matroshilov, Petrov, and Bakanov. (1999). Formation conditions and assessment of resources of organic-mineral deposits of lakes in the Lower Amur region. *Geografiya i Prirodnye Resursy* (Geography and Natural Resources). [in Russian].

Surface Water Resources of the USSR, Dal'nii Vostok, Nizhnii Amur (Far East, Lower Amur). (1970). Vol. *18*. Iss. 2. Russia: Gidrometeoizdat. [in Russian].

Ufimtsev, & Ivanov. (1984). The morphostructures of lacustrine basins of the Lower Amur region. Geomorfologiya.

Voevodin, Nikiforovskaya, & Ostapenko. (2008). Mathematical modeling of the transformation of flood waves in stream channels with floodplains. *Russian meteorology and hydrology*.

In: Floodplains
Editor: Enner Herenio Alcântara

ISBN: 978-1-62808-998-1

Chapter 7

HYDROGENIC HEAVY METALS POLLUTION OF ALLUVIAL SOILS IN THE CITY OF PERM (THE MIDDLE CIS-URALS REGION, RUSSIA)

***Yu. N. Vodyanitskii*[1*] *and A. T. Savichev*[2]**
[1]Lomonosov Moscow State University, Faculty of Soil Science
[2]Geological Institute, Russian Academy of Sciences

ABSTRACT

In 1999, the Perm region was eight among Russian regions in terms of the technogenic load per unit of area (4.4 t/km^2). The situation in the city of Perm is especially unfavorable in the ecological respect due to aerial pollution and hydrogenic pollution, because of industrial wastes entering the small rivers that are tributaries of the Kama river.

It was revealed that the alluvial soils (fluvisols) of the city of Perm are polluted by heavy metals of hydrogenic origin because of the unpurified sewage water entering them. The degree of pollution of fluvisols differed from the aerial contamination of urban soils. In the urban fluvisols, the mean Pb content is twice lower, and the content of Zn and Ni are 1.5 and 4 times higher, respectively, than in the common urban soils (urbanozems). These facts attest that the strong pollution of these alluvial soils is of hydrogenic origin.

In the fine earth of the fluvisols, a positive correlation between the content of clay particles and the contents of Sr, Zn, Ni, Cu, and Cr is absent. The strong pollution of the light textured soils attests that, nowadays, the main pollution source for the fluvisols is sewage water, rather than warp deposited. However, with time, in the floodplain of some small rivers the alluvial deposition will significantly affect the pollution of the alluvial soils. In the floodplain of other small rivers, the alluvial soils are polluted to a greater extent than the warp accumulated from the river water.

In the fluvisols, the Fe-rohrensteins are formed. Some elements are concentrated in the rohrensteins, and some others are not concentrated in them or are found in low concentrations. In Fe-rohrensteins the highly active group comprises As, Zn, Ni, Cu, Cr,

* Corresponding author: Lomonosov Moscow State University, Faculty of Soil Science, 119991, Moscow, Leninskiye gory, 1, k.12; e-mail: yu.vodyan@mail.ru.

and Pb; the moderately active one is represented by Sr, Nb, Ga, and Y; and the inert group contains Zr and Rb.

The content of some chemical elements in the Fe-rohrensteins are much greater than those in the fine earth. The Pb and Zn contents in the rohrensteins of the soil of small rivers basin are 440 and 890 mg/kg respectively. In the rohrensteins the Pb and Zn contents are 42 and 17% of their concentrations in the fine earth, respectively. Since some part of heavy metals is precipitated at the redox microbarriers around concretions (Fe-rohrensteins), it is removed from the biological cycle.

In the fluvisols, Ba is fixed in rohrensteins as barite ($BaSO_4$) of probably, technogenic nature. In the gleyed fluvisols from small rivers floodplain, Ba migrates intensively in the soil profile and fixed in ellipse-like iron-manganic nodules. The contents of La and Ce in the fine earth of polluted soils remain practically stable, and lanthanides are accumulated in the Fe-Mn- nodules: the accumulation coefficient reaches 3.5 for La and 6.9 for Ce.

In the profile of the alluvial soil on the floodplain of small rivers polluted by wastewaters from petroleum refinery, La migrates deeper into the soil profile than Ce. The latter element is more active and is sorbed in iron-manganic nodules in the topsoil. According to the average coefficients of element accumulation in the concretions (Fe-Mn-nodules and Fe-rohrensteins), the studied elements form the following sequence: Ce(3.7)>Ba(3.0)>La(2.2).

1. Introduction

Usually, technogenic pollution may be aerial or hydrogenic. Hydrogenic pollution has been studied much more poorly than aerial pollution. Nevertheless, hydrogenic pollution may be more dangerous. First, although unpurified sewage water entering small rivers contaminates limited territories, the degree of soil pollution is higher due to the pollutants being concentrated in a smaller area. Under aerial pollution, gaseous emissions are dispersed over great areas. In Perm, the industrial sewage has elevated concentrations of Cu, Pb, Zn, As, Mo, Ni, Cd, and Hg (Shchukova, 2005). Hydrogenic pollution because of the poorly purified industrial sewage is characteristic of the alluvial soils. The danger of hydrogenic pollution is explained by the fact that it occurs in limited areas with alluvial soils, where sludges having high concentrations of pollutants arrive. Second, alluvial soils are enriched with warp (silt deposits) polluted by heavy metals and metalloids (Osovetskii & Men'shikova, 2006). These include particles of charcoal, soot, nonferrous metals, ore, slag, magnetic spherules, glass, and so on. The main technogenic pollutants of the river sediments in Perm are Bi, Ag, Zn, Pb, Cu, W, and Ni (Osovetskii & Men'shikova, 2006). As a result of the circumstances mentioned, the pollution of alluvial soils by some elements in the floodplains of small rivers may be higher than that of automorphic urban soils.

The fate of heavy metals depends in many respects on soil concretions - Fe–Mn-nodules and tubular Fe-rohrensteins that are capable of removing metals from the biological cycle due to sorption and fixation of them by Fe and Mn (hydr)oxides. This binding of metals prevents their entering the soil solution; living organisms; and, hence, the biological cycle.

In 1999, the Perm region was the eighth among Russian regions in terms of the technogenic load per unit of area (4.4 t/km^2) (State National Report…, 2000). The situation in the city of Perm is especially unfavorable in the ecological respect (Osovetskii & Men'shikova, 2006; State and Protection of the Environment…, 2005) due to aerial pollution and hydrogenic pollution, because of industrial wastes entering the small rivers that are tributaries of the Kama River.

The aim of this work is to study the pollution of the alluvial soils by heavy metals within the city of Perm and to assess the role of concretions in the fixation of heavy metals.

2. Objects and Methods

The alluvial soils within the city of Perm that are flooded and covered with sediments during the high water of small rivers were studied. The soil pits were made in the zone influenced by the Perm–Krasnokamsk industrial complex on the floodplains of the left tributaries of the Kama River: the Iva, Egoshikha, Danilikha, Obva and Mulyanka and on the floodplain of the right tributary - the Las'va (figure 1). Three pits of typical gray-humus gley soils were dug on the floodplains of the Iva, Danilikha, and Mulyanka rivers. The first pit was dug at a distance of 300 m from Vosstaniya square; another one, 500 m from the Perm-2 railway station; and the last one, 200 m from the crossing of the river by the Kosmonavtov Highway bridge. The pit of fluvisols was dug on the floodplain of the Las'va River 1 km downstream of the city of Krasnokamsk. The investigations were carried out in 2006; 20 samples of 5 pits and 10 samples of nodules were analyzed.

All the soils, except for the soil in the Iva River valley, were exposed to anthropogenic impact. The lowest quality of water was recorded in the lower reaches of the Egoshikha and Danilikha rivers; the water was qualified as nonpotable. According to the report of the Department of Environment Conservation (2004), the water quality in the Iva River is of the second class (pure); at the mouth of the Egoshikha river, it is of the third class (medium polluted); in the Danilikha river, the water is of the sixth class (strongly polluted).In the Mulyanka river, of the second–third quality class, but in the upper reaches its water refers to the fourth class because of the high contents of nitrates and iron (State and Protection of the Environment…, 2005).

The reaction of the soils (pH_{water} and pH_{KCl}) was determined using a potentiometer; the particle-size composition by the sedimentation method; the organic carbon, by the Tyurin method; and the contents of mobile phosphorus and potassium, by the Kirsanov method. The total contents of macro- and microelements were analyzed using the roentgen fluorescent method with the help of a TEFA-6111 device. Instead of the traditional sample excitation by emanation from an X-ray tube, excitation by isotope source ^{241}Am with energy of 59 keV and activity of $3.7 \cdot 10^{10}$ s^{-1} was used for a TEFA-6111 analyzer in order to determine Ba, La, Ce; limit of detection is 4 -5 mg/kg. Electron scanning microscopy was used for identification of Ba-containing particles in the concretions. A Tescan MV-2300 scanning electron microscope with an INCA-200 microanalyzer was used. The detector of X-ray radiation of the microprobe had a 0.5-μm window, which made it possible to analyze oxygen and carbon. The composition of the iron compounds in some soil and sediment samples was determined using Mössbauer spectroscopy (an Ms-1104Em spectrometer) under permanent acceleration with the source being ^{57}Co in a Cr matrix at room temperature.

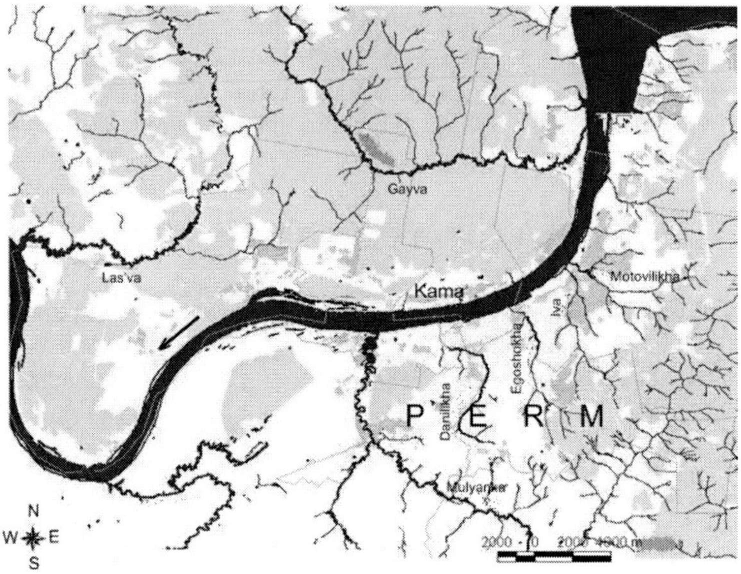

Figure 1. Map of the Kama River and its tributaries in Perm region.

3. Results and Discussion

Specific features of the hydrogenic pollution. The outskirts of the city of Perm are exposed to aerogenic and hydrogenic contamination. The aerogenic pollution of the Perm urban soils was described in detail in the work by Eremchenko & Moskvina (2005). We consider not only the mean values of the heavy metal contents but also the degree of their variation (Vodyanitskii, Vasil'yev et al., 2008; Vodyanitskii, Vasil'yev et al., 2009). According to the data of geochemists (Saet, Revich, Yanin et al., 1990), in technogenic anomalies, the variation of the contents of chemical elements is higher than in the background territories.

The urban soils of Perm (urbanozems) are contaminated by aerial pollutants (Pb, Zn, Cr, and Cu) (Eremchenko & Moskvina, 2005). The mean contents of Pb, Zn, Cu, and Cr are 71, 140, 85, and 238 mg/kg with their coefficients of variation (V) being 254, 66, 133, and 55%, respectively. Both these indices (the mean contents and high variation) indicate the technogenic aerial pollution of the soils.

The properties of the alluvial soils (fluvisols) are considered below. Some characteristics are given in Table 1. The reaction of all these soils is close to neutral: the mean value of the pH_{water} is 7.0. This value is somewhat lower than that in the urbanozems (7.6). The alluvial soils are distinguished by their low mean content of organic carbon (C_{org} = 0.46%, V = 67%). These values are much lower than in the upper horizons of the urbanozems (C_{org} =4%, V = 125%). In the lower horizons of the urbanozems, the content of C_{org} is 1.23%, which is 2.5 times higher than in the alluvial soils.

The contents of mobile phosphorus and potassium are low in the alluvial soils. Thus, alluvial soils are low fertile, and this circumstance should be taken into account upon planting trees and shrubs in the floodplain areas in the city. The ability of the alluvial soils to absorb heavy metals is low due to the small humus content in them.

Table 1. Properties of the alluvial soils

Horizon	Depth, cm	C_{org},%	pH_{water}	P_2O_5	K_2O	Clay (<1μm), %
				mg/100g		
Loamy gley fluvisol, the Danilikha river floodplain						
Warp deposit	0- 2	0,40	7,2	23,5	10,7	8,1
AJg	0-22	0,30	7,3	13,7	10,7	15,8
G~~	22-70	0,56	6,7	24,1	10,7	11,6
Light clayey gleyic fluvisol, the Las'va river floodplain						
Warp deposit	0- 2	1,64	6,6	6,0	26,6	61,8
AJ	0-10	0,51	6,1	9,7	11,4	59,0
C1g~~	10-30	0,50	6,8	21,0	16,3	53,6
C2g,h~~	30-50	0,40	6,4	12,2	7,9	35,3
C3g~~	50-80	0,17	6,6	14,0	5,4	17,0
C4g~~	80-100	0,15	7,0	15,0	7,3	26,7
Loamy-sandy gley fluvisol, the Egoshikha river floodplain						
Warp deposit	0- 2	0,57	6,9	23,5	12,9	22,2
AJg	0-15	0,37	7,4	23,5	12,1	19,1
G~~	15-30	0,34	7,7	18,3	5,4	18,6
C2g~~	30-50	0,45	7,2	20,2	9,3	36,0
Loamy gley fluvisol the Iva river floodplain						
Warp deposit	0- 2	0,44	7,4	9,6	17,0	36,6
AJg	0-15	0,28	7,6	10,8	6,0	38,2
G~~	15-30	0,15	7,39	22,2	5,4	25,3
Cg~~	30-70	0,31	7,1	21,5	7,3	25,2
Loamy gley fluvisol, the Mulyanka river floodplain						
AJg	0-10	0,53	6,9	10,2	10,0	38,5
C1g~~	10-20	0,58	7,1	11,8	4,7	25,5
G~~	30-70	0,59	7,0	6,8	13,6	32,2

The heavy metals and metalloids contents in the fine earth of the alluvial soils are given in Table 2. Table 3 presents the mean contents of heavy metals and metalloids and their variation in the fine earth of the alluvial soils and urbanozems. The Pb content is twice lower in the fluvisols than in the urbanozems (32 mg/kg). This fact is not surprising, since the main source of aerial Pb pollution is gasoline additives that are dispersed in the air.

The mean Zn content (212 mg/kg) in the fluvisols is higher by 1.5 times than in the urbanozems. These facts attest than the alluvial soils are strongly polluted by Zn of hydrogenic origin as well as by Ni, whose content (149 mg/kg, V = 86%) is higher by 4 times than in the other soils. In fluvisols the Cr content (289 mg/kg) insignificantly exceeds its concentration in the urbanozems (238 mg/kg). The high variation of the Cr content (V = 107%), as well as that of Ni, indicates their technogenic source.

No reliable differences in the contents of some chemical elements in the soils of different genesis were found. In the alluvial soils and urbanozems the mean Cu contents were 76 and 85 mg/kg, respectively. Evidently, technogenic Cu is dispersed in the air and water. The mean Sr concentration in the fluvisols (274 mg/kg) was approximately the same as in the urbanozems (257 mg/kg), although its variation was much lower in the fluvisols (14%) than in the urban soils (73%).

Table 2. The contents of heavy metals and metalloids (mg/kg) in the fine earth of the alluvial soils

Horizon	Depth, cm	Ni	Cu	Zn	Ga	As	Pb	Rb	Sr	Y	Zr	Nb	Cr
The Danilikha river floodplain													
Silt deposit	0- 2	155	67	173	7	-	48	37	250	12	104	-	300
AJg	0-22	262	121	349	17	-	90	55	267	17	241	-	600
G$^{\sim\sim}$	22-70	480	245	401	7	-	57	46	278	19	229	-	1400
The Iva river floodplain													
Silt deposit	0- 2	120	86	139	10	4	15	59	266	25	225	-	200
AJg	0-15	89	70	115	7	-	27	59	267	20	281	-	200
G$^{\sim\sim}$	15-30	60	54	87	11	-	8	49	288	21	257	-	200
Cg$^{\sim\sim}$	30-70	60	53	89	10	7	6	50	301	18	231	-	200
The Egoshikha river floodplain													
Silt deposit	0- 2	291	113	318	9	4	53	46	266	20	204	-	500
AJg	0-15	338	99	458	6	-	38	42	286	20	230	-	500
G$^{\sim\sim}$	15-30	312	88	376	13	11	16	52	281	24	263	-	400
C2g$^{\sim\sim}$	30-50	280	123	343	10	-	46	64	266	22	262	-	400
The Las'va river floodplain													
Silt deposit	0- 2	63	43	629	17	8	20	69	407	23	151	12	100
AJ	0-10	50	23	69	4	4	16	67	217	26	267	14	100
C1g$^{\sim\sim}$	10-30	59	36	137	17	13	94	64	269	29	295	15	100
C2g,h$^{\sim\sim}$	30-50	51	39	76	13	-	16	59	228	25	307	14	100
C3g$^{\sim\sim}$	50-80	56	46	61	13	7	11	53	253	27	418	11	200
C4g$^{\sim\sim}$	80-100	40	25	62	16	-	16	61	234	29	412	11	100
The Mulyanka river floodplain													
AJg	0-10	81	56	102	4	4	23	52	294	24	299	10	50
C1g$^{\sim\sim}$	10-20	63	70	114	18	7	21	54	279	26	242	11	70
G$^{\sim\sim}$	30-70	63	65	141	16	-	20	56	282	23	229	12	70
	MPC	85	55	100		2	30						
	APC*	80	132	220		10	130				350		

Note. Here and further: a dash means below the detection level; * for heavy soils with neutral reaction.

Table 3. The contents of heavy metals and metalloids (mg/kg) in the urbanozems compared to their contents in the fluvisols

Element	Urbanozems*					Fluvisols				
	n	range	average	σ	*V*,%	*n*	range	average	σ	*V*,%
Ni	69	30-300	39	24	63	20	60-480	149	129	86
Cu	70	20-1000	85	113	133	20	23-245	76	50	66
Zn	70	60-500	140	93	66	20	61-629	212	163	77
Pb	69	6-1500	71	181	254	20	6-94	32	25	78
Sr	70	0-700	257	187	73	20	217-407	274	38	14
Cr	70	20-700	238	132	55	20	50-1400	289	309	107
As		Not det.				10	4-13	7	3	46

* Data of Eremchenko & Moskvina (2005).

Note. Here and further: *n* is the number of samples, σ is the standard deviation, and *V* is coefficient of variation.

It is worth noting that some elements (Cu, Pb, and Sr) are evenly distributed in the alluvial soils: the coefficients of variation of their contents are 2–5 times lower than in the urbanozems. This fact confirms that Cu and Pb accumulate in a more concentrated pattern under the hydrogenic input of pollutants; they are dispersed under the aerial input of pollution. The heterogeneous composition of urbanozems is also of great importance. As for Sr, its distribution in fluvisols is more even than in the mixed surface layers of the urban soils.

Technogenic heavy metals in alluvial light-textured soils. It is well known that, in clay sediments, the contents of heavy metals are much higher than in sands, since they are sorbed by clays and colloids more intensely. The Zn clarke in the earth's crust is 95 mg/kg, whereas, in sand stones, it is only 16 mg/kg (Dobrovol'skii, 2003). Such differences have been determined for many heavy metals. In the background soils, the same differences in the heavy metals concentrations were observed. In fine-dispersed natural–technogenic sediments, heavy metals and metalloids are also sorbed by clay minerals. The concentrations of some heavy metals positively correlate with the content of clay particles in the warp (Osovetskii & Men'shikova, 2006).

The above mentioned allowed considering the direct relationship between the heavy metals concentration and the content of the clay fraction in the fluvisols. However, the calculation of the correlation coefficients has revealed another pattern. Only for Pb was the correlation with the clay content positive and small ($r = 0.10$). For Sr, Zn, Ni, Cu, and Cr, it was negative ($r = -0.42, -0.53, -0.55$, and -0.56, respectively). Therefore, upon pollution, these elements are concentrated in the light-textured alluvial soils. What is the reason for this phenomenon?

The decisive significance appears to belong to the filtration capacity of light-textured soils, where Sr, Zn, Ni, Cu, and Cr can migrate for long distances. In the heavy-textured deposits, hydrogenic pollutants are retained in the river bed. Thus, the high contamination of heavy-textured soils shows that the main factor of pollution of the fluvisols is the hydrogenic one but not the solid-phase factor. Nevertheless, in the alluvial soils, technogenic sediments also accumulate. Dangerous consequences of the accumulation of warp deposits are considered below.

Participation of deposited material in pollution of alluvial soils. Besides the hydrogenic pollution by sewage containing heavy metals and metalloids, the alluvial soils may be contaminated with elements accumulated in alluvium (warp), which is enriched with technogenic microelements. For the assessment of the contribution of the warp into the pollution of the alluvial soils, its chemical composition is compared to the composition of the soils. The coefficient of "solid phase heavy metals pollution of alluvial soils" $K_{i\ warp}$ is used:

$$K_{i\ warp} = C_{i\ warp} : C_{i\ soil},$$

where $C_{i\ warp}$ and $C_{i\ soil}$ are the contents of the i-heavy metal in the warp and their weighted averages in the alluvial soil. As $K_{i\ warp} > 1$, one can attribute the solid phase pollution of the alluvial soils to the warp accumulation. If this trend continues, the alluvium will be enriched in technogenic heavy metals mainly in its solid phase.

Table 4 presents the coefficients of the solid phase microelement pollution of the alluvial soils by microelements that vary from 0.26 for Cr in the soil of the Danilikha river floodplain to 3.95 for Zn in the soil of the Las'va river floodplain. The high mean K_{warp} values (>1) characterize the soils of the Iva (1.27) and Las'va (1.38) river floodplains. At the present time, these alluvial soils appear to be contaminated due to the warp deposit. In the soil of the Danilikha river floodplain, the mean K_{warp} value is 0.51. Therefore, this soil is polluted to a greater extent than the river warp.

Excess of the maximum permissible concentrations (MPC) and the approximate permissible concentrations (APC) of heavy metals in the fine earth. In Russia the values for the approximate permissible concentrations (APC) were accepted as higher for the loamy and clay soils than for the sandy and loamy-sandy ones (Bol'shakov, Vodyanitskii, Borisochkina et al., 1999 – Table 5). Table 2 shows that the highest heavy metal content was registered in the fine earth of the soils in the river floodplains studied. In the fluvisols of the Danilikha river floodplain, the contents of Ni and Zn were 480 and 400 mg/kg, respectively. The mineralogical analysis of the Danilikha river warp revealed that Ni and Zn containing minerals were absent (Osovetskii & Men'shikova, 2006). This means that the soils are polluted by these elements due to sewage.

In the soil of the Egoshikha river basin, the Ni and Zn contents reached 340 and 460 mg/kg, respectively. The Iva river is relatively pure, as was found earlier: the Ni and Zn contents in the soils of its floodplain did not exceed 90 and 129 mg/kg, respectively.

Table 4. Coefficients of the solid phase heavy metals pollution K_{warp} of the fluvisols

River floodplain	Ni	Cu	Zn	Cr	Pb	Rb	Sr	Zr	Average
Danilikha	0,42	0,32	0,45	0,26	0,71	0,55	0,91	0,45	0,51
Iva	1,81	1,51	1,48	1,0	1,38	1,14	0,91	0,91	1,27
Egoshikha	0,95	1,07	0,82	1,16	1,53	0,85	0,96	0,81	1,02
Las'va	1,22	1,19	3,95	0,77	0,66	1,16	1,67	0,42	1,38

Table 5. The values for approximate permissible concentrations (APC) of heavy metals in soils with different physical and chemical properties (Russia) (Bol'shakov, Vodyanitskii, Borisochkina et al., 1993)

Element	Soil group		
	Sandy, loamy-sandy	**Acid (clay, loamy-clay), pH_{KCl} < 5.5**	**Close to neutral, neutral (clay, loamy- clay), pH_{KCl} > 5.5**
Ni	20	40	80
Cu	33	66	132
Zn	55	110	220
As	2	5	10
Cd	0.5	1.0	2.0
Pb	32	65	130

Table 6. The excess of heavy metals and metalloids MPCs in the fine earth of the fluvisols

$$\mathbf{K_{i\,MPC} = C_{i\,sample} : C_{i\,MPC/APC}}$$

River floodplain	Horizon	Depth, cm	Ni	Cu	Zn	As*	Pb
Danilikha	Silt deposit	0- 2	1,8	1,2	1,7	-	1,6
	AJg	0-22	3,1	2,2	3,5	-	3,0
	G~~	22-70	5,6	4,5	4,0	-	1,9
Iva	Silt deposit	0- 2	1,4	1,6	1,4	0,4	0,5
	AJg	0-15	1,1	1,3	1,2	-	0,9
	G~~	15-30	0,7	1,0	0,9	-	0,3
	Cg~~	30-70	0,7	0,9	0,9	0,7	0,2
Egoshikha	Silt deposit	0- 2	3,4	2,1	3,2	0,4	1,8
	AJg	0-15	4,0	1,8	4,6	-	1,3
	G~~	15-30	3,7	1,6	3,8	1,1	0,5
	C2g~~	30-50	3,3	2,2	3,4	-	1,5
Las'va	Silt deposit	0- 2	0,7	0,8	6,3	0,8	0,6
	AJ	0-10	0,6	0,4	0,7	0,4	0,5
	C1g~~	10-30	0,7	0,6	1,4	1,3	3,1
	C2g,h~~	30-50	0,6	0,7	0,8	-	0,5
	C3g~~	50-80	0,7	0,8	0,6	0,7	0,4
	C4g~~	80-100	0,5	0,5	0,6	-	0,5
Mulyanka	AJg	0-10	0,9	1,0	1,0	0,4	0,8
	C1g~~	10-20	0,7	1,3	1,2	0,7	0,7
	G~~	30-70	0,7	1,2	1,4	-	0,7

* Excess of APC.

We compared the contents of heavy metals (Ni, Cu, Zn, and Pb) in the fine earth with their MPC and APC for As (Table 2). The fine earth of the soils in the Danilikha river turned out to be rather contaminated: the excess of the MPC for Ni, Cu, and Zn were 1.8–5.6, 1.2–4.5, and 1.7–4.0 (Table 6). In the soil of the Egoshikha river floodplain, the excesses for Ni, Cu and Zn were 3.3–4.0, 1.6–2.2, and 3.2–4.6, respectively. The excess for the As APC was insignificant (1.1). The fine earth of the soils in the Las'va and Mulyanka river floodplains was less polluted. In the soils of the Las'va river basin, the MPC were not exceeded: the K_{MPC} for Ni, Cu, and Zn were 0.5–0.7, 0.4–0.8, and 0.6–1.4, respectively. The K_{MPC} for Cu and Zn in the soil of the Mulyanka river floodplain were 1.0–1.3 and 1.0–1.4.

The contents of Fe-rohrensteins and the heavy metals concentration in them. The greatest content of Fe-rohrensteins (2% of the fine earth mass) was found in the gleyic light clayey alluvial soil of the Las'va river floodplain (Table 7). The typical loamy gley soil with a light-colored humus horizon contained the smallest amount (0.04%) of rohrensteins.

The Fe-rohrensteins in these soils were small in size. Assuming their shape to be cylinders, one can express their size via the volume when measuring their diameter and height (Table 7). The maximal size (60 mm^3) of the rohrensteins was in the loamy gley fluvisol of the Mulyanka river floodplain. The size of the rohrensteins from the gleyic light clayey fluvisol of the Las'va river floodplain was 30mm^3. The other rohrensteins had smaller volumes.

The main differences between the concretions are in the genesis of the rounded nodules and rohrensteins. The genesis of Fe-rohrensteins and that of Mn–Fe-nodules is different. Soil scientists differentiated Fe-Mn-nodules according to their morphological features. Published data on nodules are more abundant than the information about Fe-rohrensteins.

In the alluvial soils of Perm fine Fe-rohrensteins with holes are formed; there are no nonround Fe-Mn-nodules. This fact is inconsistent with idea that tubular nodules are characteristic of superaqual conditions and they are formed just above the groundwater table under the influence of organic ligands (oxalates, citrates, and so on) in the zone of root excretions: in rhizosphere of roots of some vegetables (Neubauer, Toledo-Duran et al., 2007; Kim, Owens et al., 2010).

As for the round Mn–Fe-nodules in the alluvial soils, they are mainly formed with the participation of microorganisms under the variable redox conditions. At the low redox-potential E_H values, mobile forms of Fe(II) and Mn(II) are formed in the soils; under increasing aerobic conditions, these elements are oxidized and round Fe–Mn nodules develop.

The inhibition of the Fe–Mn-nodule synthesis in the alluvial soils is explained by many reasons. One of them is the weak development of oxidogenesis, when Fe(II) does not take part in reduction processes responsible for the formation of nodules. The use of Mössbauer spectroscopy helped to reveal the different Fe(II) minerals in the warp deposited by small rivers in the city of Perm. Magnetite and Fe(II)-glauconite were found in the warp of the Las'va river; the coefficient of iron oxidation is $Ko = Fe^{3+} : (Fe^{3+} + Fe^{2+}) = 0.91$. Magnetite and Fe(II)-chlorite were present in the warp of the Egoshikha river floodplain; Ko decreased to 0.77. These values of Ko are rather low. In podzolic soils of the Russian Plain, this coefficient is higher—on average, 0.94–0.98.

Mössbauer spectroscopy revealed important differences in the types of iron hydroxides in the composition of rohrensteins and ortsteins. Feroxyhite (δFeOOH) is diagnosed in almost all rohrensteins even at room temperature. Feroxyhite exhibits magnetic hyperfine splitting with $H_{eff} = 243 - 290$ kOe. On the contrary, feroxyhite is not detected in ortsteins. Instead, they have much higher portion of hematite. The presence of ferrihydrite in the Fe-Mn-ortsteins can not be revealed at room temperature, as it is the hydroxide with a poor ordering. Cooling to liquid-helium temperature (4 K) is necessary for this purpose. Ferrihydrite containing hematite embryos was previously considered as "protohematite". The significant content of hematite in concretions indicates possible presence of ferrihydrite. Meanwhile, feroxyhite and ferrihydrite are formed under different redox-conditions (Vodaynitskii, Vasil'yev et al., 2009). Feroxyhite is formed by changing strongly reducing conditions in the weakly oxidizing. It is typical for highly water-logged soils, where plants-hydrofits supply oxygen to the roots. Micro-oxidation zones are formed in the rhizosphere of the roots, where Fe(II) is precipitated in the form of feroxyhite, which is a part of rohrensteins. Ferrihydrite is formed by changing moderately reducing conditions in highly oxidizing. Then Fe(II) and Mn(II) are able to oxidize. Ellipse-like nodules are formed in soils with totally variable redox regime.

Table 7. The content and size of Fe-rohrensteins and heavy metals and metalloids concentration in them

Horizon	Depth, cm	Fe-rohrensteins		Microelements, mg/kg											
		% of the soil mass	Diameter /length, cm	Ni	Cu	Zn	Ga	As	Pb	Rb	Sr	Y	Zr	Nb	Cr
The Iva river floodplain															
AJg	0-15	0,04	0,5/3	31	25	76	-	9	23	35	453	32	188	8	100
The Egoshikha river floodplain															
G~	15-30	0,8	1,5/10	441	114	314	-	-	65	48	199	18	185	10	500
The Danilikha river floodplain															
AJg	0-22	0,3	1,5/15	264	156	343	-	9	100	42	307	22	182	11	600
The Mulyanka river floodplain															
AJg	0-10	0,2	3/5	146	235	13	13	11	57	45	247	16	169	8	250
Clg~	10-20	0,3	3/5	95	108	260	-	14	25	49	339	21	183	11	140
G~	30-70	0,05	4/5	215	177	60	16	-	107	51	193	14	122	6	320
The Las'va river floodplain															
AJ	10-30	2,6	2/10	101	121	888	33	11	1530	87	278	28	220	6	100
C2g,h~	30-50	2,5	1,5/7	58	42	262	21	27	41	63	221	32	218	11	50
C3g~	50-80	1,3	1,5/5	73	74	677	6	42	59	63	222	34	248	14	60
C4g~	80-100	0,5	1,5/5	67	56	116	15	13	27	67	210	33	225	13	90

The concentration of some chemical elements in rohrensteins was higher than their contents in the fine earth of the alluvial soils. In the rohrensteins of the soil of the Las'va basin, the Pb and Zn contents were 1530 and 888 mg/kg, respectively. The contents of Cu and Cr in the Fe-rohrensteins of the soils of the Danilikha basin were 235 and 600 mg/kg (Table 7), respectively. The fixation of these elements in nodules is insignificant due to low content of Fe-rohrensteins (0.3%).

Table 8 presents the mean contents of heavy metals and their variation in Fe-rohrensteins. An elevated variation was characteristic of As, Pb, and Ga (V = 89, 230, and 110%, respectively). The high variation of these elements in the Fe-rohrensteins testifies to their technogenic origin. At the same time, the contents of some elements (Sr, Nb, and Zr) varied insignificantly (30, 30, and 18%, respectively). These elements appear to be of natural origin. Zirconium is a constituent of the lithogenic mineral zircon.

The Pb absorption is closely related to organic matter. Later on, this fact was clearly proved using synchrotron roentgen equipment. Lead has a chemical affinity to organic ligands; it forms strong bidentate complexes due to the chelatization by functional groups of aromatic rings (Heinrichs & Mayer, 1977; Boisset, Sarret, Hazemann et al., 1996). As a result of these reactions, Pb is stable in organic soils, where it is preserved for hundreds to thousands of years (Heinrichs & Mayer, 1977). Geochemists proved that the most part of copper was incorporated into complexes with organic matter (Linnik& Nabivanets, 1986). Cu^{2+} forms strong bonds with humus acids. Organic complexes of Cu^{2+} in soils were identified using synchrotron roentgen methods (Marcus & Tamura, 2002). The accumulation of metals–"organophils" (Cu and Pb) confirms the role of organic ligands in the formation of rohrensteins.

Table 8. The mean content of heavy metals and metalloids and their variations in rohrensteins

Element	Average	Range	σ	*V*, %
As	14	0-42	12,5	89
Zn	323	60-888	265	82
Ni	150	31-264	126	84
Cu	102	25-235	51	50
Pb	203	23-1530	467	230
Sr	267	193-453	81	30
Ga	10	0-33	11	110
Nb	10	6-14	3	30
Zr	194	122-248	35	18
Cr	221	50-600	194	88

In the soil of the Las'va river floodplain, fine Fe-rohrensteins were enriched in As. Arsenates, as well as phosphates and chromates, are bound in great amounts by iron hydroxides (Vodyanitskii, 2005). Stable complexes with As(III) are formed on the goethite αFeOOH surface due to the intraspheric absorption. The surface complex As(III)-goethite is of great importance in the binding of arsenates due to its high stability (Pedersen, Postma & Jakobsen, 2006). Thus, the relation of As to Fe-rohrensteins is explained by the formation of fine reactionary particles of iron hydroxides that absorb great amounts of As(III).

Since Ni is weakly bound in the Fe-rohrensteins of the polluted alluvial soils (concentration coefficient, $Cc_{Ni} = 1.5$), it is dispersed in the fine earth of the soils. The considerable part of the toxic elements was bound via exchange. This bond is weaker than their fixation in Fe-rohrensteins, where some toxicants (As, Cr, and so on) are strongly sorbed by Fe hydroxides.

Concentration of heavy metals in Fe-rohrensteins as related to the fine earth. The accumulation of heavy metals in soil concretions is usually compared with their contents in the fine earth using the coefficient of concentration Cc:

$$Cc = C_{nodule} : C_{fine\ earth}.$$

However, some difficulties emerge using these coefficients. Since the contents of some elements (As and Nb) are low, Cc for some rohrensteins turns into an infinite value. In order to avoid this, for the fine earth samples with the element content lower than the detection limit, we assumed the concentrations to be 3 mg/kg for As (the limit for As detection by the roentgen–fluorescence method is 4 mg/kg) and 5 mg/kg for Nb (the detection limit is 6 mg/kg). This assumption provides minimal values of Cc. The results of calculating Cc are given in Table 9; statistical indices of the distribution of Cc in the rohrensteins are given in Table 10.

Variation coefficient of the coefficient Cc for different elements varied from 14 to 112% (Table 10). The variation of Cc is related to different responses of elements to diverse physicochemical and other conditions favoring the formation of Fe-rohrensteins. As elements are sensitive to changes in the medium of the soil genetic horizons, variation of their Cc will be high and vice versa. High coefficients of variation characterized the Cc for Zn, Pb, Ga, and Cr (V > 100%). Probably, these metals are sensitive to the variation of the physicochemical conditions; they are accumulated in Fe-rohrensteins to a different extent. Such metals as Sr, Zr, Rb, and Y (V < 30%) were not sensitive to changes in the physical and chemical environments in the horizons of the alluvial soils.

According to the mean values of Cc (Table 10), the elements studied can be divided into 3 groups. Elements concentrated in Fe-rohrensteins are As, Zn, Ni, Cu, Cr, and Pb (Cc > 1). Elements not concentrated in Fe-rohrensteins are Sr, Nb, Ga, and Y (Cc ≈ 1). Rohrenstein-depleted elements are Zr and Rb (Cc < 1). Thus, soil concretions accumulate dangerous elements; in rohrensteins, they are concentrated to a different degree. The maximal As content was found in rohrensteins due to the enrichment of iron hydroxides (the mean Cc = 3.2). Sr, Nb, Ga, and Y are not precipitated on adsorption microbarriers of rhizosphere. The depletion of rohrensteins in Zr and Rb depends on their accumulation in the fine earth, probably in the coarse fractions.

Table 9. Coefficients of heavy metals and metalloids concentration Cc in Fe-rohrensteins

$$Cc = C_{rohrensteins} : C_{fine\ earth}$$

River floodplain, horizon		Ni	Cu	Zn	Ga	As	Pb	Rb	Sr	Y	Zr	Nb	Cr
Danilikha, AJg		1.0	1.3	1.0	0.0	3.0	1.1	0.8	1.1	1.3	0.7	2.2	1.0
Iva, AJg		0.3	0.4	0.7	0.0	3.0	0.8	0.7	1.7	1.6	0.7	1.6	0.5
Egoshikha, G$^{\sim}$		1.4	1.3	0.8	0.0	0.0	4.1	0.9	0.7	0.7	0.7	2.0	1.2
Las'va	AJ	1.7	3.4	6.5	1.9	0.8	16.3	1.4	1.0	1.0	0.7	0.4	1.0
	C2g,h$^{\sim}$	1.1	1.1	3.4	1.6	9.0	2.6	1.1	1.0	1.3	0.7	0.8	0.5
	C3g$^{\sim}$	1.3	1.6	11.1	0.5	6.0	5.4	1.2	0.9	1.3	0.6	1.3	0.3
	C4g$^{\sim}$	1.7	2.2	1.9	1.0	4.3	1.7	1.1	0.9	1.1	0.5	1.2	0.9
Mulyanka	AJg	1.9	2.6	2.3	3.2	2.7	2.5	0.9	0.8	0.7	0.6	0.8	5.0
	C1g$^{\sim}$	1.5	1.5	2.3	0.0	2.0	1.2	0.9	0.2	0.8	0.8	1.0	2.0
	G$^{\sim}$	3.4	2.7	0.4	1.0	1.0	5.3	0.9	0.7	0.6	0.5	0.5	4.6

Table 10. Statistical indices for the coefficients of heavy metals and metalloids concentration (Cc) in Fe-rohrensteins

Element	Average	Max	Min	σ	*V*, %
As	3.2	3.0	0.0	2.7	84
Zn	3.0	6.5	0.4	3.3	110
Ni	1.5	3.4	0.3	0.8	53
Cu	1.8	3.4	0.4	0.9	50
Pb	4.1	16.3	0.8	4.6	112
Sr	1.0	1.7	0.8	0.3	30
Nb	1.2	2.2	0.4	0.6	50
Zr	0.7	0.8	0.5	0.1	14
Ga	1.0	3.2	0.0	1.1	110
Rb	1.0	1.4	0.7	0.2	20
Y	1.0	1.6	0.0	0.3	30
Cr	1.7	5.0	0.3	1.7	100

According to Perel'man (1975) Zn, Cu, Ni, and Pb are active water migrants. They are mobile in acid and immobile in neutral and alkaline media. As a result, they are precipitated at the alkaline barrier. But almost all the soils studied have a neutral reaction. These metals and As mainly precipitate at the geochemical microbarriers around the concretions. Rohrensteins are an example of an adsorption microbarrier and oxidative microbarriers for heavy metals that are bound by organic ligands and iron hydroxides. Rhizosphere microbarriers help in Fe-rohrensteins forming.

The geochemical importance of redox-microbarriers is determined by the content of concretions and the concentration of heavy metals in them. The microbarriers were found to affect the geochemistry only of Pb and Zn in the alluvial soil of the Las'va river floodplain. In the upper horizon of this soil, the shares of Pb and Zn were 42 and 17% of their contents in the fine earth, respectively. In the other soils investigated, the effect of redox microbarriers was insignificant because of the low share of nodules.

The rohrensteins, as well as some other concretions, are not reliable depositories of toxic microelements. Under strong changes in the water and air conditions, Fe-rohrensteins and especially the fine ones may be destroyed. Upon the complete destruction of rohrensteins, only in the upper horizon of the soil in the Las'va river basin will the contents of Pb increase from 94 to 134 mg/kg (MPC~ 30 mg/kg) and the Zn concentration from 137 to 160 mg/kg (MPC~100 mg/kg). In the other fluvisoils studied, the contribution of the heavy metals deposited in the Fe-rohrensteins upon their full destruction will not influence the pollution of the fine earth.

Barium, lanthanum and cerium in alluvial soils. Meadow-bog alluvial soils were studied on floodplains of small rivers and the Kama River in Perm. Overall, 16 samples of the fine earth and iron- and iron-manganic concretions (rohrensteins and nodules) from these soils were examined (Table 11). The content of Ba in the fine earth of soils varies from 406 to 527 mg/kg, while its content in the concretions varies greatly (from 588 to 2848 mg/kg). The coefficient of Ba accumulation in the concretions also varies considerably: from 1.2 to 6.0. Light-colored formations with a size of about 20 μm were identified on internal surfaces of rohrensteins. The microanalyzer showed that these formations of composed of barite ($BaSO_4$). Barium is not typical metal of the soils of forest landscapes, so its accumulation in the iron concretions is related to the soil pollution. In the rohrensteins from the alluvial soils, Ba is mainly accumulated as barite particles precipitated on the active matrix of the concretions.

In the gleyed alluvial agrozem on the floodplain of the Mulyanka River polluted by wastewater, Ba redistribution in the soil profile takes place. In the fine earth, the Ba content is relatively stable (421–474 mg/kg). In the nodules, the concentration of Ba varies considerably and reaches its maximum (2840 mg/kg) in the C4g,t horizon at a depth of 107–137 cm. In this coarse-textured alluvial soil, Ba migrates easily to a considerable depth and is concentrated in the nodules.

The La content in the fine earth varies from 28 to 41 mg/kg. In the concretions, it varies from 14 to 108 mg/kg, and the coefficient of La accumulation in the concretions varies from 0.3 to 3.5. In the gleyed alluvial agrozem studied on the floodplain of the Mulyanka River polluted by wastewaters, including those from the petroleum refinery, the content of La in the fine earth remains practically stable (30–34 mg/kg). Its content in the nodules changes from

86 mg/kg in the deepest C5g horizon to 108 mg/kg in the C3 horizon at a depth of 75–107 cm. The technogenic lanthanum migrates to a considerable depth in this coarse-textured soil.

The Ce content in fine earth varies from 38 to 60 mg/kg. In the concretions, it varies from 16 to 324 mg/kg. The coefficients of Ce accumulation in the concretions vary from 0.3 to 6.9. In the gleyed alluvial agrozem on the floodplain of the Mulyanka River, the Ce content in the fine earth remains almost stable throughout the soil profile (45–48 mg/kg). In the Fe-Mn-nodules, it changes from 150 mg/kg in the deepest C5g horizon to 324 mg/kg in the C2 horizon at a depth of 49–75 cm. Being most active among lanthanides, the technogenic Ce migrates down the soil profile to a relatively shallow depth and is fixed in the Fe-Mn-ortsteins. This is also confirmed by other data. Thus, in the tropical laterites of Cameron, lanthanides are removed from the top iron-enriched horizons and are accumulated in the deeper layers. Cerium is deposited just below the eluvial horizon, and other lanthanides are accumulated in the deeper horizons (Braun, Viers et al., 2005). Similar differentiation of lanthanides is seen in the soil profiles on granodiorites in the New Southern Wales (Australia).

Table 11. The contents of Ba, La, and Ce (mg/kg) in the fine earth and nodules of fluvisols, coefficients of metal accumulation in the concretions (Cc), and the Ce-to-La ratios

Horizon; depth, cm	Substrate	Ba	Cc(Ba)	La	Cc(La)	Ce	Cc(Ce)	Ce/La
Humus-gley fluvisol on the Obva River floodplain								
C2g, 37-75	fine earth	527		37		58		1.6
	rohrensteins	623	1.2	25	0.7	38	0.6	1.5
G~~, 75-90	fine earth	523		41		60		1.5
	rohrensteins	673	1.3	14	0.3	16	0.3	1.1
Fluvisol on the Obva River floodplain								
C2~~, 20-27	fine earth	410		28		41		1.5
C6~~, 71-78	fine earth	406		26		38		1.5
Humus-gley fluvisol on the Kama River floodplain								
G~~, 31-55	fine earth	452		38		57		1.5
	nodules	715	1.6	56	1.5	191	3.3	3.4
Gleyed agrozem on the Mulyanka River floodplain								
C2~~, 49-75	fine earth	430		31		47		1.5
	nodules	1957	4.5	104	3.3	324	6.9	3.1
C3~~, 75-107	fine earth	421		31		46		1.5
	nodules	2120	5.0	108	3.5	302	6.6	2.8
C4g,t~~, 107-137	fine earth	474		34		45		1.3
	nodules	2840	6.0	100	2.9	243	5.4	2.4
C5g~~, >137	fine earth	441		30		48		1.6
	nodules	588	1.3	86	2.9	150	3.1	1.7
Soil clarke*		500		26		49		1.8
Average			3.0		2.2		3.7	

* according to Kabata-Pendias & Pendias (1985).

It should be noted that the contents of La and Ce in the studied soils of Perm region remain below the clarke values for the pedosphere despite the soil pollution (Taunton, Welch et al., 2000).

Different mechanisms of the formation of rohrensteins and nodules are responsible for the difference in the accumulation coefficients of the three elements. Rohrensteins formed on the floodplain of the unpolluted Obva River are characterized by the weak Ba accumulation (C_{ac} = 1.2–1.3) and the depletion of La and Ce (C_{ac} = 0.3–0.7). Organic ligands in the rohrensteins are mainly spent for iron fixation. This is reflected in the wide Fe-to-Mn ratio in them (25–100). Iron–manganic nodules are formed under conditions of alternating redox-regime and higher average E_h; they are enriched in all the three elements. They are characterized by a more even accumulation of elements and a lower Fe-to-Mn ratio (1.4–12). Cerium exhibit variable valence (III, IV) in contrast to lanthanum and majority of lanthanides. This brings it to manganese. In soils cerium behaves as manganofil. Not surprisingly, cerium is accumulated in ellipse-like Fe-Mn-nodules.

The degree of pollution of the river is also important. In the floodplain soil of the strongly polluted Mulyanka River, the accumulation coefficient of Ba in concretions varies from 1.3 to 6.0, the accumulation coefficient of La reaches 2.9–3.5, and the accumulation coefficient of Ce is as high as 3.1–6.9. Polluted waters of the Malaya Mulyanka River enter the deep Kama River and are diluted. As a result, concretions in the soils on the Kama River floodplain have lower concentrations of the studied elements; the coefficients of their accumulation are equal to 1.6 for Ba, 1.5 for La, and 3.3 for Ce.

According to the average coefficients of element accumulation in the concretions, the studied elements form the following sequence: Ce(3.7) > Ba(3.0) > La(2.2). The degree of an element accumulation in the concretion depends on its sensitivity to changes in the redox regime, sorption capacity, and the capacity of the element to form stable complexes with organic ligands. The most pronounced accumulation of Ce in the concretions is explained by its sensitivity for changes in the redox regime. Intermediate position of barium is explained by the precipitation of barite crystals on the active matrix of the concretions; Ba is a manganophilic element. A relatively low accumulation of La in the concretions is explained by its physicochemical inactivity.

Let us compare the Ce-to-La ratios in the fine earth and concretions. In the fine earth, this ratio is practically constant and averages to 1.5, which is close to the clarke value (1.9). In the concretions, it varies from 1.1 to 3.4 and averages to 2.3. A wider Ce-to-La ratio in the concretions signifies that Ce accumulation in them is more redox-active as compared with La.

Conclusion

1. The alluvial soils of the city of Perm are polluted by heavy metals of hydrogenic origin because of the unpurified sewage water entering them. The degree of pollution of these soils differed from the contamination of the urbanozems of the city. In the urban alluvial soils, the mean Pb content is twice lower, and the contents of Zn and Ni are 1.5 and 4 times higher, respectively, than in the urbanozems. These facts attest that the pollution of these alluvial soils is of hydrogenic origin.

In the fine earth of the alluvial soils, a positive correlation between the content of clay particles and the contents of Sr, Zn, Ni, Cu, and Cr is absent. The strong pollution of the light-textured soils attests that, nowadays, the main pollution source for the alluvial soils is sewage water, rather than warp deposited. However, with time, in the floodplains of some rivers, the alluvial deposition will significantly affect the pollution of the alluvial soils. In the floodplain of the Danilikha river, the alluvial soils are polluted to a greater extent than the warp accumulated from the river water.

2. In the alluvial soils, fine Fe-rohrensteins are formed. Some elements are concentrated in the rohrensteins, and some others are not concentrated in them or are found in low concentrations. The first group comprises As, Zn, Ni, Cu, Cr, and Pb; the second one is represented by Sr, Nb, Ga, and Y; and the third group contains Zr and Rb.

   The concentrations of some chemical elements in the Fe-rohrensteins are much greater than those in the fine earth. The Pb and Zn contents in the rohrensteins of the soil of the Las'va river basin are 440 and 890 mg/kg, respectively. In the upper horizon of this soil, the Pb and Zn contents are 42 and 17% of their concentration in the fine earth, respectively.

   Since some part of the microelements is precipitated at the geochemical microbarriers around concretions (rohrensteins), it is removed from the biological cycle. These microbarriers differ from permanent geochemical barriers by their temporary character. The rohrensteins may be considered as pedogeochemical microanomalies. Unlike local anomalies, whose area is one to tens of square kilometers, the microanomalies scattered throughout the fine earth are of tens to hundreds of square meters. The size of a microanomaly is determined by the content of newformations (%) and by the concentration of microelements in them as well.

3. In the alluvial soils of Perm, Ba is fixed in rohrensteins as barite ($BaSO_4$) of, probably, technogenic nature. Ba migrates intensively in the soil profile and is fixed in iron–manganic nodules. The contents of La and Ce in the fine earth of polluted soils remain practically stable, and lanthanides are accumulated in the concretions: the accumulation coefficient reaches 3.5 for La and 6.9 for Ce.

   In the profile of the fluvisols polluted by wastewaters from petroleum refinery, La migrates deeper into the soil profile than Ce. The latter element is more active and is sorbed in iron–manganic nodules in the topsoil. According to the average coefficients of element accumulation in the concretions (Fe–Mn-nodules and Fe-rohrensteins), the studied elements form the following sequence: Ce(3.7) > Ba(3.0) > La(2.2).

REFERENCES

Boisset M. C., Sarret G., Hazemann J. L. et al. (1996). Direct Determination of Lead Speciation in Contaminated Soils by EXAFS Spectroscopy. *Environ. Sci. Technol.*, V. *30*. 1540–1552.

Bol'shakov V. A., Vodyanitskii Yu. N., Borisochkina, T. I. et al. (1999). *Methodological Recommendations on the Assessment of the Contamination of Urban Soils and Snow Cover with Heavy Metals*. Moscow. 31 p. [in Russian].

Braun J. J., Viers J. & Dupre B. (2005). Solid/Liquid REE Fractionation in the Lateritic System of Goyoum, East Cameroon: The Implication for the Present Dynamics of Soil Covers of the Humid Tropical Regions. *Ceochim. Cosmochim. Acta.*, V. *62*. 273–299.

Dobrovol'skii V. V. (2003). *Fundamentals of Biogeochemistry*. Academia. Moscow. 400 p. [in Russian].

Eremchenko O. Z. & Moskvina N. V. (2005). The Properties of Soils and Technogenic Surface Formations in the Multistory Districts of Perm City. *Eur. Soil Sci.*, V. *38*. No. 7. 688–694.

Heinrichs H. & Mayer R. (1977). Distribution and Cycling of Major and Trace Elements in Two Central European Forest Ecosystems. *J. Environ. Qual.*, V. *6*. 402–407.

Kabata-Pendias A. & Pendias H. (1985). *Trace Elements in Soils and Plants*. CRC. Boca Raton.

Kim K., Owens G. & Kwon S. (2010). Influence of Indian mustard (Brassica juncea) on rhizosphere soil solution chemistry in long-term contaminated soils: A rhizobox study. *Journal of Environmental Science*. V. *22*. 98 – 105.

Linnik P. N. & Nabivanets B. I. (1986). *Migration of Metals in Fresh Surface Waters. Gidrometeoizdat.* Leningrad. 270 p. [in Russian].

Marcus M. A. & Tamura N. (2002). Quantitative Speciation of Heavy Metals in Soils and Sediments by Synchrotron X-Ray Techniques. *Applications of Synchrotron Radiation in Low-Temperature Geochemistry and Environmental Science: Reviews in Mineralogy and Geochemistry*. Washington. DC. V. *49*. 341–428.

Neubauer S., Toledo-Duran G., Emerson D. & Megonigal J. (2007). Returning to Their Roots: Iron-Oxidixing Bacteria Enhance Short-Term Plaque Formation in the Wetland-Plant Rhizosphere. *Geochemical Journal*. V. *24*. 65 – 73.

Osovetskii B. M. & Men'shikova, E. A. (2006). Natural-Technogenic Sediments. *Perm*. 208 p. [in Russian].

Pedersen H. D., Postma D. & Jakobsen R. (2006). Release of Arsenic Associated with the Reduction and Transformation Iron Oxides. *Geochim. Cosmochim. Acta.*, V. *70*. 4116–4129.

Perel'man A. I. (1975). *Geochemistry of Landscapes*. Vysshaya Shkola. Moscow. 342 p. [in Russian].

Saet Yu. N., Revich B. A., Yanin E. P. et al. (1990). *Geochemistry of the Environment*. Nedra. Moscow. 335 p. [in Russian].

Shchukova I. V. (2005). *Extended Abstract of Candidate's Dissertation in Geology and Mineralogy*. Perm. 23 p. [in Russian].

State and Protection of the Environment in the Perm Oblast in 2004. (2005). 65 p. Perm. [in Russian].

State (National) Report on the State and Use of Lands in Russian Federation in 1999. (2000). Otkrytye Sistemy. Moscow. [in Russian].

Taunton A. E., Welch S. A. & Banfield J. F. (2000). Microbial Control on Phosphate and Lanthanide Distributions during Granite Weathering and Soil Formation. *Chem. Geol.* V. *169*. 371–382.

Vodyanitskii Yu. N. (2005). *Study of Heavy Metals in Soils. Moscow*. 109 p. [in Russian].

Vodyanitskii Yu. N., Vasil'yev A. A. & Vlasov M. N. (2008). Hydrogenic Heavy Metals Pollution of Alluvial Soils of the City of Perm. *Eurasian Soil Science.*, V. *41*. No 11. 1238-1247.

Vodyanitskii Yu. N., Vasil'yev A. A., Vlasov M. N. & Korovushkin V. V. (2009). The Role of Iron Compounds in Fixing Heavy Metals and Arsenic in Alluvial and Soddy-Podzolic Soils in the Perm Area. *Eurasian Soil Science.*, V. *42*. No 7. 738 – 749.

In: Floodplains
Editor: Enner Herenio Alcântara
ISBN: 978-1-62808-998-1

Chapter 8

THE ROLE OF OVERBANK DEPOSITION IN FLOODPLAIN FORMATION AND CATASTROPHIC FLOODS IN FLOODPLAIN DESTRUCTION IN SOUTH EASTERN AUSTRALIA

Wayne D. Erskine*
School of Environmental & Life Sciences,
The University of Newcastle – Ourimbah Campus
and Environmental Research Institute of the Supervising Scientist

ABSTRACT

The transformation of mud-rich floodplains with isolated large pools, called 'chain of ponds', to low sinuosity, active sand-bed streams has occurred on many streams following upstream channel widening, bank erosion, channel incision and gully erosion in Mesozoic sedimentary basins of south eastern Australia. The present research was undertaken on Six Mile Swamp Creek and Double Swamp Creek in the Clarence-Moreton Basin and on Blaxlands Arm and Fernances Creek in the Sydney Basin. Chain of ponds generally construct mud-rich floodplains while low sinuosity, sand-bed streams build sandy floodplains dominated by floodouts, abandoned channels, natural levees and splays. Avulsions are relatively common on low sinuosity, sand-bed streams where overbank deposition has been active, producing perched channels. The main mechanism of channel avulsion is switching from elevated parts of the floodplain to connected low, vegetated depressions. Floodplain stripping by catastrophic floods removes linear strips of both muddy and sandy floodplains. Isolated floodplain stripping is one mechanism of pond excavation. Slope stability thresholds are also involved in stripping but on a much smaller scale than catastrophic floods. Discontinuous channels develop oversteepened reaches below the intersection point of a floodout which initially erode small scour pools which can enlarge into ponds and eventually into gullies. Oversteepened reaches are caused by rapid deposition with grasses and sedges stabilising the recent sediments.

---

* Corresponding author: ENVIRONMENTAL RESEARCH INSTITUTE OF THE SUPERVISING SCIENTIST, GPO Box 461, Darwin NT 0801, Australia Tel: +618 89201150 Fax: +618 89201195, Email: Wayne.Erskine@environment.gov.au.

Keywords: Chain of ponds; intermediate floodouts; sand slugs; channel avulsion; gully erosion; channel widening; bed aggradation; intersection point

1. INTRODUCTION

The most well known model of floodplain formation and destruction is the lateral accretion model. Channel migration of the outer bank of bends creates a void which is backfilled with point bars on the inside bank (Mackin, 1937; Wolman and Leopold, 1957; Leopold and Wolman, 1960). Overbank deposition produces a relatively thin surficial veneer on the floodplain because rates of lateral migration can be rapid, overbank flow velocities can be relatively high and suspended sediment concentrations in overbank flow can be low (Wolman and Leopold, 1957). We now know that this is but one of a number of potential models, many of which can occur simultaneously on the same river (Nanson and Croke, 1992). For example, concave bench accretion (Nanson and Page, 1983) or counterpoint bar deposition (Smith, 1987) also occurs on lateral migrating, meandering channels (Hickin, 1979; 1986) along with abandoned channel accretion (Page and Mowbray, 1982). However, in some cases, channel impingement on non-erodible valley sides at migrating, abrupt-angle bends is required for the development of concave benches (Carey, 1963; 1969) and this can occur on rivers without a meandering planform.

Many rivers in New South Wales exhibit floodplains with landforms produced by overbank deposition (Scholer, 1976; Nanson and Young, 1981; Erskine and Melville, 1983a; Nanson, 1986; Erskine, 2008; Erskine and Borgert, 2013). Indeed, natural levees on the Hawkesbury River have been recognised as some of the highest examples in the world (Scholer, 1976) and similar well developed levees also occur on the Macdonald and Colo rivers and Webbs Creek which are tributaries of the Hawkesbury River. Sim et al. (2010) used extensive levee deposits on the Hawkesbury River to reconstruct flood histories.

Chain of ponds are poorly defined fluvial landforms but it is generally agreed that they consist of large pools or waterholes excavated into fine-grained floodplains and were relatively common in south eastern Australia at the time of first European settlement (Eyles, 1977a; 1977b; Mack, 1983; Gallagher, 1984; Bannerman, 1987; Erskine, 1994a; 1994b; 2008; Erskine and Melville, 2008). From early descriptions, connecting channels between the ponds were usually small and well vegetated. In some cases, channels were completely absent between ponds. Concentric zones of aquatic and emergent vegetation were often present in ponds (Eyles, 1977a) although their significance has never been determined. Ponds were recognised as excellent fish habitat and were often called 'chain of mullet holes', 'chain of perch holes' and 'chain of eel holes' on historical maps and plans (Erskine, 2008). Historical gullying and channel incision have destroyed many chain of ponds either by erosion or by burial with coarse-grained sediment generated by upstream erosion (Eyles, 1977a; 1977b; Erskine, 1994a; 1994b; 2008; Erskine and Melville, 2008).

Similar features to chain of ponds have been described in other countries. For example, Bryan (1920) described in detail charcos or natural water bodies in adobe flats in the Papago Country. Charco is a Spanish word for a pool of standing or stagnant water. In other parts of America, charcos are called mudholes or mud tanks or mud tanques (Bryan, 1920). Charcos are found as single pools or as a series of pools along streams which deposit fine-grained material (Bryan, 1920). A charco or mud tank was shown in the Kevin Costner movie, 'Dances with Wolves', and so some still survive today. Tanques were also described by Leighly (1936) in New Mexico and by Mosley (1972) in Colorado. The tanques or basins of Mosley (1972) are similar to the flow-aligned scour holes which have been recorded on wet fans in south eastern Australia (Scott and Erskine, 1994). Some cienegas of the American south west also contain mud tanks or pools (Hendrickson and Minckley, 1984; Cooke and Reeves, 1976). The loss of the pools has endangered many desert fishes of the American south west (Hendrickson and Minckley, 1984). The dambos of Africa often contain ponds also.

The alternation of muddy and sandy floodplains has been recorded in the stratigraphic record of many rivers in south eastern Australia where the muddy floodplains are associated with chain of ponds and the sandy floodplains are associated with active, incised, sand-bed streams (Melville and Erskine, 1986; Erskine, 2008; Erskine and Melville, 2008). The purpose of this chapter is to present new evidence of alternating muddy and sandy floodplains since first European settlement. Like the arroyos of the American south west (Melton, 1965; Cooke and Reeves, 1976; Hendrickson and Minckley, 1984) and incised channels elsewhere in America (Schumm et al., 1984), post-settlement gully erosion and channel incision are implicated as the destroyers of muddy floodplain and the resultant sandy channels as the agent of formation of the sandy floodplains in south eastern Australia. However, the types of chain of ponds and muddy floodplains, and channel and floodplain response to historical channel erosion is far more complex and diverse than acknowledged in the literature. This chapter presents the results of four detailed case studies to illustrate this great geomorphic and sedimentologic variability.

2. Study Areas

Four floodplains in two Mesozoic sedimentary basins in New South Wales, Australia were investigated in detail. Six Mile Swamp Creek is a tributary of Bungawalbin Creek in the Richmond River basin near Casino and Double Swamp Creek is a tributary of the Clarence River near Grafton in northern NSW (Figure 1). Blaxlands Arm and Fernances Creek are tributaries of upper Wollombi Brook near Wollombi in the Hunter Valley, NSW (Figures 1 and 2).

The study reach on Six Mile Swamp Creek extends for 16.5 km from Benders Dip to Clearfield (Figure 3). The basin area at Carters Bridge is 40.4 km2 and at the Old Tenterfield Road crossing is 59.5 km^2. The study reach on Double Swamp Creek is 5 km long including a discontinuous valley-bottom gully and its downstream depositional zone (Figure 4). The total basin area is 11.3 km^2 (Figure 4). Both study areas are located in the Clarence-Moreton Basin which is an extensive intracratonic sedimentary basin up to 4000 m thick deposited on Palaeozoic basement (McElroy, 1962; Barnes and Willis, 1989; Wells and O'Brien, 1994). Both drainage basins are dominated by two formations, Kangaroo Creek Sandstone and Grafton Formation (McElroy, 1962; Barnes and Willis, 1989; Wells and O'Brien, 1994). The fluviatile late Jurassic Kangaroo Creek Sandstone outcrops in the headwaters of the Six Mile Swamp Creek basin and extensively throughout most of the Double Swamp Creek basin. It is a distinctive medium to coarse grained, white to cream, saccharoidal sandstone with minor siltstone, shale and quartz pebble sandstone (Wells and O'Brien, 1994; Willis, 1994) and exhibits planar tabular cross-bed sets with current directions indicating derivation from the south (Willis, 1994). The formation is flat lying and forms prominent cliffs and scarps. The predominantly fluviatile Grafton Formation outcrops in the lower Six Mile Swamp Creek basin and along the eastern edge of the Double Swamp Creek basin. It comprises about 100-250 m of clayey siltstone and soft sandstone, with minor claystone and coal, and is commonly highly weathered and lateritised (Wells and O'Brien, 1994; Willis, 1994). Burger (1994) concluded that the Grafton Formation is not younger than late Jurassic on palynological data.

The Six Mile Swamp Creek basin is largely part of an extensive lowland. The main land uses in the basin are grazing on freehold land, forestry in State Forests and plantation forestry on freehold land. The Double Swamp Creek basin is mainly a dissected sandstone plateau with lowlands on the eastern and lower basin. Grazing, fruit production, forestry and quarrying are the main land uses.

The basin area of upper Wollombi Brook at Wollombi is 341 km^2. Blaxlands Arm is a 27.4 km^2 tributary of upper Wollombi Brook and flows into Wollombi Brook at Yallambie (Figure 2). Fernances Creek is a 13.8 km^2 tributary of upper Wollombi Brook and flows into Wollombi Brook at Murrays Run (Figure 2). Both study areas are located in the Permo-Triassic Sydney Basin where Triassic sandstones outcrop. Both basins are comprised largely of a thick cap of Triassic Hawkesbury Sandstone with the Triassic Terrigal Formation outcropping in the valleys (Galloway, 1967; McNally, 1981; 1995). Hawkesbury Sandstone is composed of lenticular, cross-bedded and massive, quartz-rich, fine to very coarse sandstone units up to 15 m thick with sporadic, massive to well-laminated shale lenses, usually less than 5 m thick (McNally, 1981). The Terrigal Formation mainly comprises interbedded siltstones and sandstones with two persistent sandstone horizons (Hanlon *et al.*, 1953; Galloway, 1967; McNally, 1981; 1995). There is a sandstone-sandstone boundary between the two formations in the Wollombi Brook basin (Rose, 1958; McElroy, 1958). Wollombi Brook and its tributaries have dissected an extensive sandstone plateau, forming a 200 to 300 m deep valley with steep side slopes. The dominant land uses are national parks (Yengo National Park), state forest and grazing on the floodplains and lower slopes on freehold land.

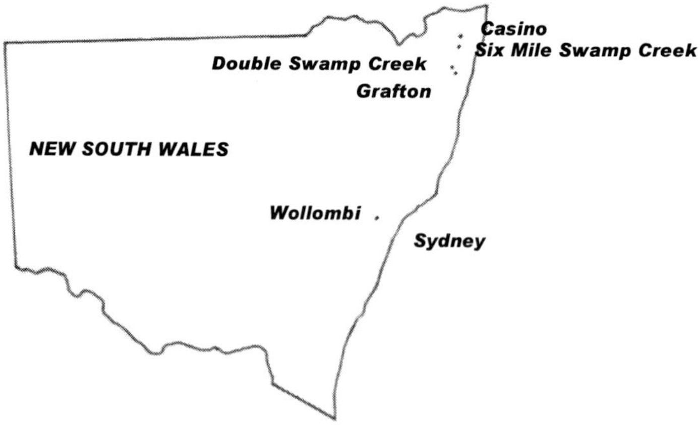

Figure 1. Location of the two Mesozoic sedimentary basins in NSW investigated in this chapter. The Clarence Moreton Basin occurs between Casino and Grafton and the Sydney Basin is centred on Sydney and Wollombi.

Figure 2. The upper Wollombi Brook drainage basin showing the two floodplains investigated here, namely Blaxlands Arm and Fernances Creek.

Mean annual rainfall is 1097 mm and median annual rainfall is 1057 mm at Casino Airport (Station No. 58063) which is the closest long-term station to Six Mile Swamp Creek. Mean and median annual rainfall at Grafton Olympic Pool (Station No. 58130), the closest long-term station to Double Swamp Creek, are 1105 and 1085 mm, respectively. The closest long-term station to Blaxlands Arm and Fernances Creek is Yallambie (Mt Auban) (Station No. 61205) and the mean and median annual rainfall are 830 and 816 mm, respectively. Monthly rainfall distribution at all three stations is similar with the highest mean and median monthly rainfall between December and March and the lowest, between July and September.

3. Six Mile Swamp Creek

According to the land owner of 'Manningville' (Mr Kinsley, personal communication, 1996), Six Mile Swamp Creek in the study reach was only a wet depression lined with 'black soil' before February 1954. Large waterholes were irregularly spaced along the valley floor and connected by a small vegetated depression. Figure 5 shows the original land survey plan of Portion 35 immediately downstream of the Old Tenterfield Road crossing of Six Mile Swamp Creek (Figure 3). It shows a 'deep lagoon' equivalent to chain of ponds which were formerly more widespread throughout south eastern Australia (Eyles, 1977a; 1977b). The deep lagoon was set in a swampy valley floor with thick tea-tree (*Melaleuca* spp.) and oak (*Casuarina* spp.) trees. A cattle and water reserve (C. & W.R. 237 in Figure 5) was notified on 15 June 1875 over a small area sandwiched between Portions 15, 34 and 35 (Figure 5). This reserve was proclaimed to enable access to reliable water in the deep lagoon for stock using the Old Tenterfield Road. Lagoons, waterholes and ponds were also shown at various locations in the study reach on early land survey plans (for example, Ewing, 1875; Wearne, 1905a; 1905b) and some are still present on the floodplain next to the present channel (Figure 3).

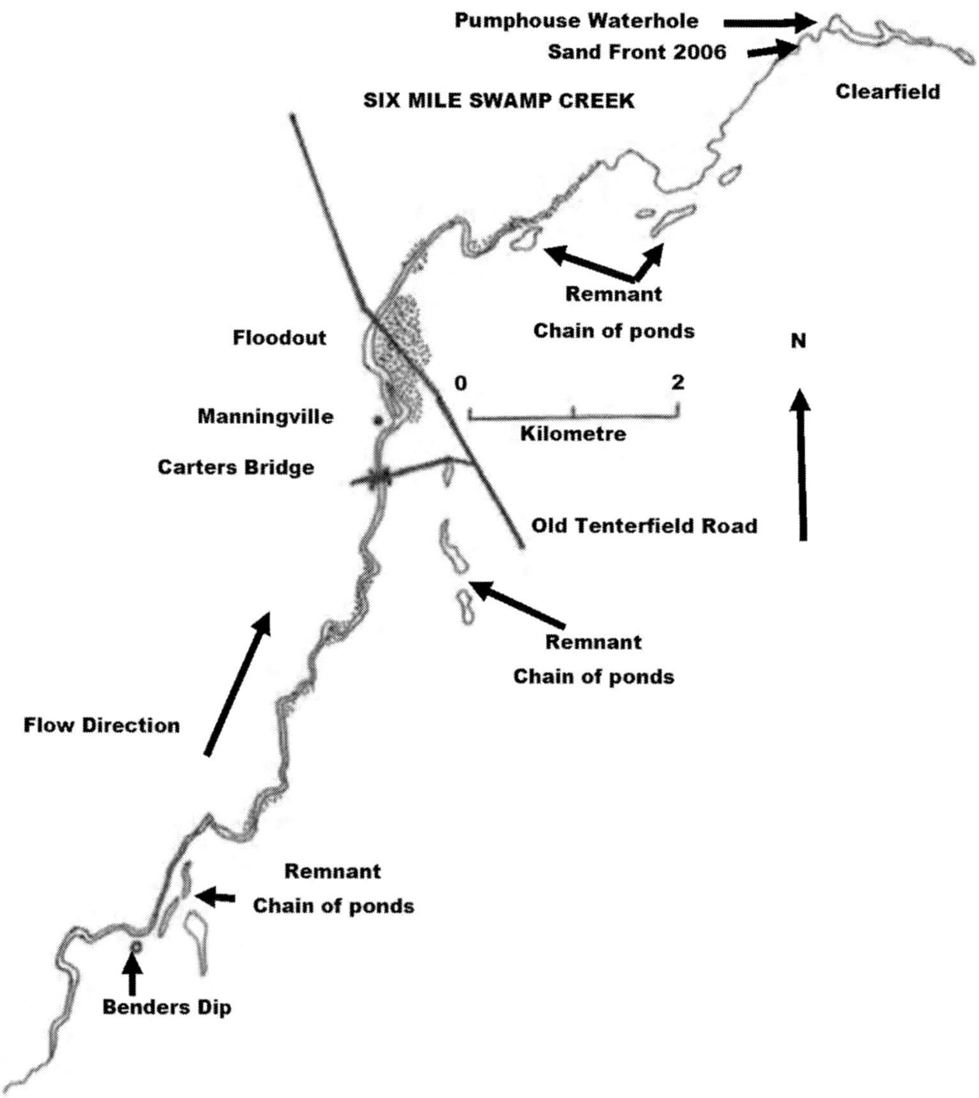

Figure 3. Study reach on Six Mile Swamp Creek.

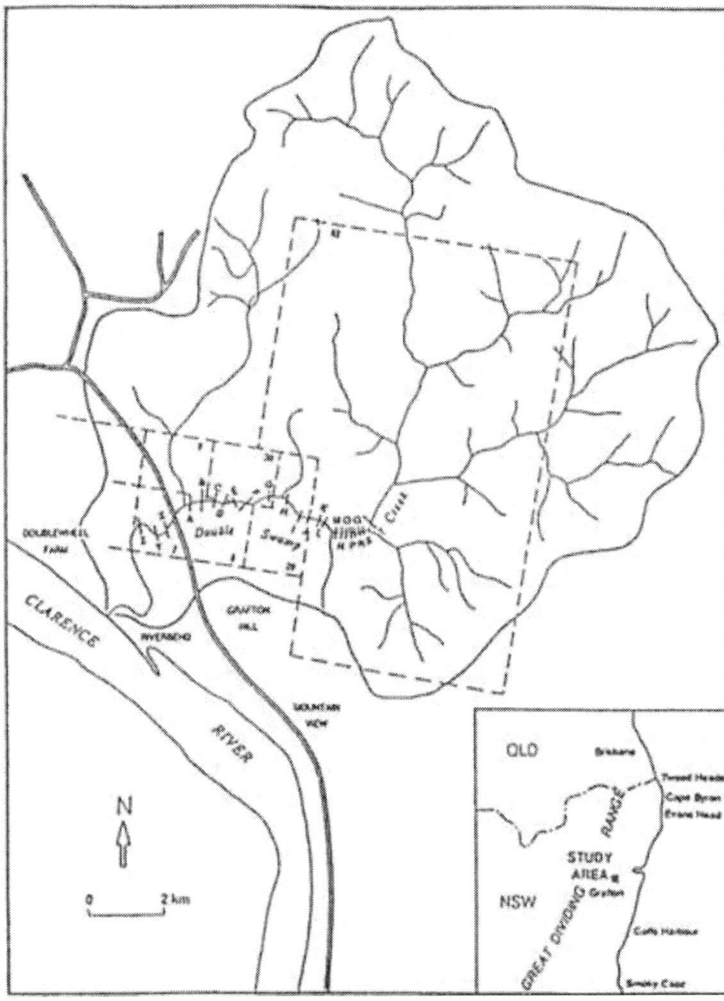

Figure 4. Double Swamp Creek basin showing location of 23 cross sections in the depositional zone of the discontinuous gully. Dashed lines show boundaries of surveyed portions whose plans were obtained for this study.

The February 1954 flood was a catastrophic event for which no rainfall and streamflow data exist for Six Mile Swamp Creek. The nearest rainfall station with a long-term record is Casino Airport (Station No. 58063). For 21-22 February 1954, 350.5 mm were recorded which is the largest 2-day rainfall on record from 1879 to 2012. However, there was a series of large floods in rapid succession between 1945 and 1956 on the Richmond River with 16 floods higher than 4.57 m recorded at Coraki (Harrison et al., 1958). The catastrophic flood of February 1954 was simply the largest of a series of large floods in a short time span. The Soil Conservation Service of NSW recorded that gully erosion commenced in the Bungawalbin Creek basin by the February 1954 flood (Harrison et al., 1958). This event, as shown below, caused massive channel widening, bank erosion, channel incision and gully erosion in the Six Mile Swamp Creek basin.

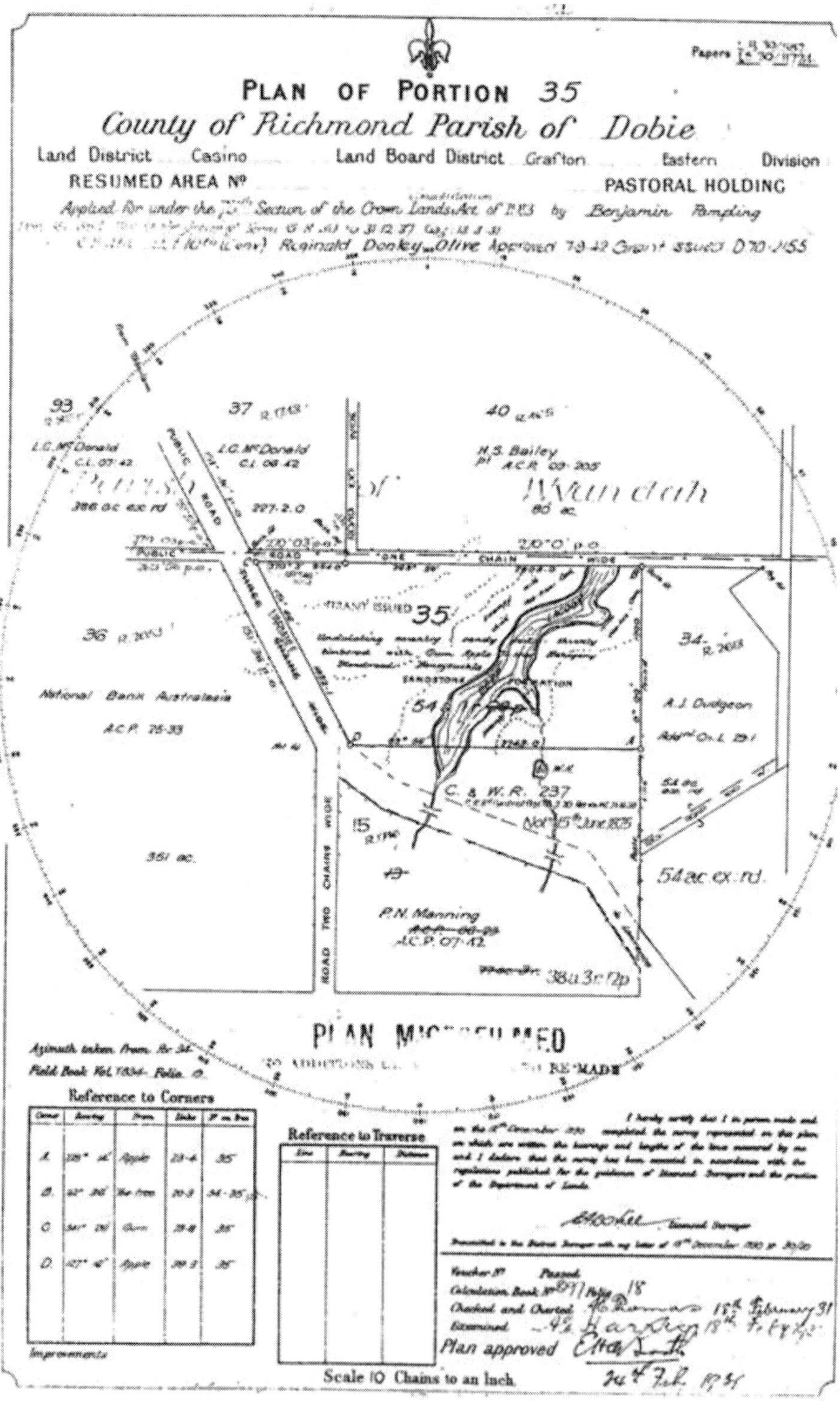

Figure 5. The 1930 survey plan of Portion 35 showing the 'deep lagoon' or chain of pond and Cattle and Water Reserve 237. This site is located immediately downstream of the Old Tenterfield Road Crossing on Figure 3.

Figure 6 shows that the 'deep lagoon' of 1930 in Portion 35 (Figure 5) had been completely buried with sand immediately after the February 1954 flood. An intermediate floodout (after Tooth, 1999) was formed by the February 1954 flood at the Old Tenterfield Road crossing. The channel continued further downstream but lost all definition between Manningville and the road crossing. These floodouts are different to those investigated in arid Australia by Tooth (1999). They are a form of channel failure where the channel changes abruptly proximal to active upstream sand sources, resulting in the deposition of the bulk of the bedload where channel capacity rapidly decreases over a short distance (Erskine and Melville, 1983b; Melville and Erskine, 1986; Erskine and Borgert, 2013). A fan-like form of deposition occurs downstream of the point where channel confinement of flow stops (Erskine and Melville, 1983b; Melville and Erskine, 1986; Erskine and Borgert, 2013). Schumm (1961) identified similar features on ephemeral channels in the USA which he called depositional or aggradational zones. The floodout extended about 2 km downstream in May 1955 and has progressively migrated further downstream since then. By 2006 sand had moved a total of 6.5 km almost to Pumphouse Waterhole at Clearfield (Figure 3). Bulk sand samples from the floodout in 1996 and 2006 had a mean size in the medium sand range (1-2 φ). A variable width, actively widening and incising channel formed during the February 1954 flood essentially on all of the alluvial length of the main channel upstream of Manningville (Figure 7). Channel erosion by catastrophic floods in south eastern Australia can produce sediment yields of 3500 to 7360 t/km^2 (Erskine and Saynor, 1996; Erskine, 1999) and the residence time of the sand usually exceeds decades (Erskine, 2008; Erskine et al., 2010). The floodout is a sand slug or bedload wave (Erskine, 1994c) which stores large amounts of sand and has now started to stabilise. Figure 8 shows the study reach in August 2006 where the floodout has contracted by substantial revegetation with *Melaleuca* spp., *Casuarina* spp. and *Acacia* spp. trees.

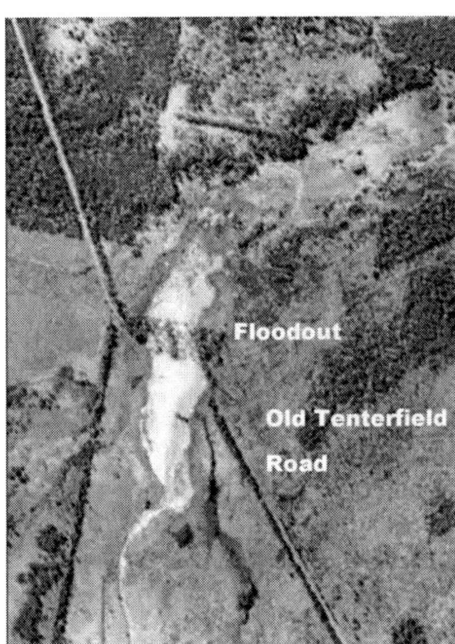

Figure 6. Air photograph of Six Mile Swamp Creek at the Old Tenterfield Road Crossing in May 1955.

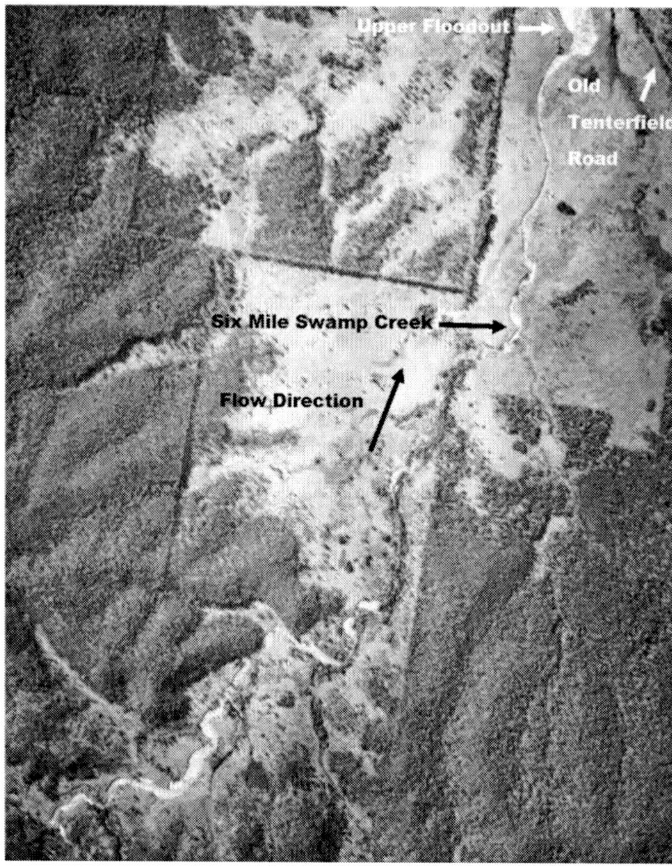

Figure 7. Air photograph of the eroded main channel of Six Mile Swamp Creek upstream of the floodout in May 1955. The land survey plans for the period before the catastrophic February 1954 flood show no integrated channel network with disconnected pools, ponds and lagoons.

According to the land owner (Mr Kinsley, 1996, personal communication), the floodout and overbank deposits on Manningville range from a minimum of 1.8 m to a maximum of 4.3 m thick and bury an organic, fine-grained swamp deposit. There are partially buried telegraph poles, fence posts and roads in the floodout. The Old Tenterfield Road was completely buried with sand so that the road was rerouted upstream to the incised section where the road crossed Six Mile Swamp Creek at Carters Bridge. Valley margin wetlands have been created where the floodout has dammed the lower reaches of tributaries. Figure 8 is an August 2006 vertical air photograph of the study reach and shows that many bends have widened by erosion since 1955.

4. Double Swamp Creek

A large discontinuous valley-bottom gully has eroded 52525 m^3 of valley-fill sediment from the middle reaches of Double Swamp Creek and tributaries upstream of cross section T and has deposited thick sand sheets on 2.1 km of floodplain downstream of cross section T (Figure 4). Land survey plans undertaken in 1874 and 1875 show swamps on the lower reaches of Double Swamp Creek where the whole valley floor is now buried by sand derived from the upstream gully. While gully erosion was certainly initiated after European settlement, the exact date of gully initiation is currently unknown. The gully was first investigated in 1986 and was very active at that time, except where sandstone was exposed in the bed and/or banks. The largest rain event before 1986 occurred in March 1974 when 274.3 mm were recorded on 11 March, 391.3 mm were recorded on 10 and 11 March, and 444.9 mm were recorded on 10-12 March at station 58130 (Grafton Olympic Pool). It is believed that this storm initiated gully erosion on Double Swamp Creek. Gully dimensions were surveyed in 1989 and assuming a sediment bulk density of 1.6 t/m^3, a sediment yield for the first 14 years of gully erosion alone is 1000 $t/km^2.yr$. This is a very high sediment yield. Channel capacity at each of the 23 cross sections of the depositional zone downstream of cross section T was determined by Manning's Equation, using estimated roughness coefficients (Cowan, 1956) and field surveyed bed slopes. The regional flood frequency method of Boyd (1978) was applied to the channel capacity data at each cross section and it was determined that bankfull discharge had an estimated return period that was usually less than 0.5 of the 1:2 year flood magnitude. This demonstrates that channel capacity is very small and that the frequency of overbank flow is very common, occurring many times each year. As a result, overbank deposition rates have exceeded 10 mm/yr since the gully was initiated on most cross sections.

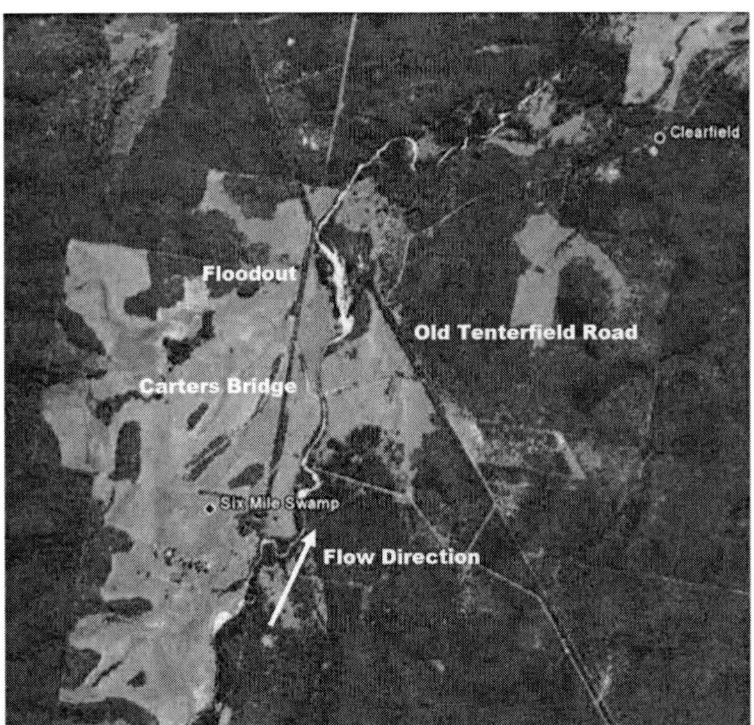

Figure 8. The floodout in August 2006 has contracted and revegetated since 1955 (compare with Figure 7). Also note that sand extends a long way downstream of the floodout.

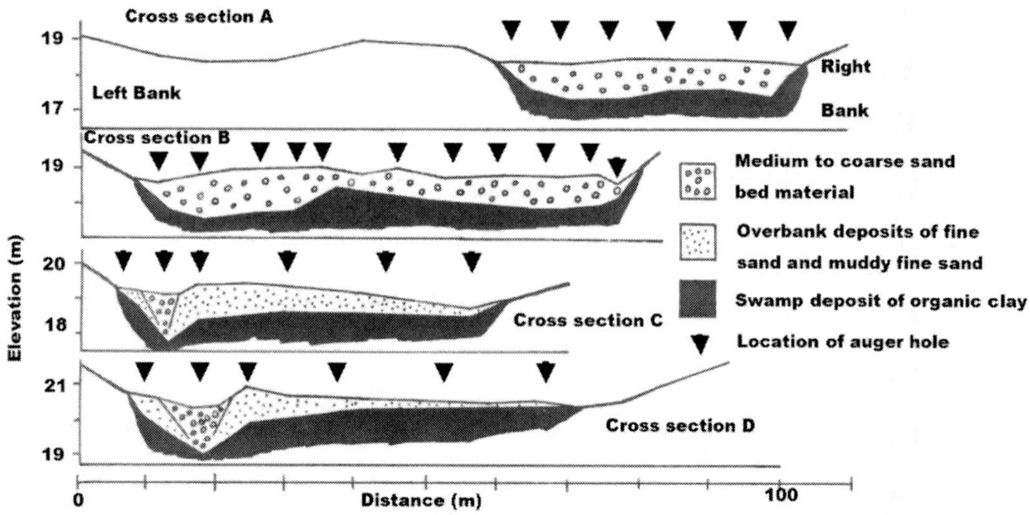

Figure 9. Floodplain stratigraphy at cross sections A to D, inclusive on Double Swamp Creek. See figures 4 and 12 for location of cross sections.

Floodplain stratigraphy in the depositional zone of the valley bottom gully is very consistent comprising a thick and extensive organic, fine-grained swamp deposit underlying surficial overbank deposits and channel sediments (Figure 9). The swamp deposit was at least 1 m thick whereas the surficial overbank deposits ranged from 0.1 up to 2.0 m thick and the channel sediments, where present, ranged from 1.0 to 2.0 m thick.

The channel flooded out immediately downstream of the culvert on the Old Copmanhurst Road in 1986. This has remained the case until the present, as shown in Figure 10. Repeated field surveys of the channel in the depositional zone documented three separate avulsions at two locations that occurred as the channel rapidly aggraded with gully-derived sand. Figure 11 shows the channel in the floodout zone in May 1986 and again in July 1987. During May 1987, the channel avulsed to a lower elevation on the left side of the floodplain. A subsequent survey in November 1989 revealed no further change. As shown in Figure 10, the channel has remained in the new location since the avulsion. The old channel is perched high above the floodplain and the new channel started aggrading as soon as it captured the bedload so that it is now also perched. However, extensive augering of the floodout (not included here) has revealed that the original course was located in a depression in the underlying swamp deposit before upstream gullying initiated rapid channel aggradation.

Two avulsions also occurred upstream of the culvert on the Old Copmanhurst Road. Field surveys of channel planform in January 1988, August 1988 and November 1989 recorded two avulsions. During a flood in April 1988 the channel avulsed to the southern side of a mound on the floodplain, as shown in Figure 12. In April 1989, the channel avulsed back to the original course (Figure 12). The cross section surveys show progressive bed aggradation at cross sections C and D (Figure 13). The floods of April 1988 and April 1989 caused a massive influx of gully-derived sediment into the channel. Figure 13 shows substantial deposition both within the channel and near overbank areas on cross sections C and D. Bed aggradation caused a reduction in bankfull mean depth from 0.27 m to 0.18 m on cross section D and from 0.19 m to 0.03 m on cross section C. Aggradation initiated the avulsion that occurred immediately downstream.

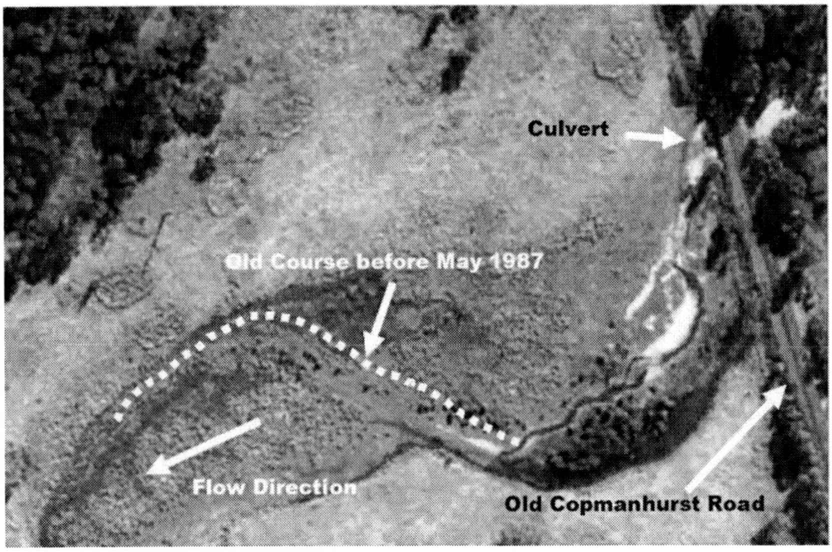

Figure 10. Air photograph of floodout on lower Double Swamp Creek downstream of the Old Copmanhurst Road in August 2006.

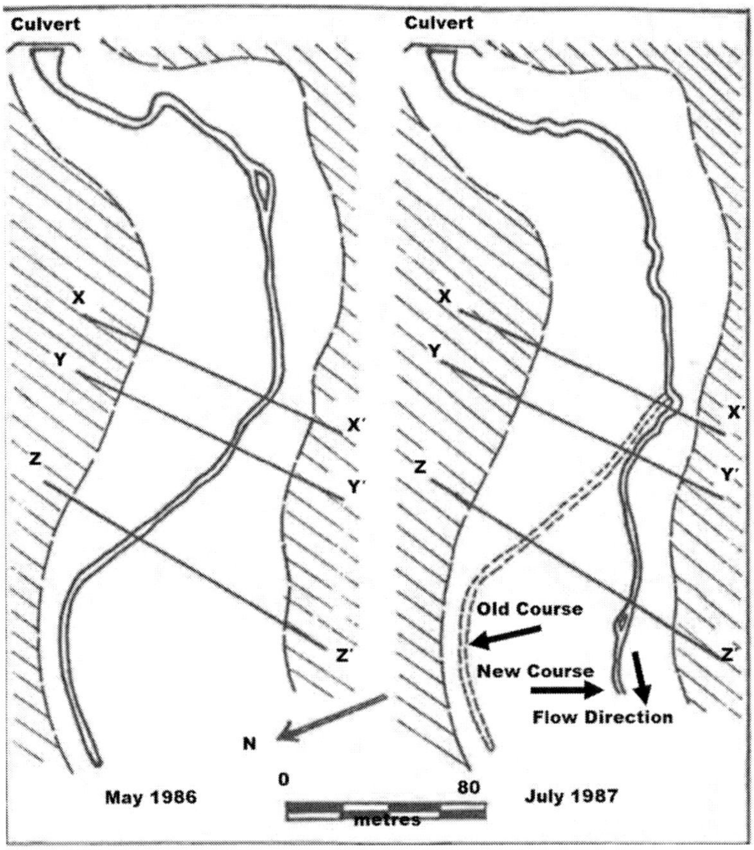

Figure 11. Channel avulsion on the floodout of Double Swamp Creek downstream of the Old Copmanhurst Road.

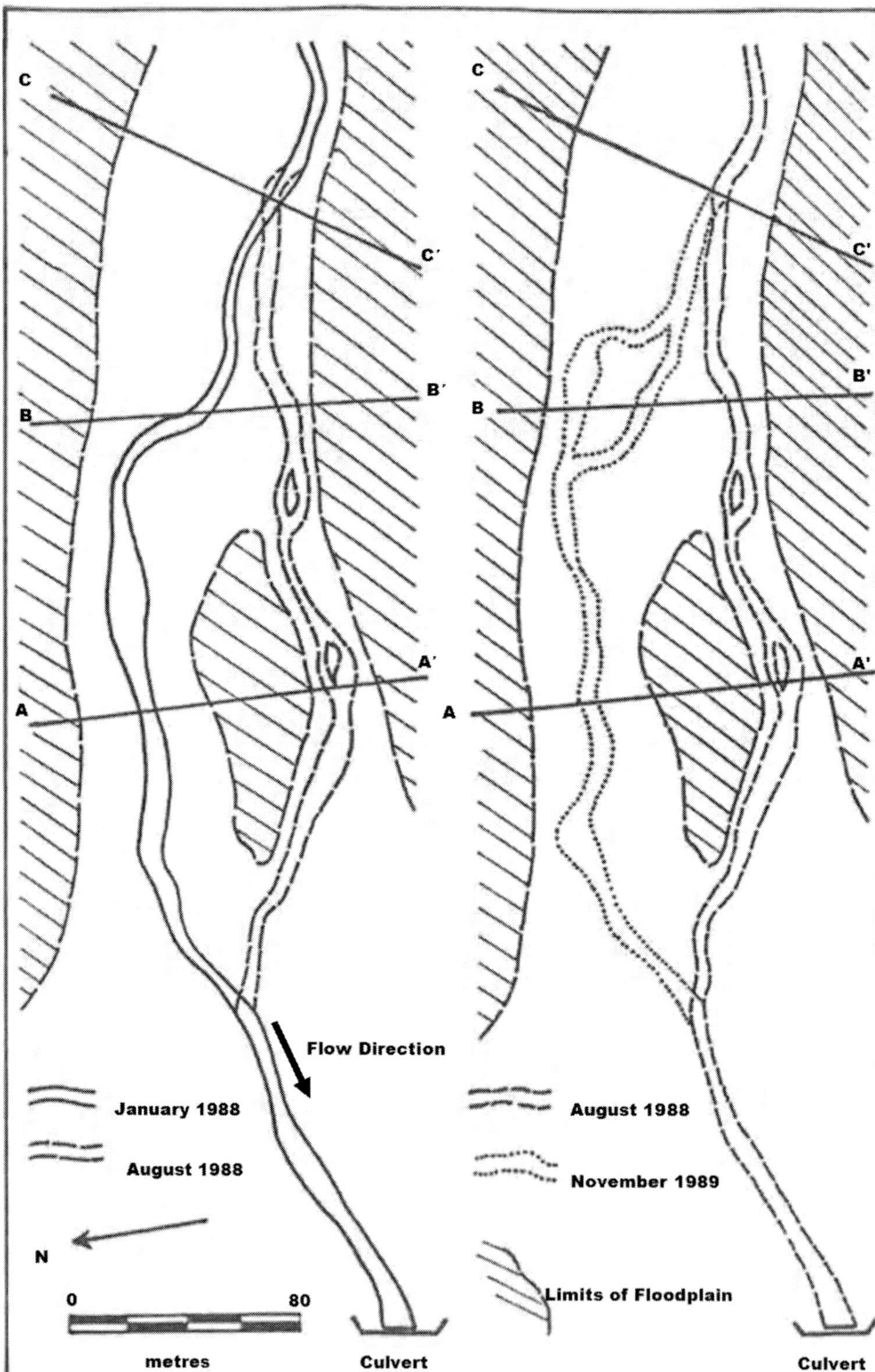

Figure 12. Channel avulsions upstream of the culvert on the Old Copmanhurst Road between January 1988 and November 1989.

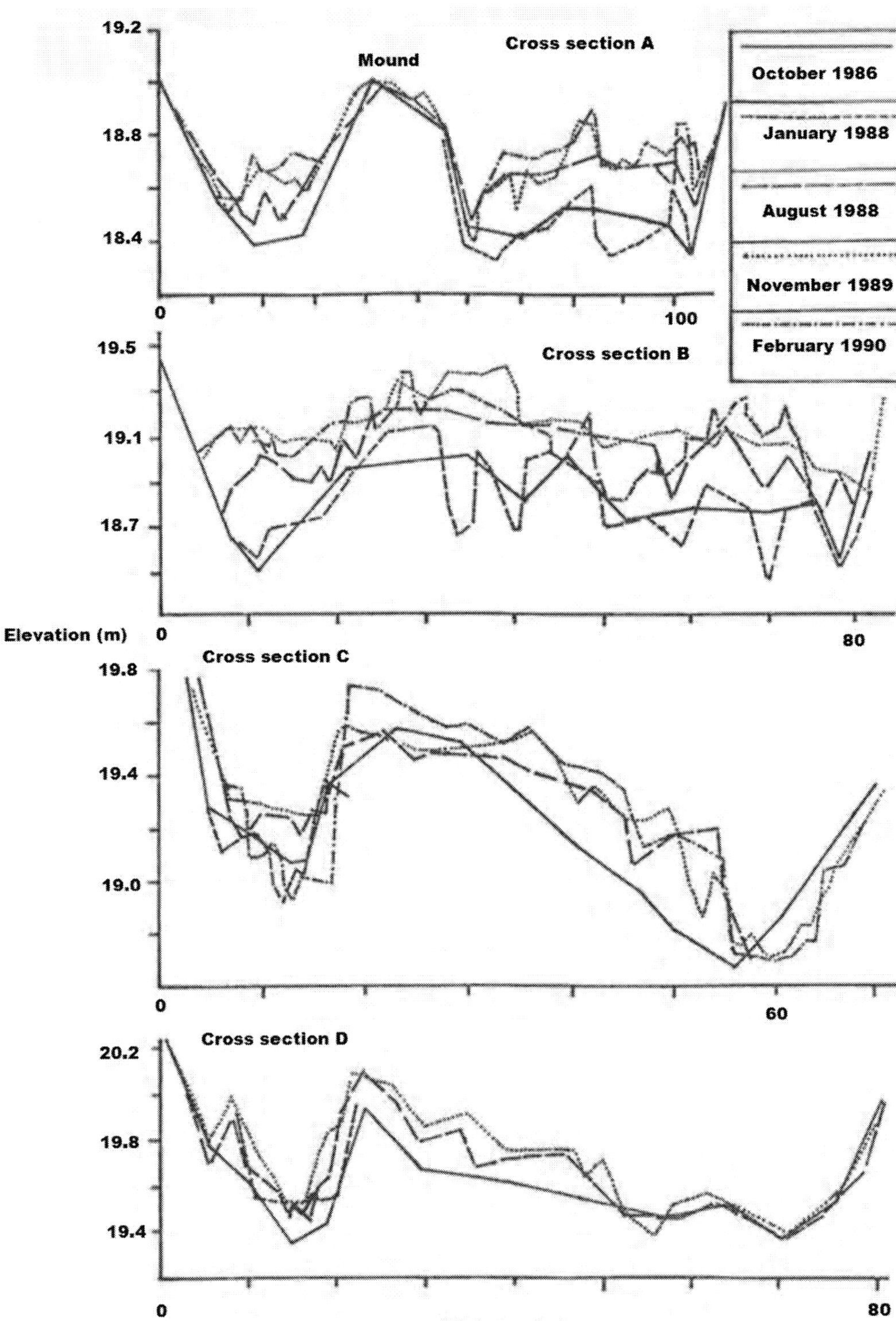

Figure 13. Repeated surveys of cross sections A, B, C and D upstream of the culvert on the Old Copmanhurst Road. See figures 4 and 12 for location of cross sections.

There is a less defined channel downstream of cross section C. As a result, gully-derived sediment splays right across the floodplain. Cross section B is located downstream of the bifurcation point of the avulsion. Cross section A is located across a vegetated mound in the centre of the floodplain which separates the pre- and post-avulsion channels. In January 1988 degradation had occurred in the centre of the floodplain on cross section B (Figure 13) and on the right bank side of cross section A. The flood of April 1988 deposited large amounts of sand across the whole valley floor. The channel at cross section B was totally infilled and mean channel depth at cross section A decreased from 0.31 m to 0.10 m. Aggradation reduced hydraulic radius, causing lower flow velocity and hydraulic efficiency. This, in turn, induced further deposition, elevating bed level above the floodplain. As the bed aggraded, the transverse slope away from the channel increased. Overbank flow was diverted away from the hydraulically inefficient channel to the left bank side of the floodplain. On flood recession the channel had avulsed to a low point on the floodplain (Figure 13).

On cross section A (Figure 13), the left bank side of the floodplain had more lateral confinement than the right side, resulting in more rapid aggradation. Deposition during the April 1989 flood reduced mean depth on cross section A from 0.17 m in August 1988 to 0.12 m in November 1989. In a similar manner to the initial avulsion, aggradation above that of the alternate course increased the transverse slope so that overbank flow was diverted away from the avulsed channel back to the original channel.

Rainfall records were analysed to determine the magnitude and frequency of the storms that caused channel avulsions because there is no river gauging station on Double Swamp Creek. The nearest rain gauge is Grafton Agricultural Research and Advisory Station (Station No. 58077) which has daily rainfall data since 1943. Rainfall frequency analysis was undertaken using the log Pearson type III distribution, as outlined by Pilgrim and Doran (1987). All data for one- and two-day annual maximum series plotted within the 5 and 95 % confidence limits. The avulsion downstream of the culvert in May 1987 occurred during a storm with return periods of 1.1 and 1.2 years on the annual maximum series for one- and two-day durations. The storms of April 1988 and April 1989 that caused the avulsions upstream of the culvert had one-day return periods of 3.8 and 4.9 years on the annual maximum series, respectively. Both events recorded the same two-day rainfall which had a return period of 2.8 years on the annual maximum series. Thus the storms which caused the three avulsions on lower Double Swamp Creek were small to moderate events which recur on average once every 1.1 to 4.9 years on the annual maximum series. Numerous larger storms have occurred in the past but did not cause channel avulsions because the discontinuous gully had not formed and hence the large amounts of sand required for the aggradation were not delivered to the channel.

5. Blaxlands Arm

The flood of 17-18 June 1949 is the largest on record in the Wollombi Brook basin and was generated by rainfall of up to 508 mm in 12 hours (Erskine, 1996). However, the 1949 flood, like the February 1954 flood on Six Mile Swamp Creek, was only one of a series of floods in a short time span (Erskine, 1996) and was greatly exceeded by at least two palaeofloods during the Holocene (Erskine and Peacock, 2002). Furthermore, two large floods occurred within 8 days in August 1952 and caused channel erosion throughout the Wollombi Brook basin (Benson, 1952). Good (1952) concluded that the large floods of 1949 to 1952 caused severe erosion on the middle and upper reaches of Blaxlands Arm. However, Good (1952) did not commence his erosion survey until 1950 and did not use any historical data sources. Therefore, it is possible that this erosion could have commenced before, although certainly reactivated by, the 1949 and 1952 floods. The present work uses many land survey plans and historical maps to determine the time and location of post-European gully erosion on Blaxlands Arm.

As outlined below, Blaxlands Arm was not only significantly impacted by the 1949 and 1952 floods but also underwent much change in the third quarter of the nineteenth century. Ogilvie (1833) showed that at the time of first European settlement, the lower reaches of Blaxlands Arm had a small channel with irregulary spaced ponds whereas the middle and upper reaches consisted of a swampy, unchanneled valley floor (Figure 14). Clearly drainage conditions at the time of first settlement were complex and diverse (Figure 14). In 1850, many isolated ponds, called 'chain of mulet (sic) holes' (Mullet refers to a number of freshwater fish species in coastal NSW), were mapped by Charlton (1850a) on the lower reaches near Noulanans Arm. By 1855, these ponds had been abandoned and replaced by a continuous incised channel on a new alignment (Rogers, 1855a). This is shown in Figure 15 where the plans of Charlton (1850a) and Rogers (1855a) have been superimposed. Gullying did not follow the exact course of the original channel because channel capacity was so small that overbank flow inundated the whole valley floor. Pond deposits of organic clay lenses at least 2 m thick buried by up to 1.3 m of stratified sand were found in 1985 at two of the locations where Charlton (1850a) depicted ponds. However, not all of the ponds on the lower reaches were abandoned. Rogers (1862) showed a number of ponds near Wollombi Brook and one of these is still there today.

A gully was initiated immediately upstream of the Noulanans Arm junction in Portion 38 on Figure 15 some time between 1850 and 1855. Rogers (1855b; 1855c; 1856) conducted many land surveys on upper Wollombi Brook in 1855 and 1856 and referred to a large flood on the neighbouring Yango Creek (Figure 2) in 1855. The 1855 flood probably cut the gully on Blaxlands Arm immediately upstream of Noulanans Arm. The area downstream of Noulanans Arm is very flat and extensive ditching to improve drainage has not accelerated channel incision over the last 150 years. The area upstream of Noulanans Arm was not only steeper but was also used for agriculture which would have increased sensitivity for gullying. Since 1855 the gully has migrated 2.8 km upstream, largely following the alignment of an inset fill. At least two cycles of post-European gullying have occurred suggesting that local deposition of gully-derived sediment within the gully increases slope predisposing the sediment deposits to new cycles of minor gullying. Secondary phases of gullying are often reported (Holland and Pickup, 1976; Cooke and Reeves, 1976; Schumm *et al.*, 1984). The sediment eroded by modern gullying has been deposited over the valley floor up to 1.7 km downstream of the Noulanans Arm junction, forming a floodout, infilling a number of ponds and straightening the channel by cutoffs. The 1949 flood not only extended and widened the gully on Blaxlands Arm upstream of Noulanans Arm but also cut a number of new discontinuous gullies upstream on the middle and upper reaches, which Good (1952) first described. The 'rich flatland' and 'rich alluvial land' described by Charlton (1850a; 1850b) on early land survey plans are now either incised by a gully or buried with deep sand (see below).

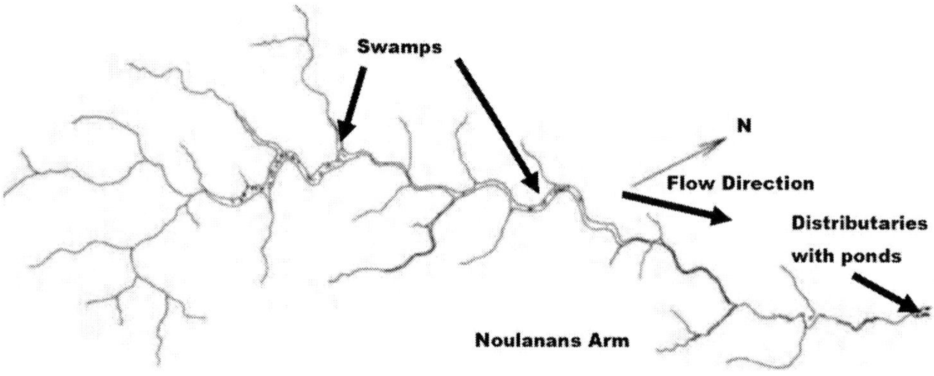

Figure 14. A redrawing of Ogilvie's (1833) map of Blaxlands Arm showing the drainage features.

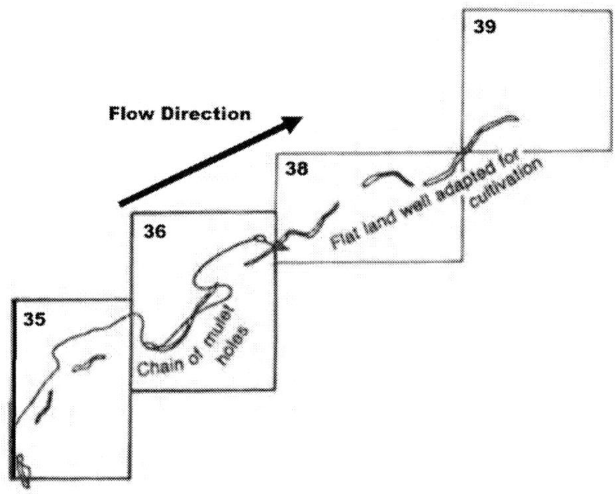

Figure 15. The chain of ponds mapped by Charlton (1850a) on portions 35, 36, 38 and 39 and the continuous incised channel mapped by Rogers (1855a) on portions 35 and 36, Parish of Blaxland, County of Northumberland. The 1855 channel alignment closely follows that of the existing incised channel. The comments on the map were written by Charlton (1850a). The Noulanans Arm junction with Blaxlands Arm occurs at the downstream end of Portion 38.

Figure 16 shows the floodplain stratigraphy at five cross sections which extend from upstream of the lower gully to the downstream floodout below the Noulanans Arm confluence. Cross section 1 is located across a floodout from a gully in the middle reaches and consists of up to 1.0 m of stratified sands. Even with a sand auger, it was not possible to penetrate the sands to the underlying sediment because of the high water table. Cross section 2 is located further downstream across the same floodout where a channel has reformed lowering the floodplain water table. As a result, it was possible to penetrate the floodout sands into the underlying organic-enriched, fine-grained swamp deposits. Cross section 3 is located across that section of the gully that was reactivated after 1949 and is near the headward limit of incision. As flow concentrates in the gully, overbank flow occurs which has partially stripped some of the floodout sand which was deposited before the gully migrated to this location. Cross section 4 is located across the floodout downstream of Noulanans Arm. A thin veneer (up to 1 m thick) of floodout sands cover the whole floodplain and the small capacity channel is formed entirely in floodout deposits. Cross section 5 is located near the basin exit where Ogilvie (1833) showed distributaries with ponds (Figure 14). The pond on the left bank is one of the two ponds shown by Ogilvie (1833). Floodout sands have blanketed the whole floodplain and overlie the organic-enriched, fine-grained chain of ponds sediments. However, sands underlying the chain of ponds sediments were also encountered at this cross section. The only other site where these sands were found were at cross section 3 at the base of the gully sidewall.

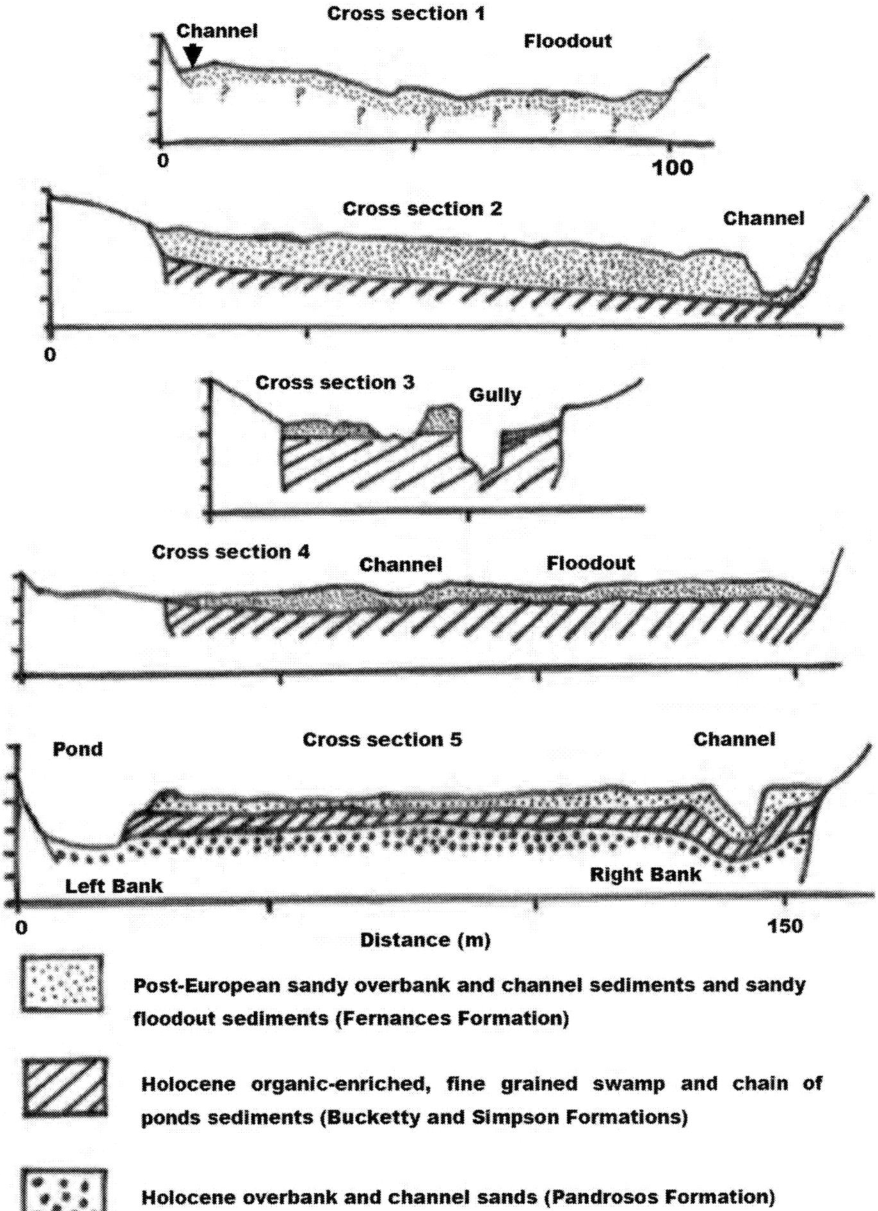

Figure 16. Floodplain stratigraphy at five cross sections on lower Blaxlands Arm arranged in downstream sequence.

Stratigraphic correlations have been made in Figure 16 with the already established lithostratigraphy for the area by Erskine and Melville (2008). The floodout facies of the Fernances Formation is a post-settlement, sandy sheet deposit that is common throughout the upper Wollombi Brook basin. The Bucketty and Simpson Formations are relatively thin sheets of fine-grained sediments that have been radiocarbon dated at up to 4070 ± 180 yBP. The Bucketty Formation overlies the Simpson Formation and is still being deposited under swampy conditions in ungullied valleys. The Pandrosos Formation is a thick sheet of sandy sediments containing discontinuous channel fill structures with granule-lined, trough sets of sand (Cullys Member).

In recent years, a series of rock chutes have been constructed where knickpoints were formerly present, such as near cross section 3. As the sediments in the floor of the gullies consisted of stratified muds and sands, the knickpoints persisted as they migrated upstream (Brush and Woman, 1960; Holland and Pickup, 1976; Gardner, 1983). Only those knickpoints initiated in sands rapidly rotated and lost their definition (Brush and Wolman, 1960). Rock chutes have successfully stabilised the knickpoints despite the occurrence of a large flood in June 2007.

6. Fernances Creek

Melville and Erskine (1986) and Erskine and Melville (2008) found that Fernances Creek was initially gullied after first European settlement at some time between 1838 and 1867. Following a series of large floods between 1856 and 1864 (Hunter River Floods Commission, 1870) and the construction by convicts of the Great North Road (Karskens, 1985), which crossed the valley floor at Fernances Crossing (Figure 17), gullying converted the swampy valley floor into an active sand-bed stream. The first land survey plans to show the gully were undertaken in 1867 and mapped extensive sand sheets downstream of Fernances Crossing (Biden, 1867a; 1867b). However, the extensive floodout on lower Fernances Creek (Figure 17) was partially gullied in 1927 when Wollombi Brook, the main stream, started to incise (Erskine, 2008) and lowered base level on the tributary (Reach 1 in Figure 17). In addition, a large flood in March 1977 initiated a short gully on a steep section of the floodout (Melville and Erskine, 1986) (Reach 2 in Figure 17).

The longitudinal profile of Fernances Creek was surveyed in March 1981. A resurvey of the profile from Fernances Crossing to downstream of cross section 11 (Figure 17) was completed in October 1990. The two profiles are shown in Figure 18. The gully bed had deepened by about 0.5 m between 1981 and 1990 from 700 to 200 m on the long profile (Figure 18). Furthermore, by October 1990, a series of deep, flow-aligned, discontinuous scour pools had formed below the intersection point. The intersection point is the point at which the bed and bank profiles intersect thus extinguishing the channel (Hooke, 1967). This is where drainage failure occurs and splays radiate from the channel. Valley-floor slope steepens greatly below the intersection point (Figure 18).

Figure 19 shows cross section 11 in March 1981 and in October 1990. Clearly, the channel is perched on top of an alluvial ridge. The cross section is located 100 m upstream of the intersection point and has been almost completely infilled by the formation of a mid-channel, longitudinal sand bar between the two surveys. The source of this sand is the degradation that occurred immediately upstream from 700 to 200 m on the long profile (Figure 18). Channel blockage diverts an increasing proportion of flow overbank, causing further sand deposition which elevates the channel higher above the floodplain (Melville and Erskine, 1986).

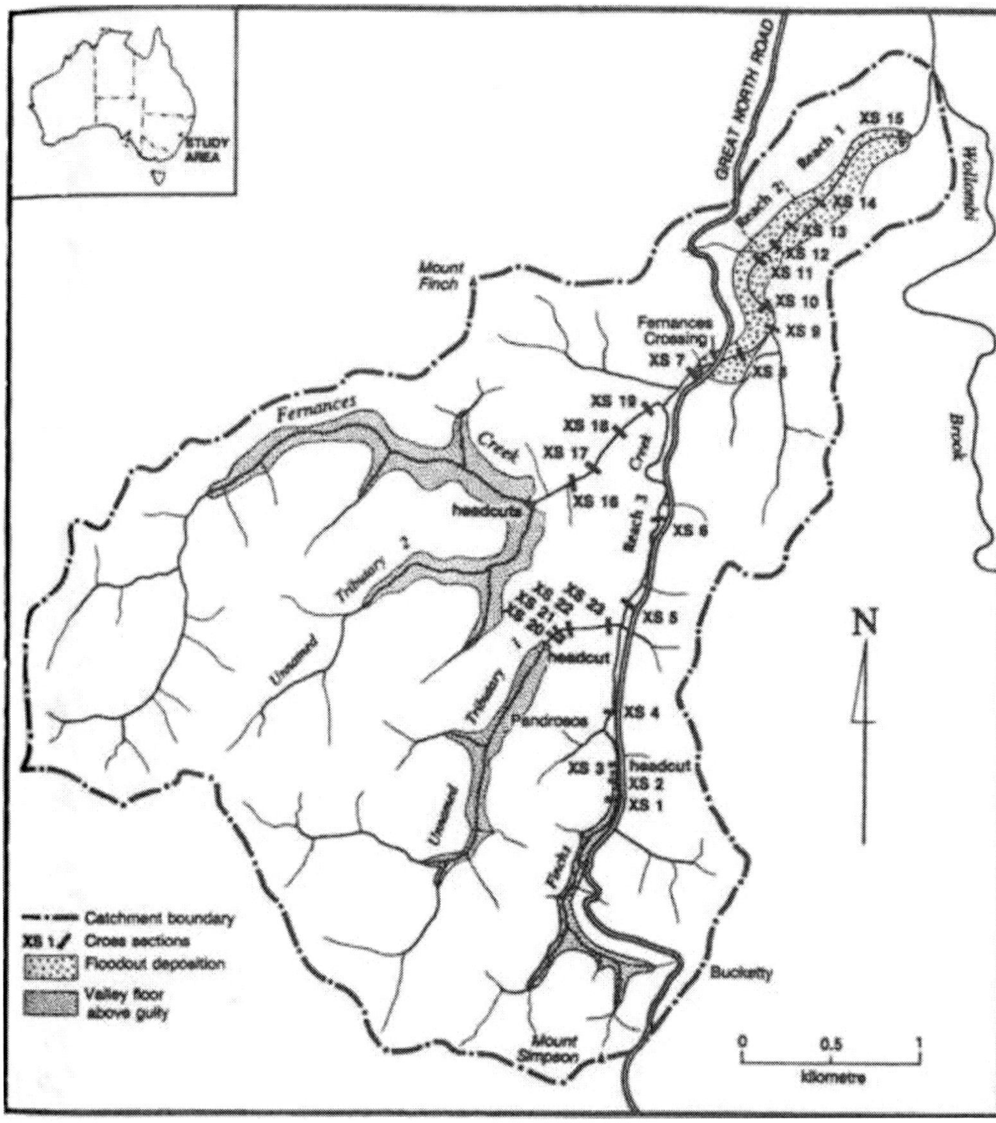

Figure 17. The Fernances Creek drainage basin showing cross section locations, floodout and valley floor above the gully. The Great North Road was originally built by convicts between 1826 and 1836 (Karskens, 1985).

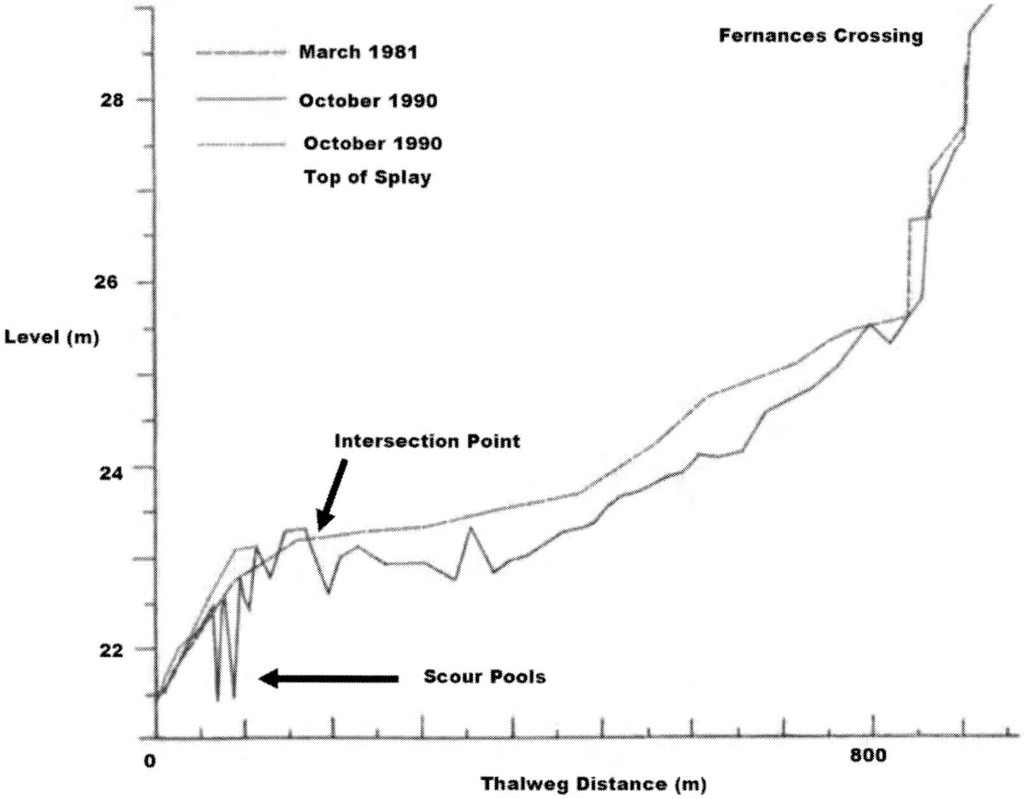

Figure 18. Field surveyed long profiles of Fernances Creek between Fernances Crossing and downstream of cross section 11.

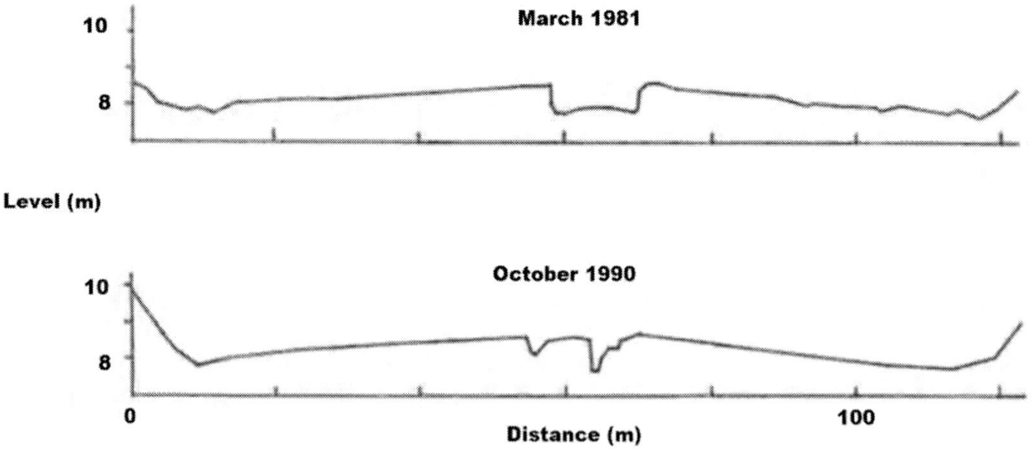

Figure 19. Changes in Cross section 11 between 1981 and 1990. Cross section 11 is located 100 m upstream of the intersection point and its location is shown in Figure 17.

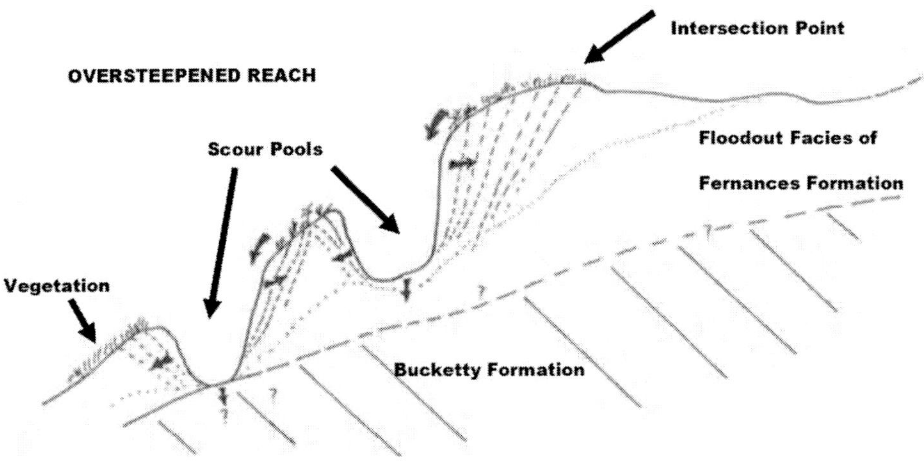

Figure 20. Model of scour pool development in the oversteepened reach below the intersection point of discontinuous, valley-bottom gullies. Scour pools can enlarge into ponds and can further enlarge into gullies.

The scour pools below the intersection point (Figure 18) have formed by localised erosion on the oversteepened reach. Scott and Erskine (1994) found identical features on nearby wet fans. Successive floods cause the pools to widen, deepen. erode headward and extend down valley. Scour pool growth can form new ponds and continued growth can eventually lead to coalescence of ponds into a continuous gully. Erskine (2005) quantified a slope threshold for gully development in the Wollombi Brook basin. Deposition at and downstream of the intersection point of discontinuous valley-bottom gullies leads to the development of oversteepened reaches which initiates scour of the grass stabilised sediment (Figure 20). Rapid deposition infills pools until reduced sand supply reinitiates scour. The development of pools into ponds (larger pools) has not been observed to date but has presumably occurred many times in the past. What is important for pool growth is the maintenance of a knickpoint as erosion extends headward. Interstratified sands and muds below the intersection point are capable of maintaining knickpoints.

Discussion and Conclusion

Both mud floodplains and sandy floodplains exist on the same river at different times in south eastern Australia and both are largely formed by overbank deposition. Channel sediments, where present, are spatially restricted. Mud floodplains can have chain of ponds with no interconnecting channels, chain of ponds with a vegetated, small capacity interconnecting channel, distributaries with and without ponds and swampy valley floors. There was a diverse range of drainage and floodplain features before European settlement. Sediment yields were low before European settlement and were mud dominated (Olive and Rieger, 1986; Wasson, 1994). Sediment yields of less than 10 t/km^{2}yr have been measured for undisturbed basins (Wasson, 1994). Floodplains functioned as channels and large floods inundated the whole

Gullying and channel incision greatly increase sediment yields. The reported sediment yield for Double Swamp Creek due solely to gully erosion was 1000 t/km^{2}yr and is two orders of magnitude greater than pre-European sediment yields. These high yields overload channels with sand, resulting in rapid and substantial aggradation. Aggradation *in situ* produces channels perched high on alluvial ridges. Relatively small floods which do not fill the floodplain enable overbank flow to flow transversely off the alluvial ridge, cutting new channels on steep slopes by avulsion. Catastrophic floods do not cause avulsions because they fill the whole floodplain with water drowning out the topography, hence negating any impact of transverse slopes.

Historical gullying and channel incision produce discontinuous channels which exhibit four channel types (Melville and Erskine, 1986; Erskine, 2008), which, in downstream sequence, are:

1. A stable upstream swampy valley floor with either no channel or with a very small capacity channel and chain of ponds which have not been eroded by historical gullying.
2. A gully or incised channel where either the swamp has been gullied or the small capacity channel has been greatly incised.
3. An active, small capacity sand-bed channel overloaded with sand from upstream channel erosion. Such channels usually floodout and completely loose definition. Deposition greatly reduces sand yields so that a new channel forms further downstream.
4. Either a sinuous small capacity channel or a swampy valley floor with chain of ponds downstream of the area where gully-derived sand has been deposited.

The channel of upper Wollombi Brook is an excellent example of these four channel types (Erskine, 2008). Six Mile Swamp Creek did not exhibit reach 1 whereas Blaxlands Arm and Fernances Creek did not exhibit reach 4. Only Double Swamp Creek exhibited all four reaches.

Floodouts are not confined to arid zone rivers which lose discharge due to high transmission losses downstream of the point where tributaries join the main stream. Internal hydraulic adjustments can occur over short distances and prevent channels from transporting the sand load. In south eastern Australia, intermediate floodouts are found in humid areas and form by rapid internal hydraulic adjustments.

The location of intermediate floodouts is dependent on large-scale, rapid sand deposition which completely obliterates the channel. Substantial sand storage occurs in the floodout. Melville and Erskine (1986) found that of the sediment generated by historical gully erosion on Fernances Creek, 90 % was deposited in either the floodplain in the gully or in the floodout downstream of the gully. Erskine (2008) proposed that discontinuous gullies and incised channels in the Wollombi Brook basin form a well-defined *sand compartment* where all the sand derived by historical gullying and incision is completely retained in the gully/incised channel and downstream floodout. This has not been recognised before but is very common in south eastern Australia. River managers stopped a dam being built on Wollombi Brook because of the presumed high sand load when, in fact, the sand was largely being deposited upstream of the dam.

Although many areas of south eastern Australia are characterised by both high flow and flood variability (Finlayson and McMahon, 1988; Erskine and Saynor, 1996), chain of ponds are one form of morphological adjustment to this high variability. The whole floodplain conveys the flow by catastrophic floods under conditions of chain of ponds and swampy valley floors. As a result, flood power is dissipated across a broad area and flood erosion is relatively minor. When incised or large capacity channels form they confine most of the flood energy to the channel and, as a result, erode catastrophically (Nanson, 1986; Erskine, 1994; Erskine and Saynor, 1996; Erskine and Livingstone, 1999). Natural levees can confine high discharges to the channel or channel incision can so deepen channels that no overbank flow occurs. High flood variability predisposes channels to catastrophic flood response (Erskine, 1996).

The catastrophic flood of June 1949 certainly caused gully erosion in the upper Wollombi Brook basin. However, some streams had gullied before 1949, as shown here for Blaxlands Arm and Fernances Creek. A slope threshold has been shown by Erskine (2005) to control gully initiation in the Wollombi Brook basin. Threshold exceedance can occur during a catastrophic flood but can also occur during small and moderate floods (Schumm, 1973).

Catastrophic floods are not as important for eroding mud and sand floodplains as other geomorphic processes in the Mesozoic sedimentary basins of south eastern Australia. While this seems counter-intuiative, these events are so large that inundate the whole floodplain and must be confined to channels by natural levees to concentrate flood energy. While Nanson (1986) emphasised the significance of catastrophic floods in causing floodplain stripping, Warner (1997) also pointed out that catastrophic floods cause large-scale channel adjustments and the macrochannel can include the whole valley floor, including the floodplain, in narrow bedrock-confined valleys.

ACKNOWLEDGMENTS

For their assistance with various aspects of this work, I thank the late Darcy Erskine, the late Timothy Erskine, Vanessa Erskine, Christine Erskine, Mike Melville, Ingrid McCarthy, Anthony Acret, Mike Saynor, Anita Chalmers, Pam Scott, Robert Tucker, James Purtill and Ali Akhoond-Ali. The University of New South Wales and the University of Newcastle financially supported this research. Dr Anita Chalmers, Dr Mike Saynor and Dr Rick van Dam constructively criticised a draft of the chapter.

REFERENCES

Bannerman, S. M. (1987). *An investigation of chains of ponds as a distinctive hydrogeomorphic form (a case study of Killarney Chain of Ponds)* B.A. (Hons) Thesis, University of Sydney.

Barnes, R. G. & Willis, I. L. (1989). *The geology of the Grafton and Maclean 1:250000 sheet areas*. Geological Survey of NSW Report No. GS 1989/17.

Benson, R. (1952). *Preliminary Report on Floods in Hunter River*, August 1952. Water Conservation and Irrigation Commission, North Sydney.

Biden, W. D. (1867a). *Plan of Two Portions of Land in the Parish of Hay*, County of Northumberland. NSW Department of Lands Plan No. N727 1501.

Biden, W. D. (1867b). *Plan One Portion of Land in the Parish of Blaxland*, County of Northumberland Comprising 100 Acres and Road 12 Perches No. 52 Measured for Auction. NSW Department of Lands Plan No. N728 501.

Boyd, M. J. (1978). Regional flood frequency data for NSW streams. *Civil Engineering Transactions of the Institution of Engineers Australia* CE20, 88-95.

Brush, L. M. & Wolman, M. G. (1960). Knickpoint behaviour in noncohesive material: a laboratory study. *Geological Society of America Bulletin*, *71*, 59-74.

Bryan, K. (1920). Origin of rock tanks and charcos. *American Journal of Science*, 4th Ser. 50, 188-206.

Burger, D. (1994). Palynology of the uppermost Walloon Coal Measures, Kangaroo Sandstone, and Grafton Formation, Clarence-Moreton Basin. In: *The Geology and Petroleum Potential of the Clarence-Moreton Basin*. (eds A.T. Wells and P.E. O'Brien), 181-188, Australian Geological Survey Organisation Bulletin 241.

Carey, W. C. (1963). The mechanisms of turns in alluvial streams. *Military Engineer*, Jan-Feb 1963, 14-6.

Carey, W. C. (1969). Formation of flood plain lands. *Journal of Hydraulics Division*, Proceedings of American Society of Civil Engineers, *95*, 981-94.

Charlton, H. (1850a). *Plan of Three Allotments of Land containing each thirty acres in Blaxlands Valley near Wollombi*, County of Northumberland. NSW Department of Lands Plan No. H669 663.

Charlton, H. (1850b). *Plan of thirty acres of land at Blaxlands Valley in the County of Northumberland*. NSW Department of Lands Plan No. H685 663.

Cooke, R. U. & Reeves R. W. (1976). *Arroyos and Environmental Change in the American South-West*. Clarendon Press, Oxford.

Cowan, W. L. (1956). Estimating hydraulic roughness coefficients. *Agricultural Engineering*, *37*, 473-5.

Erskine, W. D. (1994a). River response to accelerated soil erosion in the Glenelg River catchment, Victoria. *Australian Journal of Soil and Water Conservation*, *7*(2), 39-47.

Erskine, W. D. (1994b). Late Quaternary alluvial history of Nowlands Creek, Hunter Valley, NSW. *Australian Geographer*, *25*, 50-60.

Erskine, W. D. (1994c). Sand slugs generated by catastrophic floods on the Goulburn River, NSW. In: *Variability in Stream Erosion and Sediment Transport*. (ed. by L.J. Olive, R.J. Loughran and J.A. Kesby), 143-51, International Association of Hydrological Sciences Publ. No. 224.

Erskine, W. D. (1996). Response and recovery of a sand-bed stream to a catastrophic flood. *Zeitschrift fur Geomorpologie N.F.*, *40*, 359-83.

Erskine, W. D. (1999). Oscillatory response versus progressive degradation of incised channels in southeastern Australia. In: *Incised River Channels*. (ed. by S.E. Darby and A. Simons), 67-95, Wiley, Chichester.

Erskine, W. D. (2005). Gully erosion. In: *Water Encyclopedia Surface and Agricultural Water*. (ed. by J. Lehr, J. Keeley, J. Lehr and T.B. Kingery III), 183-188, Wiley-Interscience Hoboken.

Erskine, W. D. (2008). Channel incision and sand compartmentalization in an Australian sandstone basin subject to high flood variability. In: *Sediment Dynamics in Changing Environments*. (ed. by J. Schmidt, T. Cochrane, C. Phillips, S. Elliott, T. Davies and L. Basher) 283-290, International Association of Hydrological Sciences Publ. No. 325.

Erskine, W. D. & Borgert, R. (2013). Geomorphology and sedimentology of the Mogo Creek fluvial delta, NSW, Australia. In: *Deltas: Landforms, ecosystems and human activities*, (ed. by G. Young and G. Perillo), in press, International Association of Hydrological Sciences Publ. 358.

Erskine, W. D., Chalmers, A. C. & Townley-Jones, M. (2010). The importance of sediment control for recovery of incised channels. In: *Sediment Dynamics for a Changing Future* (ed. by K. Banasik, A.J. Horowitz, P.N. Owen, M. Stone and D.E. Walling), 211-219, International Association of Hydrological Sciences Publ. 337.

Erskine, W. D. & Livingstone, E. A. (1999). In-channel benches: the role of floods in their formation and destruction on bedrock-confined rivers. In: *Varieties of Fluvial Form*. (ed. by A.J. Miller and A. Gupta), 446-475, John Wiley and Sons, Chichester.

Erskine, W. D. & Melville, M. D. (1983a). Impact of the 1978 floods on the channel and floodplain of the lower Macdonald River, NSW. *Australian Geographer*, *15*, 284-92.

Erskine, W. D. & Melville, M. D. (1983b). Sedimentary properties and processes in a sandstone valley: Fernances Creek, Hunter Valley, New South Wales. In: *Aspects of Australian Sandstone Landscapes* (ed. by R.W. Young and G.C. Nanson), 94-105, Australian and New Zealand Geomorphology Group Spec. Publ. No. 1.

Erskine, W. D. & Melville, M. D. (2008). Geomorphic and stratigraphic complexity: Holocene alluvial history of upper Wollombi Brook, Australia. *Geografiska Annaler*, *90A*, 19-35.

Erskine, W. D. & Peacock, C. T. (2002). Late Holocene flood plain development following a cataclysmic flood. In: *The Structure, Function and Management Implications of Fluvial Sedimentary Systems*. (ed. by F.J. Dyer, M.C. Thoms and J.M. Olley) 177-184, International Association of Hydrological Sciences Publ. No. 276.

Erskine, W. D. & Saynor, M. J. (1996). Effects of catastrophic floods on sediment yields in southeastern Australia. In: *Erosion and Sediment Yield: Global and Regional Perspectives*. (ed. by D.E. Walling and B.W. Webb) 381-388, International Association of Hydrological Sciences Publ. No. 236.

Ewing, T. (1878). *Plan showing 7 Portions No. 1. 2. 3. 14. 15. 16 and 17, Parish of Dobie, County of Richmond measured for sale on application by Henry Bell*. NSW Department of Lands Plan No. R496 1744.

Eyles, R. J. (1977a). Birchams Creek: the transition from a chain of ponds to a gully. *Australian Geographical Studies, 15*, 146-57.

Eyles, R. J. (1977b). Changes in drainage networks since 1820, Southern Tablelands, NSW. *Australian Geographer*, *13*, 377-86.

Finlayson, B. L. & McMahon, T. A. (1988). Australia v the World: a comparative analysis of streamflow characteristics. In: *Fluvial Geomorphology of Australia*. (ed. by R.F. Warner), 17-40, Academic Press, Sydney.

Gallagher, W. (1984). Changes in drainage forms on the Cumberland Plain since 1788. B.Sc. (Hons). *Thesis*, University of Sydney.

Galloway, M. C. (1967). The stratigraphy of the Putty-Upper Colo area, Sydney Basin, NSW. *Journal and Proceedings of Royal Society of NSW*, *101*, 23-36.

Gardner, T. W. (1983). Experimental study of knickpoint and longitudinal profile evolution in cohesive homogeneous material. *Geological Society of America Bulletin*, *94*, 664-72.

Good, A. W. (1952). *Warkworth Catchment Area. General Report (Investigations 1950-1952).* Soil Conservation Service of NSW, Scone.

Hanlon, F. N., Osborne, G. D. & Raggatt, H. G. (1953). Narrabeen Group: its subdivisions and correlations between the South Coast and Narrabeen-Wyong districts. *Journal and Proceedings of Royal Society of NSW*, *87*, 106-120.

Harrison, E. W., Murray J. W., Powles, R. & Middleton, C. S. (1958). *Report of the Richmond River Valley Flood Mitigation Committee*. Department of Public Works, Sydney.

Hendrickson, D. A. & Minckley, W. L. (1984). Cienegas – vanishing climax communities of the American Southwest. *Desert Plants*, *6*, 131-75.

Hickin, E. J. (1979). Concave-bank benches on the Squamish River, British Columbia, Canada. *Canadian Journal of Earth Sciences, 16*, 200-3.

Hickin, E. J. (1986). Concave-benches in the floodplains of the Muswa and Fort Nelson Rivers, British Columbia. *Canadian Geographer*, *30*, 111-22.

Holland, W. N. & Pickup G. (1976). Flume study of knickpoint development in stratified sediment. *Geological Society of America Bulletin*, *87*, 76-82.

Hooke, R. L. (1967). Processes on arid region alluvial fans. *Journal of Geology, 75*, 438-60.

Hunter River Floods Commission (1870). *Report of Commission Appointed to Enquire into and Report Respecting Floods in the District of the Hunter River*. NSW Government Printer, Sydney.

Karskens, G. (1985). The construction of the Great North Road, NSW 1826-1836. *Multi-Disciplinary Engineering Transactions, Institution of Engineers Australia*, *GE9*, 102-11.

Leighly, J. (1936). Meandering arroyos of the dry southwest. *Geographical Review*, *26*, 270-82.

Leopold, L. B. & Wolman, M. G. (1960). River meanders. *Geological Society of America Bulletin*, *71*, 769-94.

Mack, J. J. (1983). *Chain of Ponds. A Narrative of a Victorian Pioneer*. Neptune Press, Newtown.

Mackin, J. H. (1937). Erosional history of the Big Horn Basin, Wyoming. *Geological Society of America Bulletin*, *48*, 813-94.

McElroy, C. T. (1958). Notes on the field use of heavy mineral studies in the Wollombi-Broke district. *NSW Department of Mines Technical Reports*, *6*, 99-100.

McElroy, C. T. (1962). The Geology of the Clarence-Moreton Basin. *Memoirs of the Geological Survey of NSW*, *Geology No. 9*.

McNally, G. H. (1981). Valley bulging, Mangrove Creek Dam. *Quarterly Notes Geological Survey of NSW*, *42*, 4-11.

McNally. G. H. (1995). Engineering geology of the Mangrove Creek Dam. In: *Engineering Geology of the Newcastle-Gosford region* (ed. by S.W. Sloan and M.A. Allman), 208-220, Australian Geomechanics Society, Springwood.

Melton, M. A. (1965). The geomorphic and palaeoclimatic significance of alluvial deposits in southern Arizona. *Journal of Geology*, *73*, 1-38.

Melville, M. D. & Erskine, W. D. (1986). Sediment remobilization and storage by discontinuous gullying in humid southeastern Australia. In: *Drainage Basin Sediment Delivery*. (ed. by R.F. Hadley), 277-86, International Association of Hydrological Sciences Publ. No. 159.

Mosley, M. P. (1972). Evolution of a discontinuous gully system. *Annals of the Association of American Geographers*, *62*, 655-63.

Nanson, G. C. (1986). Episodes of vertical accretion and catastrophic stripping: a model of disequilibrium flood-plain development. *Geological Society of America Bulletin*, *97*, 497-504.

Nanson, G. C. & Croke, J. C. (1992). A genetic classification of floodplains. *Geomorphology*, *4*, 459-86.

Nanson, G. C. & Page, K. J. (1983). Lateral accretion of fine-grained concave bank benches in meandering rivers. In: *Modern and Ancient Fluvial Systems*, (ed. by J.D. Collinson and J. Lewin), 135-43, International Association of Sedimentologists Spec. Pub. 6.

Nanson, G. C. & Young, R. W. (1981). Overbank deposition and floodplain formation on small coastal streams of New South Wales. *Zeitschrift fur Geomorphologie*, *25*, 332-47.

Ogilvie, P. G. (1833). *Survey of Blaxlands Valley near Richard Wiseman's Inn on the Wollombi Brook*. State Archives Authority of NSW Map No. 3048.

Olive, L. J. & Rieger, W. (1986). Low Australian sediment yields – a question of inefficient sediment delivery? In: *Drainage Basin Sediment Delivery*. (ed. by R.F. Hadley), 355-364, International Association of Hydrological Sciences Publ. 159.

Page, K. J. & Mowbray, P. D. (1982). Cutoff and oxbow lake (b) on a meandering river. *Australian Geographer, 15*, 177-80.

Pilgrim, D. H. & Doran, D. G. (1987). Flood frequency analysis. In: *Australian Rainfall and Runoff Volume 1*. (ed. by D.H. Pilgrim), 195-236, Institution of Engineers Australia, Barton.

Rogers, J. (1855a). *Plan shewing (sic). Position of Country Lots of Land, Blaxlands Arm, Wolombi (sic).* Brook, County of Northumberland, Applied to Purchase by W. Sternbeck, Moore and others. NSW Department of Lands Plan No. N111 1501.

Rogers, J. (1855b). *Plan of Four Small Farms Situate at Yango Creek, Wolombi (sic) Brook, Parish of Yango*, County of Northumberland, Applied to Purchase by D. Davis and A. Forbes. NSW Department of Lands Plan No. N88 1501.

Rogers, J. (1855c). *Plan Shewing (sic) Position of Three Portions of Land, Yango Creek, Wolombi (sic)*, County of Northumberland, Applied to Purchase by Goodwin, Welling and Others. NSW Department of Lands Plan No. N90 1501.

Rogers, J. (1856). *Plan Shewing (sic) Position of Three Portions of Land, Yango Creek, Wolombi (sic)*, County of Northumberland, Applied to Purchase by T. Polson and Others. NSW Department of Lands Plan No. N164 1501.

Rogers, J. (1862). *Plan Shewing (sic) Position of Two 40 Acres of Land Blaxlands Arm, Wolombi (sic)*, Northumberland, Applied for at the Office of the Land Agent for the Wolombi (sic) under the Regulations by W.R. Wiseman and Andrew Wiseman. NSW Department of Lands Plan No. N754 1501.

Rose, G. (1958). The geology of the Triassic rocks in the southern section of the Hunter River catchment: preliminary report. *NSW Department of Mines Technical Reports*, *6*, 97-8.

Scholer, H. A. (1976). Discharge of sands by sandy bed streams and regime of leveed rivers in coastal flood plains with special reference to rivers in eastern New South Wales. *Journal and Proceedings of Royal Society of NSW*, *109*, 81-9.

Schumm, S. A. (1961). Effect of sediment characteristics on erosion and deposition in ephemeral stream channels. *U.S. Geological Survey Professional Paper 352-C*.

Schumm, S. A. (1973). Geomorphic thresholds and complex response. In: *Fluvial Geomorphology*. (ed by M. Morisawa), 299-310, George Allen and Unwin, London.

Schumm, S. A., Harvey, M. D. & Watson, C. C. (1984). *Incised Channels. Morphology, Dynamics and Control.* Water Resources Publications, Littleton.

Scott P. F. & Erskine W. D. (1994). Geomorphic effects of a large flood on fluvial fans. *Earth Surface Processes and Landforms*, *19*, 95-108.

Sim, A., Erskine, W. D. & Drysdale, R. (2010). Application of sediment studies to the management and planning of water resources in the Sydney region. In: *Sediment Dynamics in a Changing Future* (ed. by K. Banasik, A. Horowitz, P.N. Owens, M. Stone and D. Walling) 229-237, International Association of Hydrological Sciences Publ. 337.

Smith, S. A. (1987). Gravel counterpoint bars: examples from the River Tywi, South Wales. In: *Recent Developments in Fluvial Sedimentology*. (ed. by F.G. Etheridge, R.M. Flores, M.D. Harvey and J.N. Weaver), 5-82, Society for Economic Palentology and Mineralogy Spec. Publ. 39.

Tooth, S. (1999). Floodouts in central Australia. In: *Varieties of Fluvial Form*. (eds A.J. Miller and A. Gupta), 219-247, Wiley, Chichester.

Warner, R. F. (1997). Floodplain stripping: another form of adjustment to secular hydrologic regime change in southeast Australia. *Catena*, *30*, 263-82.

Wasson, R. J. (1994). Annual and decadal variation of sediment yield in Australia, and some global comparisons. In: *Variability in Stream Erosion and Sediment Transport*. (ed. by L.J. Olive, R.J. Loughran and J.A. Kesby)., 269-79, International Association of Hydrological Sciences Publ. No. 224.

Wearne, F. (1905a). *Plan of Portion 24, County of Richmond, Parish of Dobie*. NSW Department of Lands Plan No. R1634 1744.

Wearne, F. (1905b). *Plan of Portion 25, County of Richmond, Parish of Dobie*. NSW Department of Lands Plan No. R1633 1744.

Wells, A. T. & O'Brien, P. E. (1994). Lithostratigraphic framework of the Clarence-Moreton Basin. In: *The Geology and Petroleum Potential of the Clarence-Moreton Basin*. (eds A.T. Wells and P.E. O'Brien), 4-47, Australian Geological Survey Organisation Bulletin 241.

Willis, I. L. (1994). Stratigraphic implications of regional reconnaissance observations in the southern Clarence-Moreton Basin, New South Wales. In: *The Geology and Petroleum Potential of the Clarence-Moreton Basin*. (eds A.T. Wells and P.E. O'Brien), 48-71, Australian Geological Survey Organisation Bulletin 241.

Wolman, M. G. & Leopold, L. B. (1957). River floodplains: some observations on their formation. *U.S. Geological Survey Professional Paper 282-C*.

In: Floodplains
Editor: Enner Herenio Alcântara
ISBN: 978-1-62808-998-1

Chapter 9

EFFECT OF HYDRODYNAMICS ON THE WATER QUALITY OF LAKES AND RESERVOIRS

Javier Vidal and Xavier Casamitjana*
Environmental Physics Group Physics Department
University of Girona

ABSTRACT

The aim of this work is to see how different hydrodynamic processes which have their origin in the interaction of meteorological variables (wind and solar radiation) with a stratified water body of a lake or reservoir determine the circulation patterns of the water and can significantly influence its quality. To accomplish this we will present the results of different studies conducted in different Spanish reservoirs. These studies have been published in prestigious journals (Vidal et al, 2005; Serra et al., 2007; Vidal et al., 2007; Vidal and Casamitjana, 2008; Vidal et al., 2012).

1. STANDING WAVES

1.1. The Music of Lakes

The standing waves produced in musical instruments, including string and wind, are responsible for the music that we can hear. The musicians, blowing or hitting with the hand excite these waves that propagate through the air to reach our ears. Recent investigations in lakes and reservoirs show that its behavior is very similar to a musical instrument. In spring and summer, solar radiation stratifies the water column that can be considered formed by two distinct layers: the so-called surface mixed layer (sometimes also epilimnion), where the effects of wind mixing are present and the bottom layer or hypolimnion, relatively isolated from the outside, both layers are separated by a strong thermal gradient called thermocline. Many Mediterranean reservoirs are away from this pattern, mainly due to the strong water

* Email: xavier.casamitjana @udg.edu.

withdrawals that occur; this creates an additional mixing of the water column, which is now formed by more layers, each with a different temperature. The wind excites waves in these layers (see Figure 1), creating what we might call the music of the lakes.

The mechanism of excitation of standing waves in a lake works as follows. The wind blows and piles up the water at one end of the lake. This creates an excess of pressure that is balanced with an inclination of the first layer or epilimnion (see Figure 1). When the wind stops blowing a surface current in the opposite direction of the wind tend to bring the water back to its equilibrium position, but due to inertia the water is being piled up at the other end of the Lake and the inclination of the epilimnion is reversed. As a result a standing wave, also called a seiche, has been created. Moreover, in the deeper layers different currents are formed, the direction of which alternate, and they are getting weaker due to the viscosity effects. The oscillation of the water surface is named barotropic oscillation while the oscillations of the internal layers are named baroclinic oscillations. In a two-dimensional approximation the standing waves are classified according to the number of nodes and its vertical or horizontal position using the notation ViHj where i indicate the number of vertical nodes and j the number of horizontal nodes.

As an example we see in Figure 2 the modes V1H1, V2H1 V3H1 on the right column and the modes V1H2, V2H2 and V3H2 on the left column. Each of these modes have a characteristic frequency (or period) of vibration which depends largely on the stratification, i.e. the number of layers, and the temperature. When one of these frequencies coincides with the frequency of the wind, the forcing agent, there is resonance and the lake vibrates as a whole at this frequency.

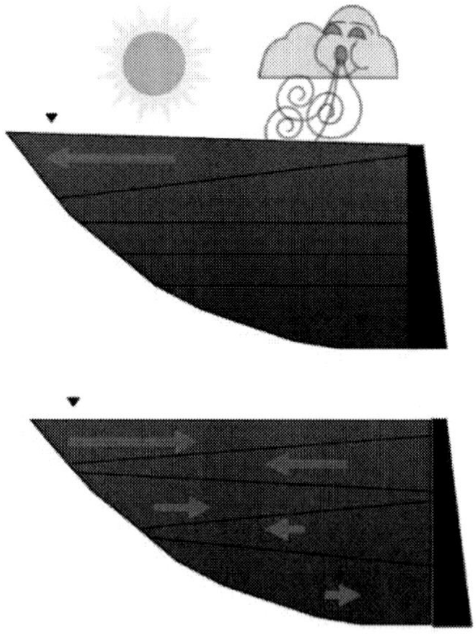

Figure 1. This figure explains how the "music of the lakes" is produced. Solar radiation stratifies the water column and the wind mixes it. In the Mediterranean reservoirs the volume water withdrawal is high. As a result, the water column is composed by several layers of different temperatures and therefore different densities. The wind blows and piles up the water at one end of the lake. When the wind stops blowing a surface current in the opposite direction is created, which tends to bring the water back to its equilibrium position. The internal layers also oscillate, but the speed is increasingly smaller.

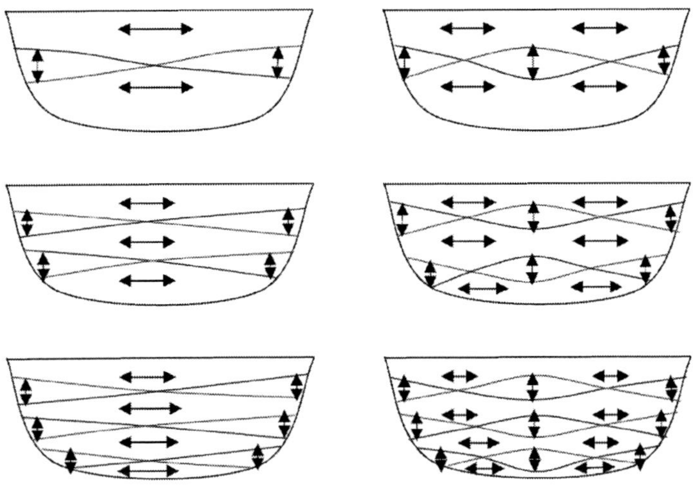

Figure 2. Schematic view of various oscillation modes in lakes. In the left column the modes: V1H1 (one vertical node and one horizontal node), V2H1 (two vertical nodes and one horizontal node) and V3H1 (three vertical nodes and one horizontal node). In the right column the modes: V1H2 (1 one vertical node and two horizontal nodes), V2H2 (two vertical nodes and two horizontal node) and V3H3 (three vertical nodes and two horizontal node).

While V1H1 mode has been observed many times the other modes have been much less. This is probably due to the fact that in many European lakes, studied since many years, the two-layer approximation is valid, while in Mediterranean reservoirs the strong water extraction causes a more continuous stratification (see for example Casamitjana et al., 2003). In reservoirs formed by two layers with a regular bathymetry the V1H1 mode is expected. Horizontal modes tend to occur in lakes or reservoirs with sudden changes of its width. Münnich et al (1992) observed a dominant V2H1 mode in Alpnach See, a side arm of Lake Lucerne. Perez Losada et al (2003) observed a mode V3 in Boadella reservoir.

1.2. Seiches in Sau Reservoir

Sau reservoir is located 426 m above the sea level with a maximum volume storage of 168 Hm^3 and a maximum area of 580 ha. Extraction nozzles are located at 21 m, 50 m 35 m from the base of the dam. One of its most significant features is its canyon shaped morphology, encased in the bottom of the valley carved out by river Ter in its passage through the Pre-Pyrenees mountains (Armengol et al. 1999). This river provides a high nutrient load, ammonia and soluble reactive phosphorus. While the nutrient load has been reduced in recent years as a result of the coming into operation of wastewater treatment plants, nutrient load is still quite high and generates an abundant proliferation of phytoplankton. With the exception of some peaks of chlorophyll-a as high as 200 $\mu g\ l^{-1}$, the annual variability of chlorophyll-a does not exceed 26.5 $\mu\ gl^{-1}$.

Mediterranean reservoirs, such as Sau and Béznar (see Figure 3) have an elongated shape and are placed at the bottom of the valley that the river has been excavating. They usually have a regular wind regime with a periodicity of 24 hours flowing in the direction of the main axis of the valley.

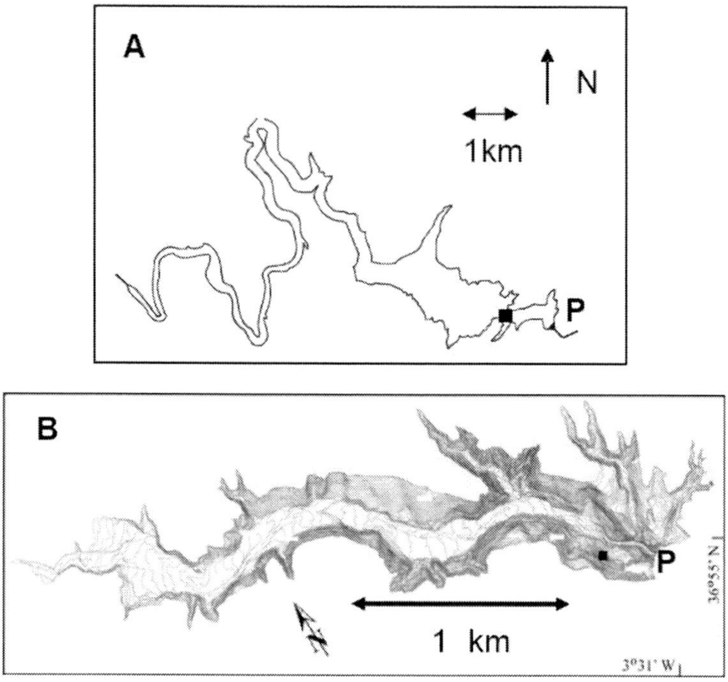

Figure 3. Schematic maps of Sau (A) and Béznar (B) reservoirs. Black squares indicate measurement points and P indicates the dam position.

Sau reservoir is 18.2 km and together with Susqueda and Pasteral forms a series of three dams on the middle of the river Ter, which is 200 km long; when full, can have a length of 3.6 km. During the morning, wind blows towards the dam (positive direction in Figure 4), but in the afternoon the wind changes direction and blows upstream. At the time of the change, the higher values of wind speed are recorded. Also, it has to be considered that the passage of meteorological fronts change the pattern of winds described above.

Now we want to compare the pattern of the wind speed, which is the forcing agent of the oscillations in the reservoir, with the oscillations occurring in the reservoir. The system reminds us the forced harmonic oscillator, however it is much more complicated because the reservoir is producing many types of movements that mask the typically hydrodynamic oscillatory response. To study the fluctuations in the reservoir we used thermistors chains that recorded the temperature in different parts of the water column from the surface to the bottom; we have also measured the water currents with an ADCP (Acoustic Doppler current profiler). The ADCP is placed on the surface of the water and sends sound waves towards the bottom. These waves collide with particles in suspension in the water and are reflected back to the ADCP that calculates its speed using the Doppler effect. Since the particles are suspended, the speed of the particles is equal to the speed of the water. The study of time series of temperature and water velocity is what will establish the oscillatory response of the reservoir.

In Figure 5A, we see the evolution of the power spectral density (PED) of the wind speed projected in the principal axis of the Sau reservoir during the summer of 2004. This measurement is based in the Fourier decomposition of a periodic time series and shows the predominant frequencies (or periods) of these time series.

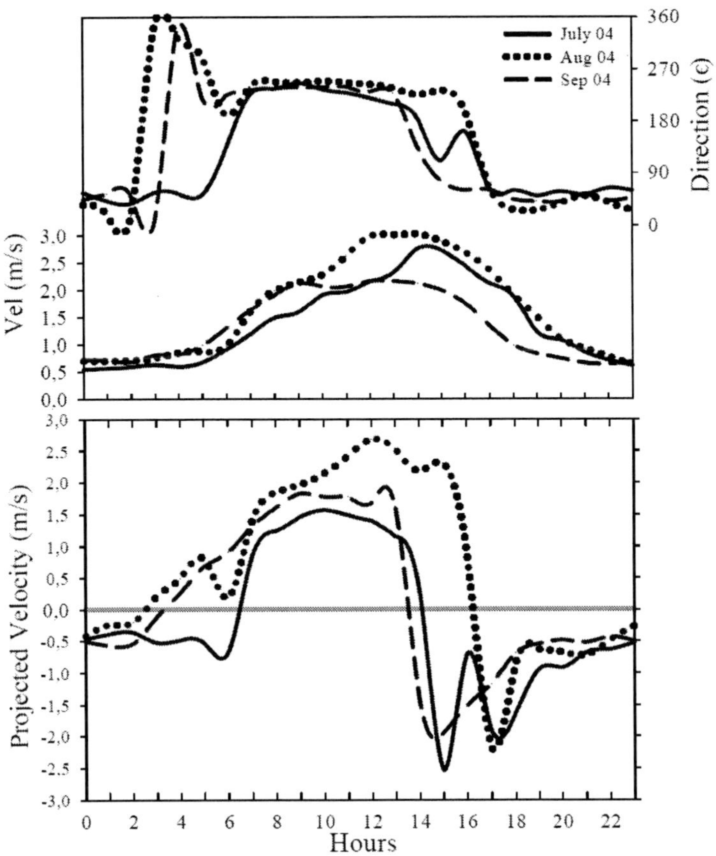

Figure 4. (A) July, August and September, 2004, averaged hourly values for the wind velocity and for the wind direction. (B) July, August and September, 2004, averaged hourly values for the wind velocity projected following the main axis of the reservoir.

In Figure 5A, we have calculated the PED every two days, taking intervals of 10 days series. The PED has maximums (red areas in Figure 5A) coinciding with the 24 hour period, as expected, since this this is the period of the wind, but also 12 hours.

Both periods correspond to the first harmonics in the Fourier decomposition of a square wave of period 24 hours, therefore we can say, that the wind forces our reservoir with a periodicity of 24 hours, the most important, but also with another of 12 hours. Figure 5B gives us an idea of the response of the reservoir to the pulses of the wind; we have represented the PED series of potential energy per unit volume (PE) of the water column. The PE of the water column has been calculated using the expression:

$$PE(t) = \frac{1}{2} N^2(t)\rho_0(t)\xi^2(t)\,, \tag{1}$$

where

$$N^2 = -\frac{g}{\rho_0}\frac{d\rho}{dz}$$

N^2 is the square of the frequency of stratification, g is the acceleration of gravity, ρ is the density of water, t is the time and ξ is the displacement of the isotherm calculated where the displacements are maximum. Values of N^2 and ρ_0 values are averaged throughout the period. To measure ξ we used data from the thermistors chain that we left on the measurement point shown in Figure 3A. We see here also dominant modes of 12 hours in mid-July and 24 hours from mid-August, coinciding with the maximum of PED series of wind. But in early August, when the wind has also a dominant period of 24 hours, we found no remarkable period in the response of the reservoir.

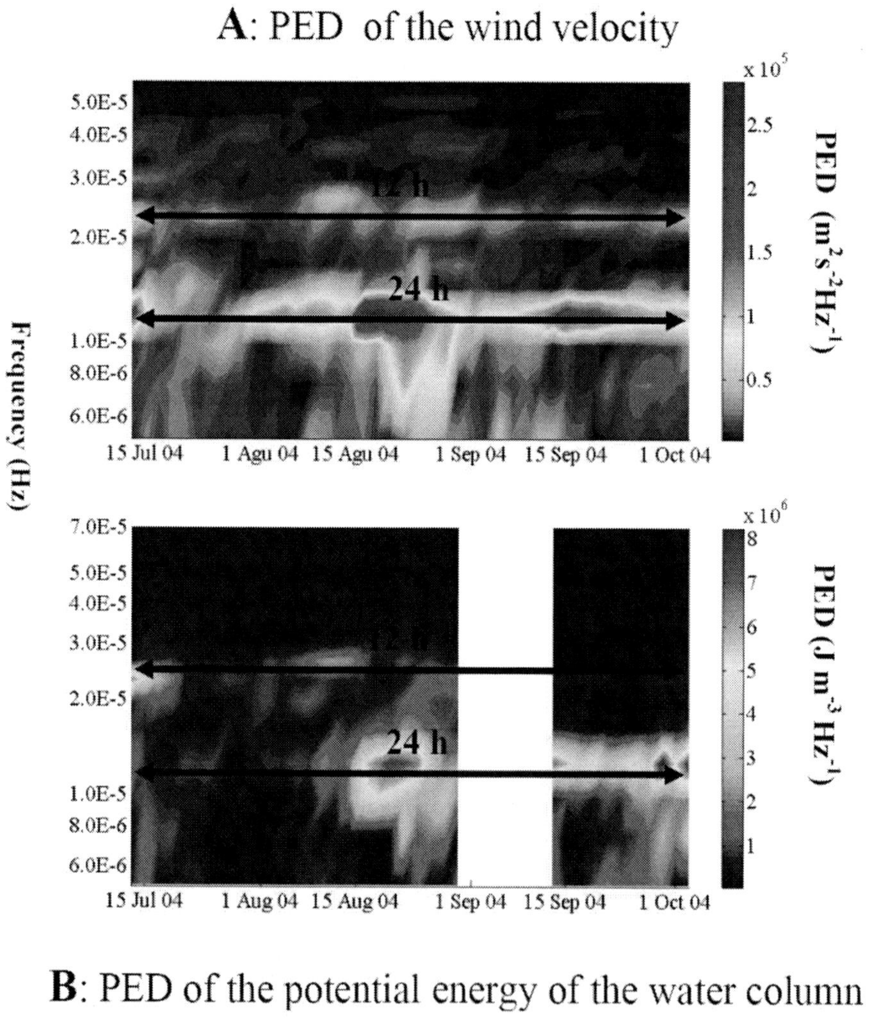

Figure 5. (A) Power Spectral density (PED) for the wind velocity projected following the main axis of the reservoir and (B) for the potential energy per unit volume of the water column from 15th July till 1st October 2004 in Sau reservoir. Periods of 12 h and 24 hours (showed in the figure) give maximum values for the wind and for the potential energy.

Now we wonder if we can attribute the modes of 12 and 24 hours we have seen in Figure 5B with the standing waves described in Figure 2. We will see that the 12-hour mode corresponds to the vertical mode V2 mode and the 24 hours mode to the vertical mode V3. In early August we note that although there is a clear wind forcing with a period of 24 hours, the PED of the PE has no significant peaks, indicating that no characteristic vibrational modes are selected in the reservoir .This is due to the fact that the stratification does not allow any vibration mode having a characteristic period close to 24 hours. Later on we will find out which horizontal modes correspond to these vertical V2 and V3 modes.

In Figure 6A we can see the velocity measured with the ADCP at the point indicated by the Figure 3A. To see more clearly the structure of three layers characteristic of the V2 mode we filtered the velocity values of Figure 6A with a 12 hours filter (Figure 6B). Notice that the maximum and the minimum values of the isotherms, which are represented in the figure, occur when the current changes its direction, as expected in the structure of an oscillating mode. In Figure 6C we have represented the average thermal profile obtained on 20 July 2004. We see a nearly continuous temperature change with depth and hence the stratification is different form the two layer structure.

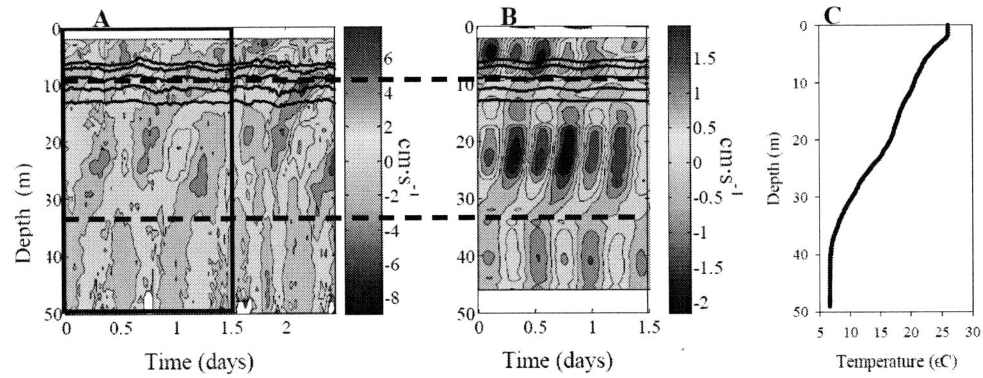

Figure 6. (A) Projected velocity following the main axis of the reservoir measured with the ADCP at Station 1 in Sau reservoir and (B) 12-hours filtered velocity. Day 0 corresponds to 27 July 2004. Dotted lines are used to illustrate the 3-layer characteristic of V2 mode; continuous black lines show the evolution of the isotherms. C shows the average temperature profile at station 1 on 20 July 2004.

If in the last paragraph we have corroborated the presence of the V2 mode with velocity profiles now we will demonstrate the existence of the V3 mode using temperature profiles. In Figure 7 the vertical displacement of three isotherms (22, 13.5, 8.5°C), clearly shows an oscillation of 24 hours, this is especially clear from the 19 September 2004 (day 9 in Figure 7). Furthermore, the 22°C isotherm oscillation is in the opposite phase with the 13.5°C isotherm and in phase with the 8.5°C isotherm (see dashed lines in Figure 7A) as is characteristic mode V3.

In Figure 7B we have represented two temperature profiles measured at the times indicated by the two discontinuous lines in Figure 7A which are separated by 12 hours, it can also be appreciated the characteristic structure of the V3 mode with 3 nodes, or fixed points, and three anti-nodes, or points of maximum oscillation.

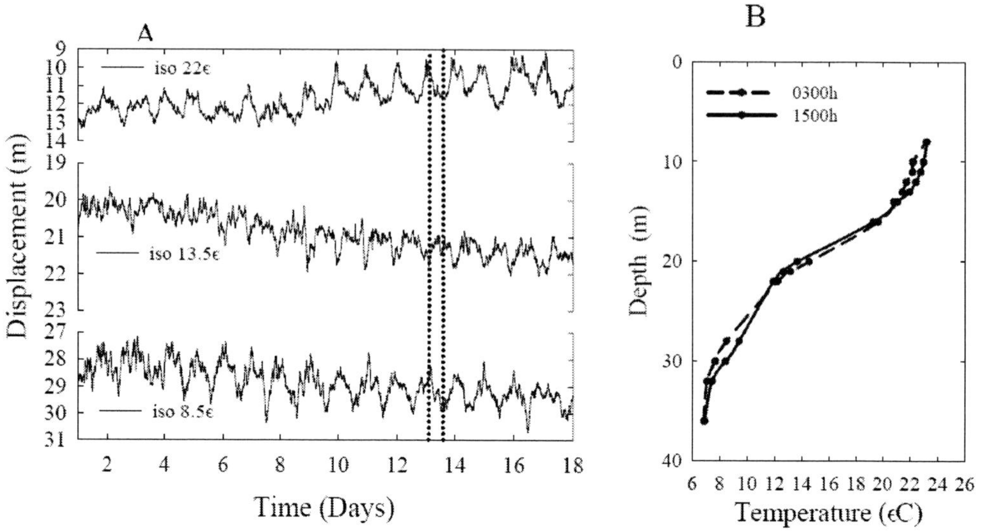

Figure 7. (A): Isotherm vertical displacements at station 1. Day 0 corresponds to 10 September 2004. Dotted lines show the experimental evidence for the mode V3. (B): Temperature profiles at two different hours measured with the thermistor string at station 1 on 22 September.

1.3. Using Three Dimensional Models

In Figure 8 we present the simulations obtained with the ELCOM model during a 24-hour cycle for three isotherms (22 ° C, 16 ° C and 8.5 ° C). This model solves the Navier Stokes equations with the Boussinesq approximation together with the transport equations for scalar quantities . Among the simulated processes there are the barotropic and baroclinic responses in the lakes, the effects of the rotation of the earth, the wind stresses and all the thermal processes that occur on the surface of lakes : shortwave, longwave , latent and sensible heat; the model also simulate the evolution of the inflows along the lake and the effects of withdrawals and other water outlets. For a more detailed description of the model see: Hodges et al (2000) and Hodges and Imberger (2001). The numerical model ELCOM allows us to know the thermal structure of the reservoir at any point, and it is a useful tool because we only have experimental data of temperature in one place (Figure 3).

Figure 8A corresponds to 20 September 2004 at 15.00 hours and the interval of time between figures A-D is 6 hours. The model is initialized on 16 September 2000 and used the wind data measured at the meteorological station near the dam. We see clearly the structure of the V3 mode with 4 layers fluctuating, as we outlined in Figure 8E. Notice how in Figure 8A and 8C water displacements are maximum and opposite.

We also see that the mode is V3H2 because besides the three vertical nodes are two horizontal nodes, represented by the two dashed lines in Figure 8E. In addition, vertical displacements are smaller away from the dam (to the left in Figures 8A-E), since this part is more sheltered from the wind and internal waves originate closer to the dam. The use of the model ELCOM during the middle of July, will let us to know that the mode V2 corresponds with a mode H2 and therefore we had a V2H2 mode.

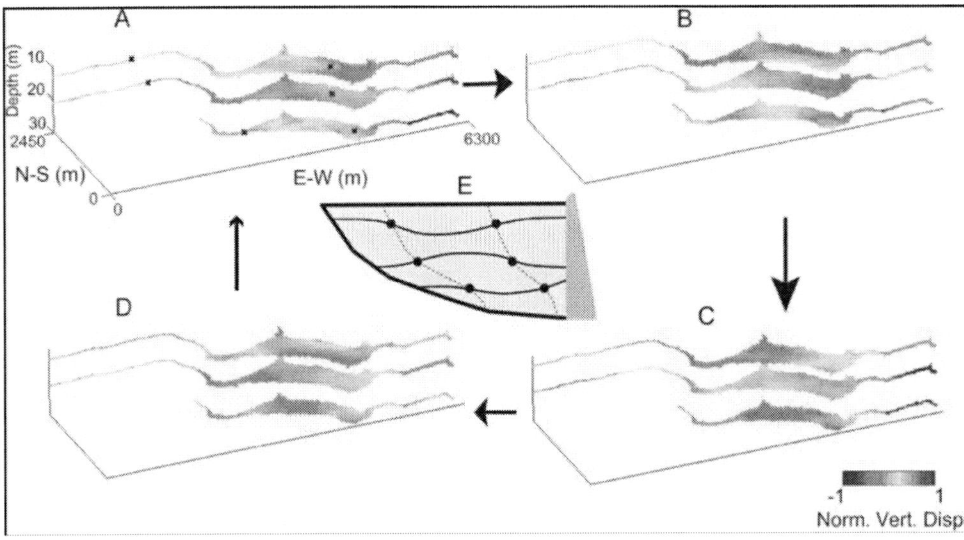

Figure 8. (A-D). Evolution of the ELCOM predicted vertical displacement of the 22°, 16° and 8.5°C isotherms at intervals of 6h, corresponding to 20 September 2004. Crosses in (A) show the approximate position of the nodal points in the horizontal structure. E: Sketch of the vertical structure corresponding to (C), showing the nodal points.

Béznar reservoir, in Andalucía, (Figure 3B) has many similarities with Sau reservoir. Beznar reservoir collects water from Sierra Nevada and has the maximum water volume in winter and spring, coinciding with the melting snow. When full, covers 107.26 ha and its maximum depth is 102 m. In this reservoir we also have obtained wind data in a weather station and temperature data from chain of thermistors located at the point indicated in Figure 3. Our aim is also to compare the response of the reservoir to the periodic forcing exerted by the wind. To do this, we analyzed the time series using the continuous transform "wavelet" of the square of the wind speed. The higher the real part of the transform coefficients represented in Figure 9A, the most important is the period in the time series. As in the case of Sau reservoir, we see a very clear predominance of the 24 hours period.

Figure 9B examines the response of the reservoir to the wind pulses, also by using the continuous transform "wavelet" of the potential energy per unit volume (PE) of the water column, defined by the expression (1). Here we see a strong response from the reservoir at the 24 hours mode. As before, we wonder which stationary mode of those in Figure 2 corresponds with that mode. Given the two-dimensional structure of the reservoir, we will use a model that solves the generalized eigenvalue equation of the stream function ϕ, where x and z are the horizontal and vertical dimensions and ω is the angular frequency of the standing wave

$$\frac{\partial^2 \phi}{\partial x^2} = \frac{\omega^2}{N^2} \frac{\partial^2 \phi}{\partial z^2} \tag{3}$$

Solving this equation for different values of the stratification represented by N^2 allows us to find the oscillatory modes of the dam, we have represented the Figure 9B; For this we used the finite element technique as described in Münnich et al (1992) which consists of taking the x direction along the main axis of the lake, discretize the domain and solve (3) using centered finite differences. Among all the possible modes of oscillation that appear in the model in Figure 9B we represented the vertical modes V1H1 to V5H1, which are the ones that we are able to find in the reservoir.

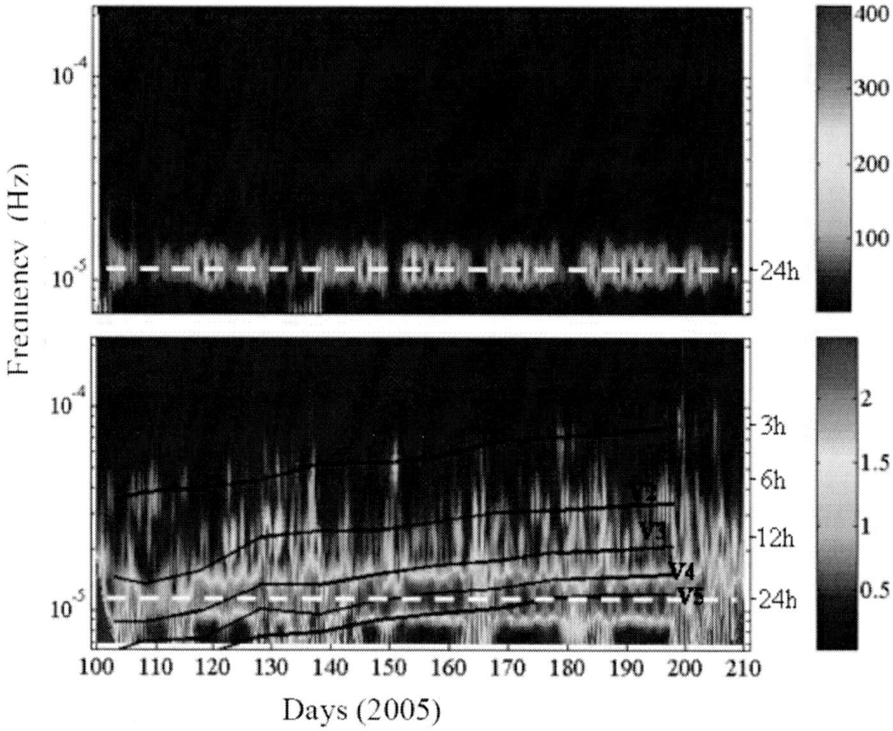

Figure 9. (A) Real part of the continuous wavelet coefficients in absolute values of square of the wind speed and (B) real part of the integrated signal of the continuous wavelet coefficients of discretized PE along the water column (see text). Circles connected by black lines indicate the evolution of the frequency of the V1H1, V2H1, V3H1, V4H1, and V5H1 mode, as obtained with the 2D model.

We have also seen how low modes are important when wind patterns are still the usual, especially when weather conditions are altered with the passage of fronts. As the stratification changes in time, frequency of the modes, which depends on stratification, also will change; Figure 9B shows how it slightly increases in time. It is interesting to notice that the modes V1H1 and V2H1 are only excited when the period of the main wind is no longer the 24 hours one; however usually excited modes are as high or V4H1 V5H1. We interpret this as follows. The wind forces our dam with a 24-hour period. When some of the modes characteristic of the reservoir is close to 24 hours this is the one that is excited and the others are damped. When the wind does not force our system the oscillation is the V1H1 which is the most commonly excited lakes.

There are several factors controlling the standing waves that occur in lakes: the morphology, the stratification and the forcing agent (the wind in most cases). Generally the lower modes are predominant modes (V1H1). This happens especially in lakes that can be considered composed of two layers (the epilimnion and hypolimnion). However, in reservoirs where the stratification is almost continuum, like many Mediterranean reservoirs mainly due to water extraction, and which are formed for many layers, the spectrum of internal waves is much longer. This is the case of Sau, and where we found V2H2 V3H3 modes and Béznar where vertical modes are observed to V5H1.

Although the presence of high modes in reservoirs had not been described until now, this study shows that these modes are not so uncommon. Reservoirs behave like oscillating multi modal systems in which the modes close to the forcing agent are selected from the entire spectrum of possible modes of oscillation. We have also seen how low modes are important when wind patterns are different from usual, especially because of the passage of fronts.

1.4. Water Quality Consequences

The influence of all the oscillation modes in the water quality evolution of a Lake is still not very well understood. In Figure 10 we present results of a study carried out in Sau reservoir (Serra et al., 2007). In this study, together with the temperature and currents, fluorescence profiles of different phytoplankton communities were carried out using an in situ submersible probe.

This probe determines four different communities (green algae, diatoms, cyanophyceae, and cryptophyceae), which have been previously calibrated by the manufacturer by considering particular algae populations corresponding to each algae group. Figure 10A show the results of the July 2004 survey, when the mode V2 was predominant and in Figure 10B for the September survey when the predominant mode was V3.

The picture presented in Figure 10 can be described as follow. Mixing caused by surface cooling causes the surface layer to deepen and a mixed distribution of phytoplankton to form at the surface. Mixing caused by surface cooling generally dominates at night, whereas during the day there are only weak mixing and low turbulence levels because of the weak winds that prevail in this time. Because of this fact, layering is possible mainly during the day. In contrast, these layers quickly disappear when water parcels mix because of a medium external forcing (convection) leading from low to moderate turbulence levels. During the day phytoplankton populations become shallower because of the rise of isotherms caused by both solar heating and internal waves. Therefore, organisms that do not sink or swim will move passively because of all of the above mentioned processes.

Horizontal advection caused by the wind-driven current at the surface of the water column also plays an important role in the transport of phytoplankton. This indicates that the phytoplankton is distributed in patches along the reservoir, probably because of horizontal differences in habitat conditions. Diatoms and cryptophyceae seem to be heterogeneously distributed, whereas the horizontal distribution of green algae seems to be more homogenous. More work should be done to investigate the reason for the different distributions of these communities throughout the reservoir. In addition, the wind excites internal waves or seiches in the reservoir. Seiches and the wind-driven current, together with solar heating and the above mentioned convection, determine phytoplankton distribution in the water column in the reservoir from the surface down to deeper layers.

To a lesser degree, community migrations to accommodate cells to their preferred light requirements can also be dominant when both convection and advection of water parcels stop.

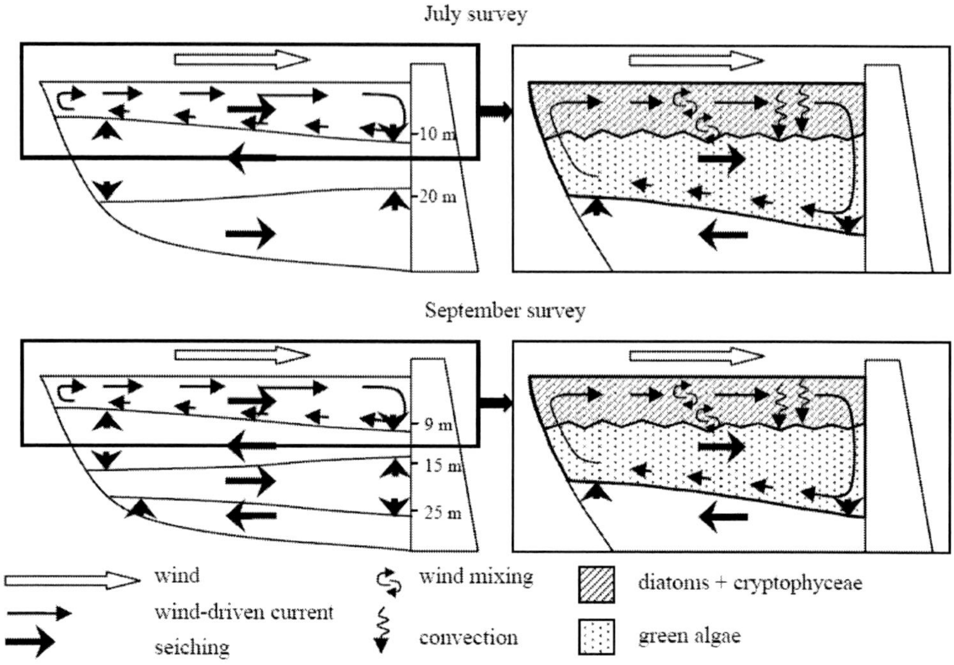

Figure 10. Schematic diagram of the seiching, the wind-driven current, the mixing caused by wind and the surface cooling for both (A) July and (B) September surveys. The extension of green algae, diatoms, and cryptophyceae is also indicated in the enlarged diagrams located in the right panels of each figure.

All of these results demonstrate that the phytoplankton dynamics depend on several external factors, and to understand their behavior it is necessary to determine all of the external variables, such as the heat balance, wind velocity, and the presence of internal wave oscillations, i.e., the different modes explained before.

2. Phytoplankton Response to Fluctuating River Inflow

River valley reservoirs, in Mediterranean regions, are often large and narrow, receiving water from a single river inflow. These reservoirs have important longitudinal changes controlled by the river intrusions across them. That progressive transformation from a river to a lake system is a key point to understand the water quality variations in the reservoir. The quick response of the river temperature to daily and short term variations (days to weeks) make the river-reservoir interaction a highly dynamic system. Likewise, the fate and transport of the river-born incoming nutrients, and therefore the phytoplankton evolution, will depend on the inflow dynamics. Now, we will show how algae blooms in zones with high nutrient concentration are related to the hydrodynamics of the river inflows.

In general, a reservoir can be divided along the longitudinal axis into three zones (see Figure 11) the riverine zone, the transition zone and the lacustrine zone. The riverine zone is characterized by higher flow, short residence time, and high values of nutrients and suspended solids. The transition zone where the river meets the reservoir is characterized by high phytoplankton productivity, decreasing flow velocity, increased water residence and large sedimentation. Finally, the lacustrine zone consists of the area near the dam with longer residence time, lower nutrient availability and lower suspended matter.

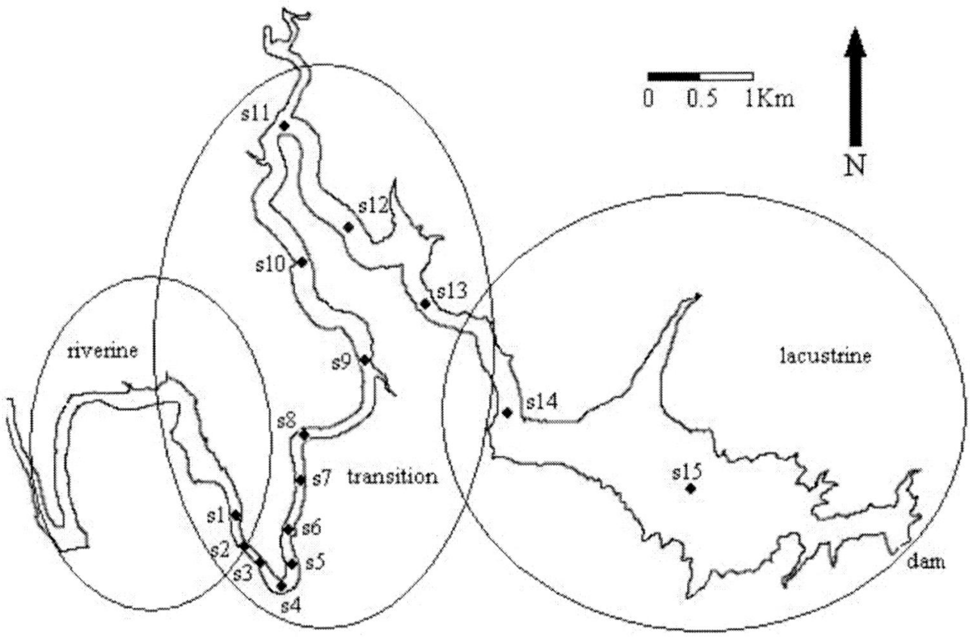

Figure 11. Location of the measuring stations on Sau Reservoir. The reservoir was divided into riverine, transition and lacustrine zones.

The boundaries between the three zones are not well defined and can be highly variable, in response to inflow characteristics. The interaction of river inflows in aquatic systems is based on the density difference between the inflow and the main water body, which generates overflows, interflows or underflows. The interaction between the river and the water body is essential determining the fate not only of nutrients or sediments, but also of possible contaminants. Armengol et al. (1999) studied the long term longitudinal processes associated with the river circulation in Sau reservoir. They described that, in winter, the inflow temperature is usually lower than that of the reservoir and thus the river intrudes the reservoir forming a density current or underflow. This deep flow continues until February, when the river temperature rises faster than the surface water of the reservoir, resulting in an overflow. From February to April-May, the surface flow corresponds to the start of the spring phytoplankton bloom due to the introduction of nutrients into the photic zone. The transition between spring and summer is characterized by an interflow that sinks progressively until mid November when it reaches the bottom as an underflow.

2.1. Inflow Dynamics of a River in a Reservoir

The pathways of river water distribution in stratified reservoirs, are mainly governed by the river inflow rate and the density difference between the river and the reservoir. In freshwater systems, temperature is the main factor controlling density of both river and reservoir.

The temperature of river Ter, the river discharging to Sau reservoir (see 1.2), is highly dependent on the meteorological conditions. The last days of July 2005 were quite warm (see Figure 12A), fluctuating at diurnal and synoptic scales; while the river temperature responded quite fast to changes in air temperature at both scales. The river inflow rate, during that survey, was around 2 to 3 $m^3 s^{-1}$ (Figure 12B) and the outflow, which was intermittent, was normally higher than the inflow so the water level was decreasing. Nevertheless, the outflow structure was located more than 10 km away (see Figure 11), deeper than the intrusion and below the thermocline and thus the effect of such water withdrawal into the inflow dynamics may be neglected. The conductivity of the river (Figure 12C), despite of showing pronounced peaks at different times during the study period, was always higher that the ambient reservoir conductivity, and then, it can be used as a tracer of the inflow. Moreover, the river Ter passes through a farming catchment area what increases significantly the nutrient concentration of the river. That is evident if we see the high phosphate concentration obtained in the automatic station (Figure 12D). Note that, phosphorous, is usually the limiting factor of the algal growth in Sau (Armengol et al. 1999).

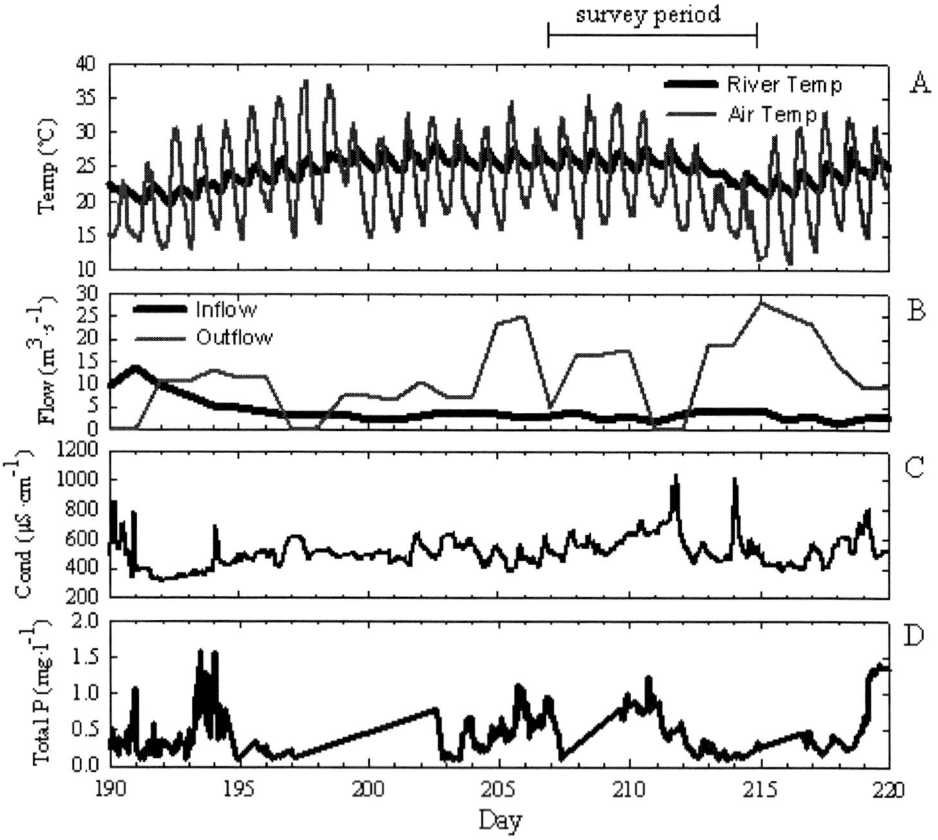

Figure 12. (A) Time series of river temperature and air temperature. (B) Inflow and outflow discharges; (C) river conductivity, and (D) concentration of SRP in the river Ter during the period of our study (days 207–215).

The river responds faster than the reservoir to the solar heating. As the day progresses the river became warmer, eventually reaching temperatures similar to those of the surface water of the reservoir, thus generating an overflow in the afternoon and early in the night. However, during the night, the river became colder than the water surface and eventually, late in the night and during the early morning an underflow intruding at 6- 7 meters depth occurred. Such dynamics can be observed by looking at ADCP and temperature string data (Fig 13a), at station s6 (Figure 11). During the survey, the intersection of the river with the reservoir, where the plunge of the river took place (plunge point), was detected by visual inspection to occur between stations s2 and s3 (Figure 11). Thus, s6 was located around 900m downstream from the plunge point. Figure 13b shows the velocity profiles during Julian days 208 – 210; here, positive values indicate that the river flowed towards the dam. From Fig 13 it is clear how the inflow (positive values) changes from an underflow to an overflow. Black lines in Figures 13a and 13b show the central position of the inflow, as estimated from the ADCP measurements. During the underflow episodes, cold dense water from the river plunges into the warmer water of the reservoir. During the overflow episodes, warm water from the river displaces the surface water of the reservoir, as shown in Figure 13a. In both processes, an upstream current is generated (see the negative values in Figure 13 b) and water from the reservoir is entrained in the inflow, especially at the plunge point where the mixing is greater. That is, during the overflow events, cold water from lower layers travelled upstream and during the underflow events, warm water from the surface of the reservoir travelled upstream.

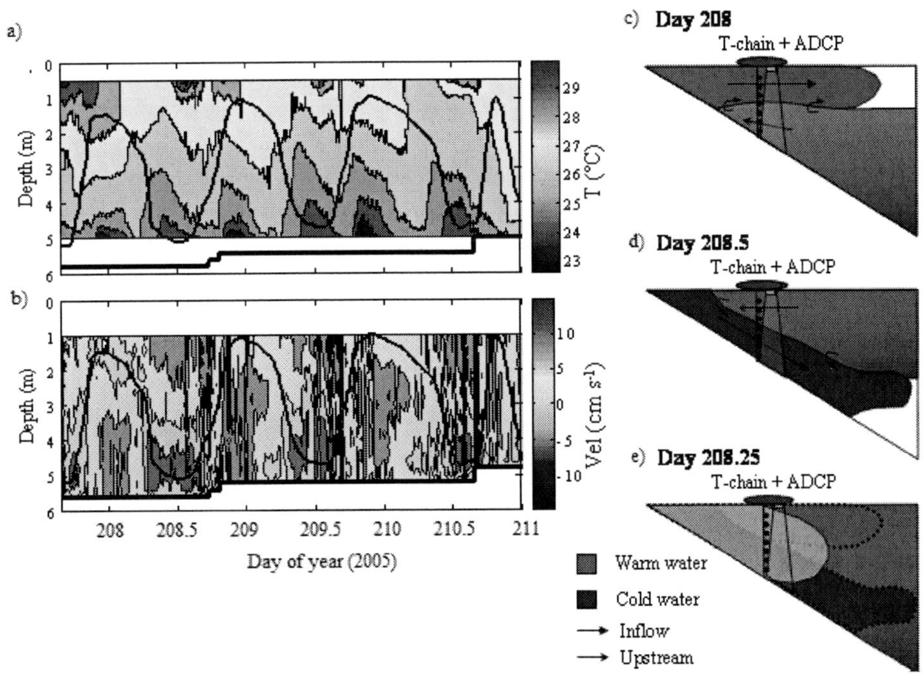

Figure 13. Observations at stations s6 from day 208 to 210 in Sau Reservoir: (A) time series of temperature and (B) velocity profiles. The black line represents the depth of neutral buoyancy estimated from temperature records, presuming that lake and river water do not mix. Figures (C), (D) and (E) represent the overflow, interflow and transition events occurring at day 208.0, 208.5 and 208.25, respectively. Color shading indicates the temperatures, and the lengths of arrows are proportional to velocity.

It can also be seen that the stratification decreases during the transition of such events (Figure 13a) suggesting that the inflow mixes vertically in the water column. In Figures 13c to 13e, a schematic representation of these situations is presented. Note that the time delay between an overflow (Figure 13c) and an underflow (Figure 13d) is about 12h. Eventually, underflow events will end up as interflow after s6 as later will be seen.

The CTD measurements, together with the ADCP measurements carried out during the morning of 31 Jul 05 (Julian day 212) when an interflow situation occurred, are presented in Figure 14. Unfortunately, we do not have CTD measurements in the afternoons and nights, when overflows occurred. Based on the velocity profiles, represented by the arrows in Figure 14, the inflow evolution (positive velocities) can be represented (see the green contours in the figure). When the interflow occurred, the temperature of the river was lower than the surface temperature and the inflow plunged at about 1000 m length from station s1 and between s2 and s3 (Figure 14A). The inflow behaved as an underflow afterwards going down till a depth of aprox. 6m and 3000 m length from s1, when it reached the insertion point. High values of turbidity (Figure 14B) in the bottom coinciding with the underflow suggest shear induced resuspension; however, after insertion, great part of the resuspended material sedimented and turbidity decreases rapidly after the insertion point. Inflow conductivity (Figure 14C), despite being quite variable, was always higher than the conductivity of the reservoir water and can be used as a natural tracer of the river pathways. High conductivity values in the bottom (inside the green area) can be attributed to the present measured inflow. However, the high conductivity values found at the surface in layers up to ≈ 3000m long (outside the green area), likely correspond to recent past overflow events that have been injecting inflowing water into the surface in the fluctuation produced due to river density variation.

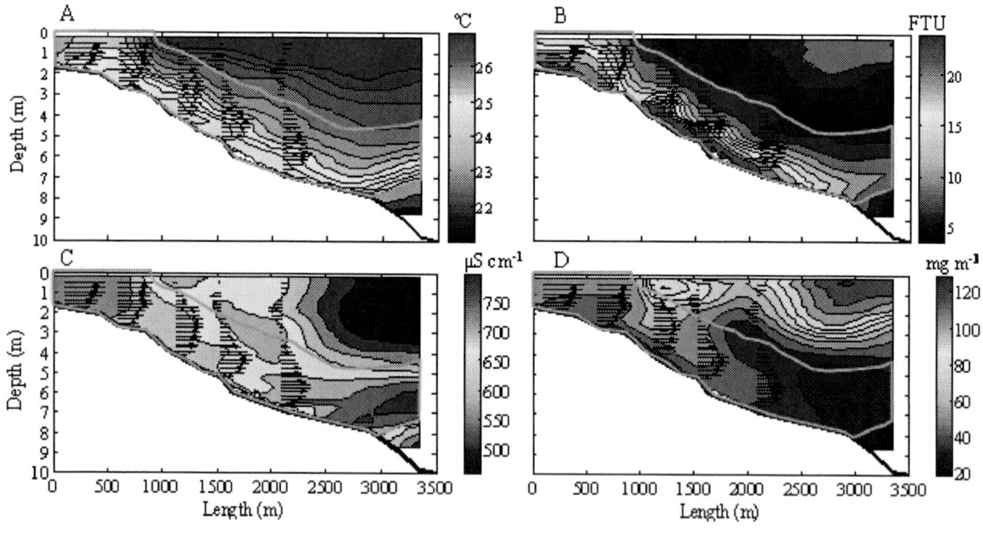

Figure 14. Longitudinal contours following the thalweg during day 212,starting at s1 at 0900 h and finishing at 1100 h, of (A) temperature, (B) turbidity, (C) conductivity, and (D) chlorophyll a. The zero position corresponds to station s1 (Figure 11). Black arrows show velocity profiles while green contour shows the estimated interflow position based on the velocity profiles and turbidity (that also agrees with temperature and conductivity).

The Chlorophyll-a values (Figure 14D) are characterized by two main peaks. The first one is clearly located in the plunge zone where the river converges with the ambient water; here an accumulation of matter coming from both the river and the reservoir, together with a phytoplankton peak, can be expected. Such peak was probably produced during the interflow and had a short duration, since overflows would likely disperse the peak. Part of this Chlorophyll entrained into the inflow and was dragged to the bottom, as can be seen in Figure 14D. The second, higher peak, with values of more than 100 mg/m3, is located at around 3000-3500m distance coinciding with the head of the overflow events. This peak will be explained in detail in the next paragraph.

2.2. Biological Response

From a water quality point of view, it is important to know how the river nutrients will be distributed along the reservoir and the effect of such nutrients over the phytoplankton populations. The input of nutrients from the River Ter in Sau reservoir is high, and thus, plays an important role in the phytoplankton evolution. The oscillating dynamics showed by the inflow during the field experiment varying between an overflow and an interflow means that new nutrients are injected everyday into the reservoir surface. When the injected water, rich in nutrients but poor in phytoplankton, encounters the eutrophic quiescent reservoir water, conditions for phytoplankton growth are excellent. Conductivity can be used as a tracer of the river water into the reservoir, and help to explain the progression or evolution of the river and the head of the injected overflows (Figure 15a).

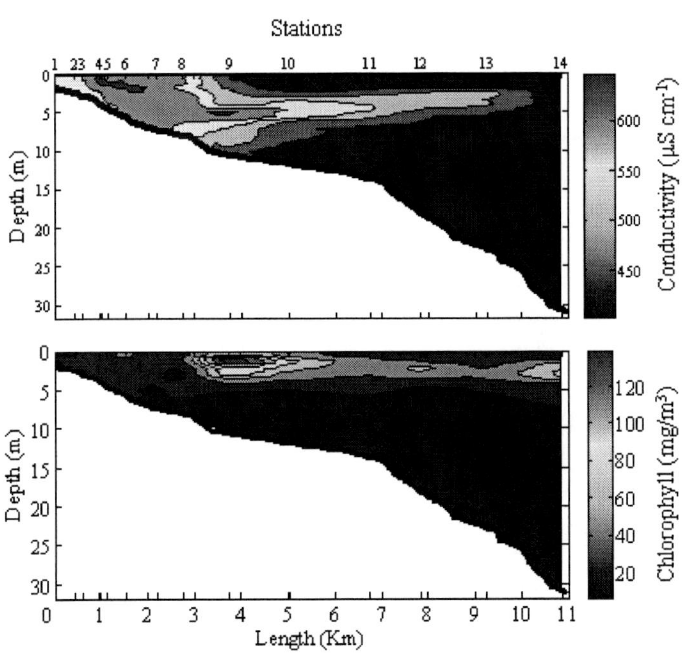

Figure 15. Longitudinal contour of (A) conductivity and (B) chlorophyll a following the thalweg on day 209. River conductivity, during the 2-dayperiod prior to day 209, averaged 600 μS cm^{-1}. Areas with conductivity values above 450 μS $cm^{-1}$are interpreted here as influenced by the river.

A bloom of phytoplankton located around station s9 (4 km), where the head of the injected overflowing water meets the reservoir water, can be appreciated in Figure 15b. At the time of the field experiment the reservoir was strongly stratified and solar radiation was high, thus near s9 where nutrients from the river mixes with the reservoir water in the euphotic zone, the conditions are favourable to the formation of blooms.

Summarizing, river inflows are usually a source of contaminants or nutrients into reservoirs. The fate and transport of the inflowing water is directly influenced by the density difference of the inflow and reservoir and the hydrodynamics of the reservoir. Such mechanisms evolve at different scales, though, in Sau reservoir, the diurnal scale has shown to be determinant in the pathways of the inflow. By one hand, the temperature of the river Ter responses to meteorological daily evolution faster than the reservoir temperature, modifying its density and therefore its intrusion depth at sub-daily scales (diurnal pattern). By the other hand, the daily periodic wind force basin-scale internal waves having the same period. In the riverine part of the reservoir the internal waves are in phase with the oscillation of the intrusion depth The combined effect of both mechanisms increase the mixing of the inflow and therefore the residence time of the river in the reservoir. Furthermore, during most of the spring and summer, at certain hours of the day, the temperature of the river is slightly greater than the temperature of the reservoir surface, and overflows are generated injecting the rich nutrient water into the euphotic zone; such injection favours the spread of algae. Under strong stratification cyanobacterias tend to dominate and generate located blooms in the intersection of the inflow water with the reservoir water.

REFERENCES

Armengol J, Garcia JC, Comerma M, Romero M, Dolz J, Roura M, Han BH, Vidal A, Simek K (1999) *Longitudinal processes in Canyon type reservoir: the case of Sau* (N.E. Spain). In: Tundisi JG, Straskraba M (eds) *Theoretical reservoir ecology and its applications*. Backhuys Publishers, Leiden, pp 313–345.

Casamitjana, X., T. Serra, C. Baserba, I J. Pérez. (2003). Effects of the water withdrawal in the stratification patterns of a reservoir. *Hydrobiologia* 504: 21–28.

Hodges BR, Imberger J (2001) Simple curvilinear method for numerical methods of open channels. *J. Hydr Eng-ASCE* 127:949–958.

Hodges BR, Imberger J, Saggio A, Winters KB (2000) Modeling basin-scale internal waves in a stratified lake. *Limnol Oceanogr 45*:1603–1620.

Münnich, M., Wüest, A. I Imboden, D. M. (1992). Observations of the second vertical mode of the internal seiche in an alpine lake. *Limnol. Oceanogr.*, 37(8), 1705–1719.

Perez-Losada, J., E. Roget, and X. Casamitjana. (2003). Evidence of high vertical wave-number behaviour in a continuously stratified reservoir. *Journal of Hydraulic Engineering,* 129: 9,734-737.

Serra T, Vidal J, Colomer J, Casamitjana X, Soler M (2007) The role of surface vertical mixing in phytoplankton distribution in a stratified reservoir. *Limnol Oceanogr* 52: 620–634.

Vidal, J., Rueda, F. I Casamitjana, X. (2007).The seasonal evolution of high vertical–mode internal waves in a deep reservoir. *Limnol. Oceanogr.,* 52(6), 2656–2667.

Vidal J, Casamitjana X (2008). Forced resonant oscillations as response to periodic winds in a stratified reservoir. J Hydr Eng 134:416–425.

Vidal J, Casamitjana X, Colomer J, Serra T (2005) The internal wave field in Sau Reservoir: observation and modeling of a third vertical mode. *Limnol Oceanogr* 50:1326–1333.

Vidal, J., Marcé, R. Serrra, T., Colomer, J., Rueda, F., Casamitjana, X., (2012). Localized algal blooms induced by river inflows in a canyon type reservoir. *Aquatic Sciences*, 74, 315-327.

EDITOR CONTACT INFORMATION

Dr. Enner Herenio de Alcântara
Sao Paulo State University - UNESP
Department of Cartography
649, Djalma Dutra Street, apto 44
Presidente Prudente, São Paulo State,
Brazil 19015-040
enner@fct.unesp.br

INDEX

E

F

G

H

I

O

P

Q

R

S

T

U

V